Laser Light Scattering in Biochemistry

Laser Light Scattering in Biochemistry

Edited by

S.E. Harding
University of Nottingham

D.B. Sattelle
University of Cambridge

V.A. Bloomfield
University of Minnesota

Based on papers presented at a UK Biochemical Society Meeting. University of Cambridge, 13–15 September 1990

Special Publication No. 99

ISBN 0-85186-486-4

A catalogue record of this book is available from the British Library

Published by The Royal Society of Chemistry,
Thomas Graham House, Science Park, Cambridge CB4 4WF

Printed in England by Redwood Press Ltd, Melksham, Wiltshire

Preface

This book is based on a U.K. Biochemical Society Meeting held at the University of Cambridge from the 13th to the 15th September, 1990: the senior authors of most of the chapters each gave a lecture at this meeting.

Historically, light scattering and the University of Cambridge are inextricably linked. J. Tyndall, who posed the two great enigmas of 19th-century meteorology ("Why is the sky blue?" and "Why is skylight plane polarized?") was a Rede lecturer there. J.C. Maxwell, whose electromagnetic theory provided the basis for solving Tyndall's enigmas, was the first Cavendish Professor at the Cavendish Laboratory. J.W. Strutt (Lord Rayleigh), who used Maxwell's theory to explain the scattering of light by small particles, was the second Cavendish Professor. G.G. Stokes, widely referred to by the 'Dynamic' or 'Quasi-elastic' light scattering fraternity was Professor of Applied Mathematics and Theoretical Physics, and W.L. Bragg was also a Cavendish Professor. So, Cambridge provided an appropriate venue for the first international meeting in the U.K. dedicated to light scattering in biochemistry for almost a decade.

In this book we have sought to cover the state of the art in both classical (or 'Total Intensity') as well as dynamic (or 'Quasi-elastic') light scattering. Although the last two decades have seen a great deal of attention paid towards the contribution that dynamic light scattering is making towards biomolecular science, classical light scattering is now also making significant inroads, particularly with the appearance of so-called 'Maxwell Demon' techniques involving the coupling of classical light scattering photometers to a gel permeation or size exclusion chromatography system. Part I ("Techniques") of this book seeks to address all these developments. Part II ("Macromolecules") deals with specific applications to systems of biological macromolecules, from proteins and glycopolymers right through to nucleic acids. The final part of the book ("Macromolecular Assemblies") deals with the application of light scattering techniques to the study of large biomolecular assemblies, including membrane systems, whole micro-organisms and other cellular systems, and finally it addresses polymer systems for drug delivery.

We are extremely grateful to all who helped in the organisation of the meeting, particularly to Margaret Clements and Sue Scott of the Department of Zoology at Cambridge, to Robert Dale and Marion Roberts of the Biochemical Society, and to Dr. John Horton whose skills with the Apple Macintosh were invaluable in the production of this book. We are also grateful to the following organisations who provided generous financial support: Academic Press, British Sugar Ltd., Brookhaven Instrument Corporation (U.S.A.), Ciba-Geigy (Horsham) Ltd., Coherent Ltd., Coulter Electronics Ltd., Fisons Pharmaceuticals Ltd., Glaxo Ltd., LDC Analytical, ICI Ltd., Malvern Instruments Ltd., Nestlé Ltd. (Switzerland), Newport Ltd., Oriel Scientific Ltd., Oros Instruments Biotage Ltd., Polymer Laboratories, Polymer Standards Service (Germany), Photon Control Ltd., Schlumberger Ltd., Unilever Ltd., Wyatt Technology Ltd. (U.S.A.) and Yamanouchi Ltd. In this field, academics and manufacturers often work closely together - the advances of each are intimately interwoven.

It is hoped that the reader finds the book as lively and informative as the 130 participants found the meeting.

Stephen Harding

David Sattelle

Victor Bloomfield

Nottingham, Cambridge & St.Paul

August 1991

J. Tyndall

J.C. Maxwell

J.W. Strutt
(Lord Rayleigh)

G.G. Stokes

W.L. Bragg

Contents

Part I: Techniques

Part I: Techniques

1

Static and Dynamic Light Scattering Approaches to Structure Determination of Biopolymers

By Walther Burchard

INSTITUTE OF MACROMOLECULAR CHEMISTRY, UNIVERSITY OF FREIBURG, GERMANY

1. INTRODUCTION

Biological macromolecules are in principle polymers like synthetic long chain molecules: thus at first sight there may be no need to treat biopolymers differently with respect to structure characterisation in solution. However, biopolymers do display features which impose serious constraints on the applicability of several commonly used techniques of polymer characterisation. Some of these characteristics peculiar to biopolymers are as follows.

(i) Biologically-active macromolecules are characterised by a unique structural conformation in solution. Change in temperature or in the solvent composition (*i.e.* salt concentration or pH) can cause changes in this structure, which may be associated with a loss of biological activity; the biopolymer becomes denatured. Thus, care must be taken not to destroy the biologically-active structure.

(ii) The characterisation of synthetic polymers is facilitated by the possibility of preparing a series of homologous chains of different lengths, but of the same chemical composition: such variation of the chain length is often not possible with biopolymers.

(iii) The well-defined, tertiary structure of biopolymers is stabilized by special interactions between the many, very polar side groups. Not all of them will be fully saturated by *intra*molecular interactions, and a pronounced *inter*molecular interaction will become effective as the concentration of polymer is increased. Association - which can be non-specific - will occur, and will then lead to randomly branched clusters. Often, however, the interactions *are* specific, and will induce well-organized supramolecular structures. These non-specific interactions and specific association phenomena are especially difficult to investigate.

2. BASIC RELATIONSHIPS

In static light scattering (LS) the intensity is averaged over a fairly long time (~2 s), and this is in most cases long enough to smooth out all internal mobility. The time average equals the ensemble average; *equilibrium* or *static* properties are measured.

Nowadays we are able to record the scattering intensity from tiny scattering volumes in time intervals as short as 50 ns. This enables us to follow the motion of macromolecules in solution. This is achieved by the construction of a time correlation function[1].

$$G_2(t) = \langle i(0)\, i(t) \rangle \qquad (1)$$

where the scattering intensity at time zero is compared with that at a delayed time $t \approx 10^{-6}$ s - 10^{-3} s. This intensity correlation function is difficult to interpret, but can under most conditions be converted into a correlation function $g_1(t)$ of the scattered electric field so that

$$G_2(t) = A + (Bg_1(t))^2 \qquad (2)$$

with

$$g_1(t) = S(q,t)/S(q) \qquad (3)$$

where $S(q,t)$ is the dynamic structure factor, which is given by

$$S(q,t) = \sum \sum \langle \exp (iq|r_j(0) - r_k(t)|) \rangle \qquad (4)$$

Here, q is the magnitude of the scattering vector

$$q = (4\pi/\lambda) \sin \Theta/2 \qquad (5)$$

and $r_j(0)$ is the position of the jth scattering element at time zero while $r_k(t)$ is that of the kth element at time t. $S(q)$ is time independent and is the static structure factor

$$S(q) = \langle \sum\sum \exp (iq|r_j(0) - r_k(0)|) \rangle \qquad (6)$$

The two equations for the static and dynamic structure factors appear very similar but differ essentially in one point: in static LS the average must be taken only over all spatial fluctuations of the ensemble, while in dynamic LS the average has to be performed with respect to space *and* time. In most cases we have a very limited knowledge of the required space-time distribution functions, and only the simplest cases are known. One example is that of small particles with a narrow weight distribution, and for monodisperse, large, hard spheres. Then we can write

$$g_1(t) = B \exp(-\Gamma t) \quad (7)$$

with B a constant close to unity and a decay constant Γ, that is related to the translational diffusion coefficient D_{trans} by the equation

$$\Gamma = D_{trans}\, q^2 \quad (8)$$

In most cases simple single exponential behaviour is not observed, and deviations from the straight line of log $g_1(t)$ against t are apparent (see Fig. 1). These deviations may result from a degree of polydispersity, because in this case each species contributes its own exponential, according its specific diffusion coefficient. Fluctuations resulting from internal molecular flexibility can also contribute to departures from single exponential behaviour.

In these cases a cumulant expansion is made[2]

$$\ln g_1(t) = -\Gamma_1 t + (\Gamma_2/2!)\, t^2 - (\Gamma_3/3!)\, t^3 + \ldots \quad (9)$$

Most interesting is the first cumulant Γ_1 which is related to the translational diffusion coefficient by[3]

$$\Gamma_1/q^2 = D_c\,(1 + C\, R_g^2 q^2 - \cdots) \quad (10)$$

where C is a structure-dependent coefficient, that arises from internal flexibility but one which will not be discussed further in this context[3-5]. Eq. (10) is rigorous only when q is not too large, *i.e.* $R_g q < 2$ (where $R_g = (\langle S^2\rangle_z)^{1/2}$ is the root of the z-average means square radius of gyration and D_c denotes the translational diffusion coefficient at concentration c). D_c can often be well approximated by the linear equation

$$D_c = D_{trans}\,(1 + k_D c) \quad (11)$$

Comparing the two equations (10) and (11) with the common static LS equation

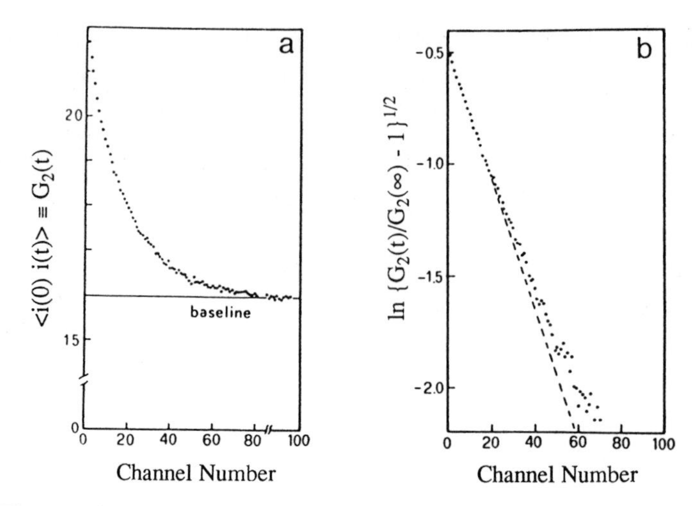

Figure 1. Typical example (using PMMA in acetone as the scattering material) of the time correlation functions of (a) the scattering intensity $G_2(t) = \langle i(0)i(t) \rangle$, and (b) the scattered electric field $g_1(t)$. Note in (b) the deviations from the straight line relationship which corresponds to single exponential behaviour predicted for a monodisperse sample.

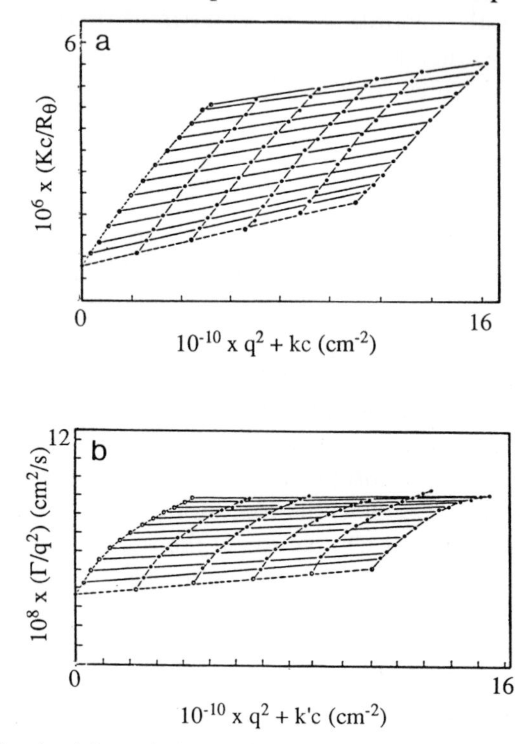

Figure 2. Static (a) and dynamic (b) Zimm plots of the *Rhizobium trifolii* exopolysaccharide, strain TA-1, in 0.1 M NaCl[6].

$$K_c/R_\Theta = (1/M_w) + (1/3)\, q^2 R_g^2/M_w + 2A_2c + \ldots \quad (12)$$

one notices a striking similarity. Thus, both static and dynamic LS measurements may in a first instance be evaluated by static and dynamic Zimm plots[3]. Fig. 2 gives an example of this using a microbial polysaccharide sample which will be discussed in detail later in this Chapter. The only difference between these Zimm plots is that Kc/R_Θ is plotted against q^2+kc in the static plot and $\Gamma_1/q^2 = D_{app}(q)$ is plotted against q^2+kc in the dynamic one. It therefore follows that the six parameters given in Table 1 *viz* M_w, R_g, A_2, D_{trans}, C and k_D can all, in principle, be obtained from such a pair of Zimm plots.

Table 1: The six quantities which can be obtained from the intercepts and slopes of static and dynamic Zimm plots.

Type of Experiment	Intercept	Slope of c = 0 Line	Slope of Θ = 0 Line
Static LS	M_w	R_g^2	A_2
Dynamic LS	D_{trans}	CR_g^2	k_D

3. SMALL PARTICLES

It is natural to combine the quantities referred to in Table 1 to provide useful information on the biological particles under investigation.

Often the biological particles are rather small in size. In this case no angular dependence can be observed, and R_g remains unobtainable, unless the smaller wavelengths of X-rays and thermal neutrons can be applied. However some useful information can be obtained; namely the average radii from D_{trans} and the second virial coefficient A_2. Applying the well known Stokes-Einstein relationship we find

$$D_{trans} = \frac{kT}{(6\pi\eta_0 R_h)} \quad (13)$$

(where k is the Boltzmann constant and η_0 the viscosity of the medium) which defines the effective hydrodynamic radius of the macromolecule. Another estimate for an effective radius may be obtained from[7]

$$A_2 = (16\pi/3)\, N_A\, (R_{eq}^3/M_w^2) \tag{14}$$

or

$$R_{eq} = \left\{\frac{3A_2\, M_w^2}{16\pi\, N_A}\right\}^{1/3} \tag{15}$$

Eqs. (13) and (15) are exact for hard spheres and should give the same value, *i.e.* R_{eq}=R_h. Such behaviour is indeed observed for some enzymes; for instance for β-galactosidase we found

R_h = 8.3 nm ± 3% R_{eq}= 8.7 nm ± 8%

(These data lie a little outside the experimental error but can easily be explained in terms of an ellipsoidal structure of the enzyme.)

These two radii do not coincide in all cases, and often R_{eq} is smaller than R_h or can even become negative. For structures which are built up from linear or cross-linked chains, the more general relation holds

$$A_2\, M_w = 4\pi^{3/2}\, N_A\, (R_g^3/M_w)\, \Psi(z) \tag{16}$$

where $\Psi(z)$ is the so-called interpenetration function which reaches a constant value if the interaction parameter $z>1$ ($z \sim \beta N^{1/2}$, where β is the excluded volume of one segment and N the number of segments). This plateau value corresponds to repulsion between the particles, but A_2 becomes small or even negative in poor solvents. In the limit of large z, the ratio X_R=R_{eq}/R_h becomes close to unity, but for small z values $R_{eq}/R_h<1$. The precise relationship is known so far only for linear and flexible chains, and more work is necessary in order to represent the behaviour of loosely cross-linked structures or cyclically closed macromolecular structures. Further differentiation can be made by calculating the segment density (d), *via*

$$d = M_w/(N_A(4\pi/3)\, R_x^3) \tag{17}$$

where x can represent either "h" or "eq".

4. PARTICLES WITH $R_g q \approx 1$

For this case, an angular dependence is observed, and R_g can be determined from the initial slope of the static Zimm plot. The condition of $R_g q \approx 1$ can, of course, be fulfilled, either by fairly large particles (if visible light is used (q is small) or by small particles if the short wavelength of thermal neutrons is used (when q is large)). In my laboratory, we have been in a rather rare situation of being able to measure R_g from mussel glycogen by both techniques[8,9], and have found the following result shown in Table 2.

Table 2. Radii R_g, R_{eq}, and R_h of mussel glycogen obtained by static (LS) and dynamic (DLS) light scattering, small angle neutron scattering (SANS) respectively and by electron microscopy (R_{EM}).

R_g (LS)	R_g (SANS)	R_h (DLS)	R_{eq} (LS)	R_{EM}
18.9 nm	18.8 nm	19.3 nm	14.6 nm	16.3 nm
$M_w = 1.89 \times 10^6$ g/mol			$A_2 = 0.83 \times 10^{-5}$ ml.mol.g^{-2}	
$\rho = R_g/R_h = 0.977$			$X_R = R_{eq}/R_h = 0.756$	

What can this set of information tell us of the macromolecular structure? Several years ago in our laboratory, we calculated the ρ-parameter ($=R_g/R_h$) for different molecular architectures[3] and found the data shown in Table 3. The experimental value for mussel glycogen is close to that of a hard sphere, but is even closer to that of a star-branched macromolecule with many arms, and this structure gives equivalent results to an "AB_2"-polycondensate, *i.e.* a monomer with one A-group and two B-groups, where only A can react with B. This is in fact the condition for a glucose (Fig. 3), where B_1 (the OH in C6 position) and B_2 (the OH in C4 position) have different reactivities.

For glycogen, $X_R \sim 0.76$, and such $X_R < 1$ values indicate that the conformation is not spherical. It has been known for a long time that X_R can be correlated with the concentration coefficient k_D in the equation

$$D_c = D_{trans}(1 + k_D c) \tag{18}$$

Recently Selzer[10] has found for flexible chains the semi-empirical equation

$$k_D^* = 4.8\, X_R^3 - 2 \qquad \rightarrow 0.242 \quad \text{from experiment} \tag{19}$$

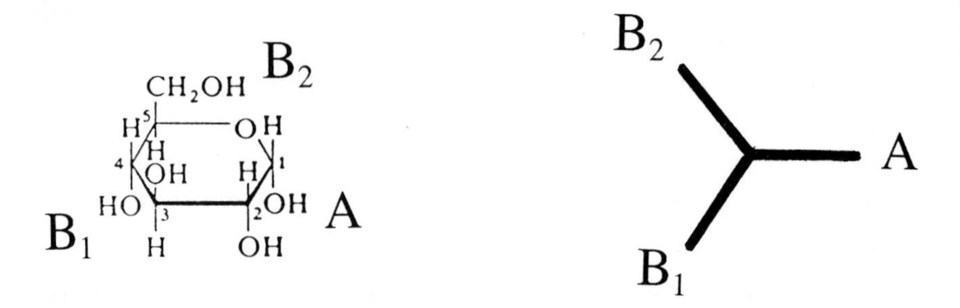

Figure 3. Chemical structure of glucose. The OH-group in the C1 position (A-functional group) forms, together with the OH-groups in C4 and C6 positions (B_1 and B_2 functional groups), an AB_2-type monomer that builds up glycogen, where A can react only with either B_1 or B_2.

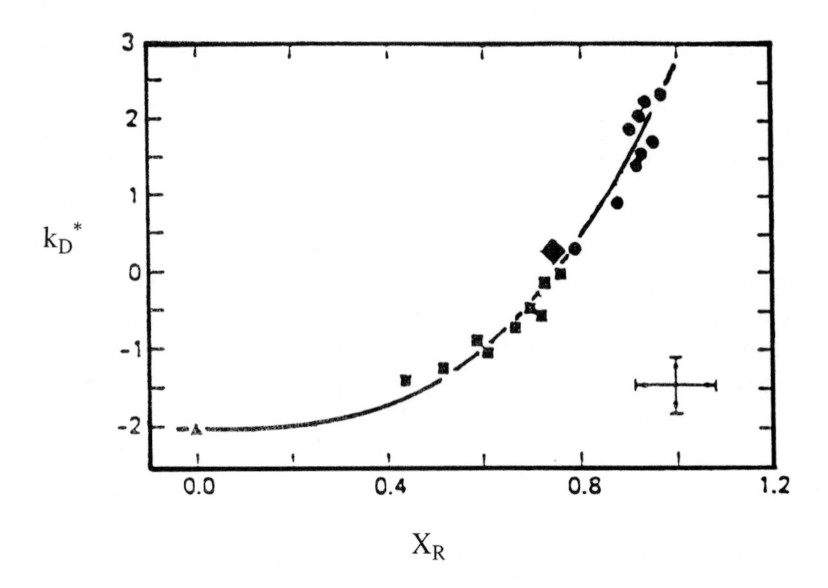

Figure 4. Plot of k_D^* against X_R based on measurements with poly-α-methylstyrene and the semi-empirical relationship eq. 19. The large filled squares correspond to glycogen.

with

$$k_D^* = k_D (N_A R_h^3/M_w) \quad \rightarrow .074 \qquad \text{from } X_R \qquad (20)$$

Table 3: The theoretical ρ-parameters and C-parameters for some selected structures.

Architecture	ρ	C
Homogeneous sphere	0.778	0
Random coil, monodisperse		
Θ-conditions	1.50	0.173
good solvent	1.78	-
Random coil, polydisperse, z = 1		
Θ-conditions	1.73	0.200
good solvent	2.05	-
Regular stars		
Θ-conditions, f= 4	1.33	0.148
Θ-conditions, f » 1	1.079	0.098
Rigid rod		
monodisperse	> 2.0	0.042
polydisperse, z = 1	> 2.0	0.16

Fig. 4 shows the plot of k_D^* *vs.* X_R according to eq. 19, together with results from a synthetic polymer (poly-α-methylstyrene) and our point calculated from A_2 of glycogen. Hard spheres should give a k_D value close to +2, while for fully interpenetrable "soft" spheres a value of -2 is predicted. Evidently, we have a particle which shows behaviour almost exactly between these two limits. The experimental value lies clearly above the value predicted by Selzer's equation. This might be a result of a highly branched structure and polydispersity, but the experimental error is still too high: more measurements need to be performed in the future to clarify this point.

5. ASYMPTOTIC BEHAVIOUR OF THE ANGULAR DEPENDENCE (R_gq»1)

Consideration of the three radii R_g, R_h, and R_{eq} and their two characteristic ratios alone is not sufficient for a clear description of these particles, and other information has to be applied for a more

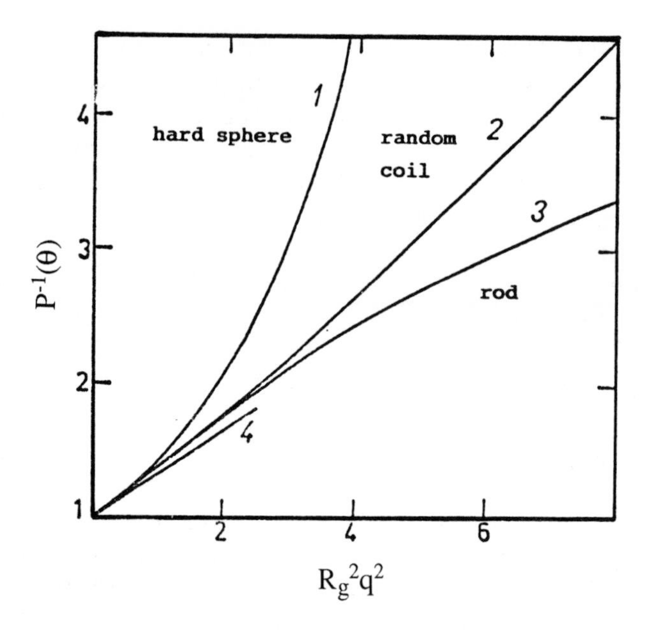

Figure 5. Plot of the inverse particle scattering factor $1/P(\theta)$ against $R_g^2q^2$ for three typical architectures: (1) hard spheres; (2) random coil of flexible chains; (3) stiff rods.

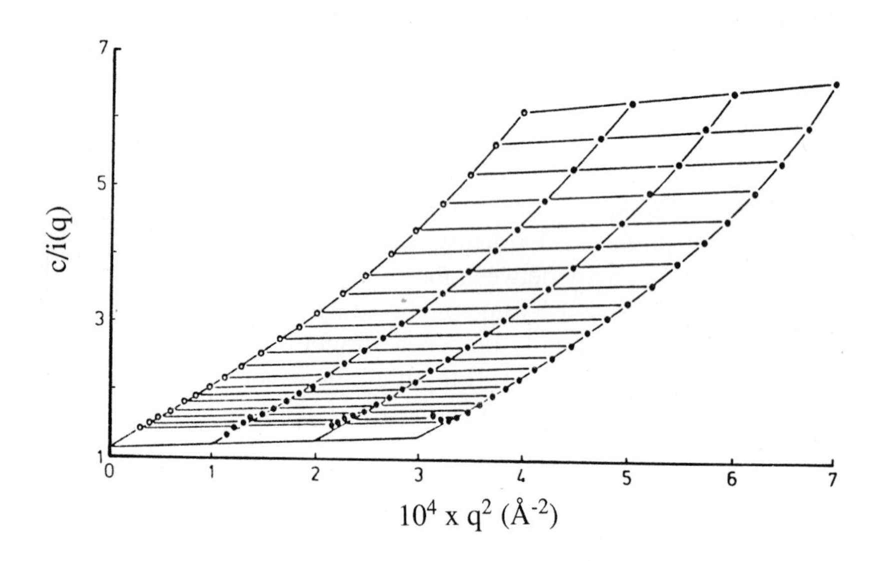

Figure 6. Common Zimm plot from the SANS data of glycogen[8].

specific answer. This information could come from, for example, viscometry or electron microscopy (neither of which are the subject of this contribution), and it could come from a specific analysis of the asymptotic angular dependence of the scattering envelope *i.e.* where $R_g q \gg 1$. This region could be reached with this example by small angle neutron scattering (SANS).

Nevertheless, a first impression can be obtained from plots of the reciprocal scattering intensity as a function of q^2. Fig. 5 shows a schematic plot of the inverse particle scattering factor for three selected structures, a sphere (1), a random coil (2), and a rod (3).

A Zimm plot of the LS data from mussel glycogen is shown in Fig. 6: a clear upturn is observed. This could be indicative of spheres. An even better insight is gained from a Kratky plot (*i.e.* $q^2P(q)$ *vs.* q) of the data at $c \rightarrow 0$. (Fig. 7). We found a pronounced maximum, and an increase at rather large values of qR_g. The dotted line represents the theoretical curve for an AB_2 type polycondensate, assuming full flexibility of the highly branched chains and no excluded volume effect[8]: a rather good description is obtained with this model. The minimum and the subsequent weak increase of the experimental curve have recently been predicted by Monte Carlo simulations[11], being the result of (a) overcrowding of the outer chains (Fig. 8) and (b) their chain stiffness. Thus the final conclusion is that glycogen is a highly branched particle of spherical symmetry, but it is not a hard sphere.

A very similar result was obtained with β-casein micelles in a phosphate-free medium. Fig. 9 shows a Berry plot and Fig. 10 the corresponding Kratky plot of the curve at $c \rightarrow 0$. In a Berry plot $(Kc/R_\Theta)^{1/2}$ is used, by means of which a linearization of a curve showing strong upward curvature is often achieved. The dotted line represents the theoretical curve (no excluded volume) for a star-branched macromolecule with about 36 Gaussian chains as arms. This value was easily obtained from the measured micelle molar mass and the known molar mass of the individual chain. Fig. 11 shows our model suggestions for the β- and the α-casein, though the latter is not discussed further in the present context.

The result of this model can be cross-checked *via* the $\rho = R_g/R_h$ parameters. The result is shown in Fig. 12, where the results for the α_s-casein are also shown.

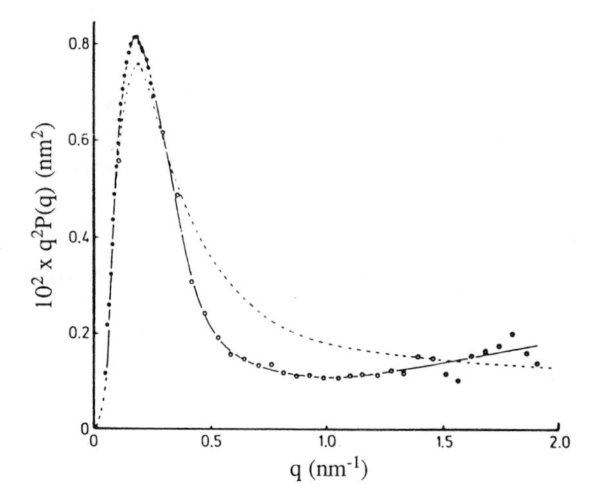

Figure 7. Kratky plot of the same data as shown in Fig. 6 at c=0.

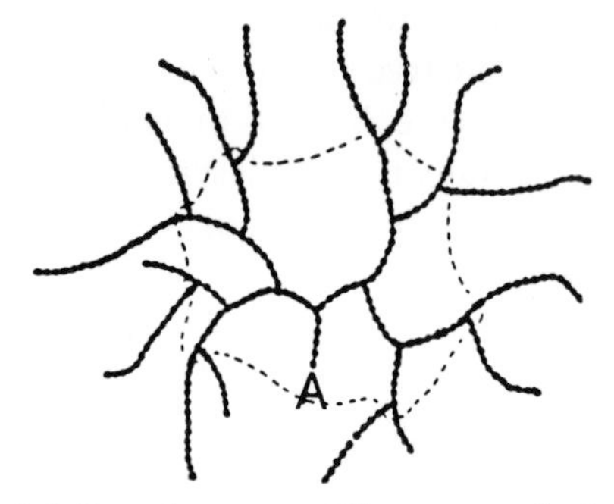

Figure 8. Model for glycogen. The outer chains are so densely packed that a stretching out is observed. The chain stiffness is increased and causes the asymptotic linear increase depicted in Fig. 7.

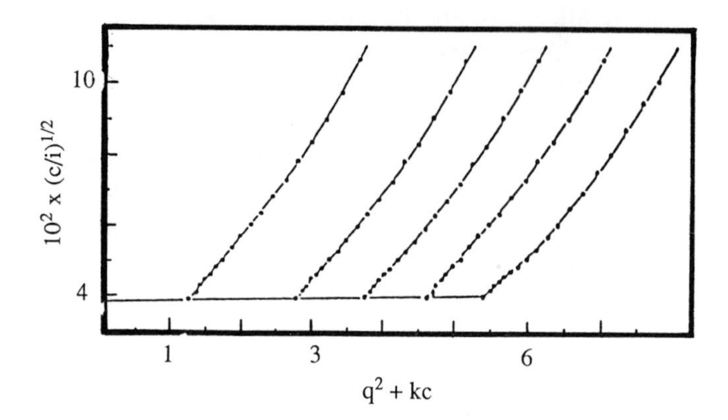

Figure 9. Berry plot from SANS data of β-casein[12].

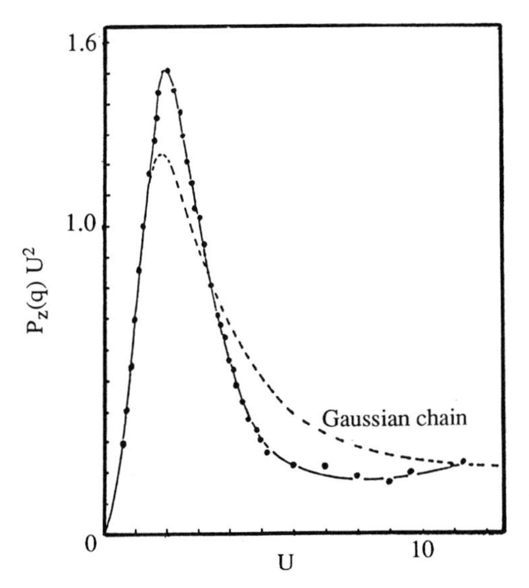

Figure 10. Kratky plot of the data from Fig. 9, where c=0 [12].

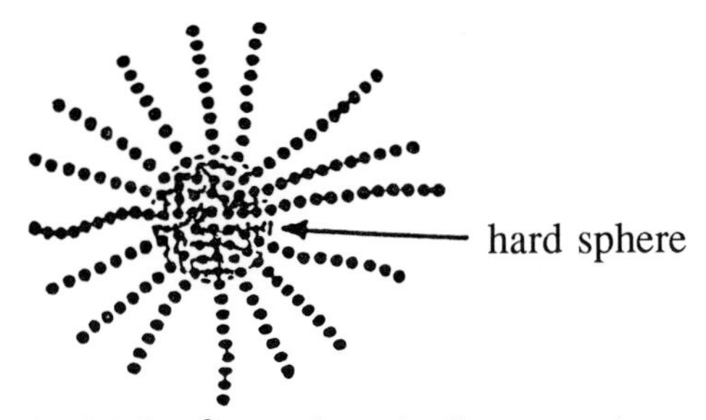

Figure 11. Model for β-casein micelle.

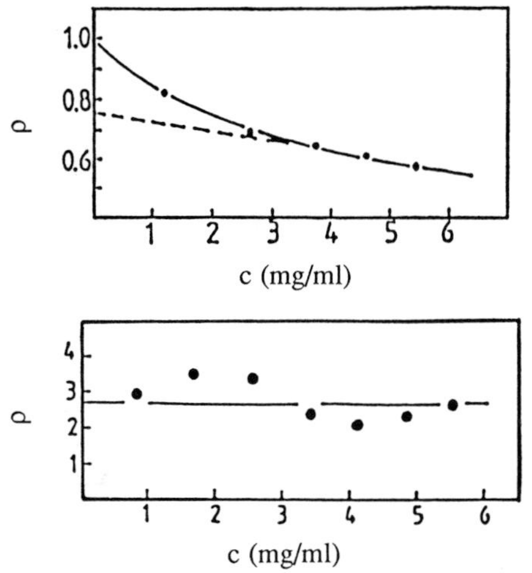

Figure 12. The ρ-parameter for β- (upper) and α_s-caseins (lower) as a function of concentration c[12].

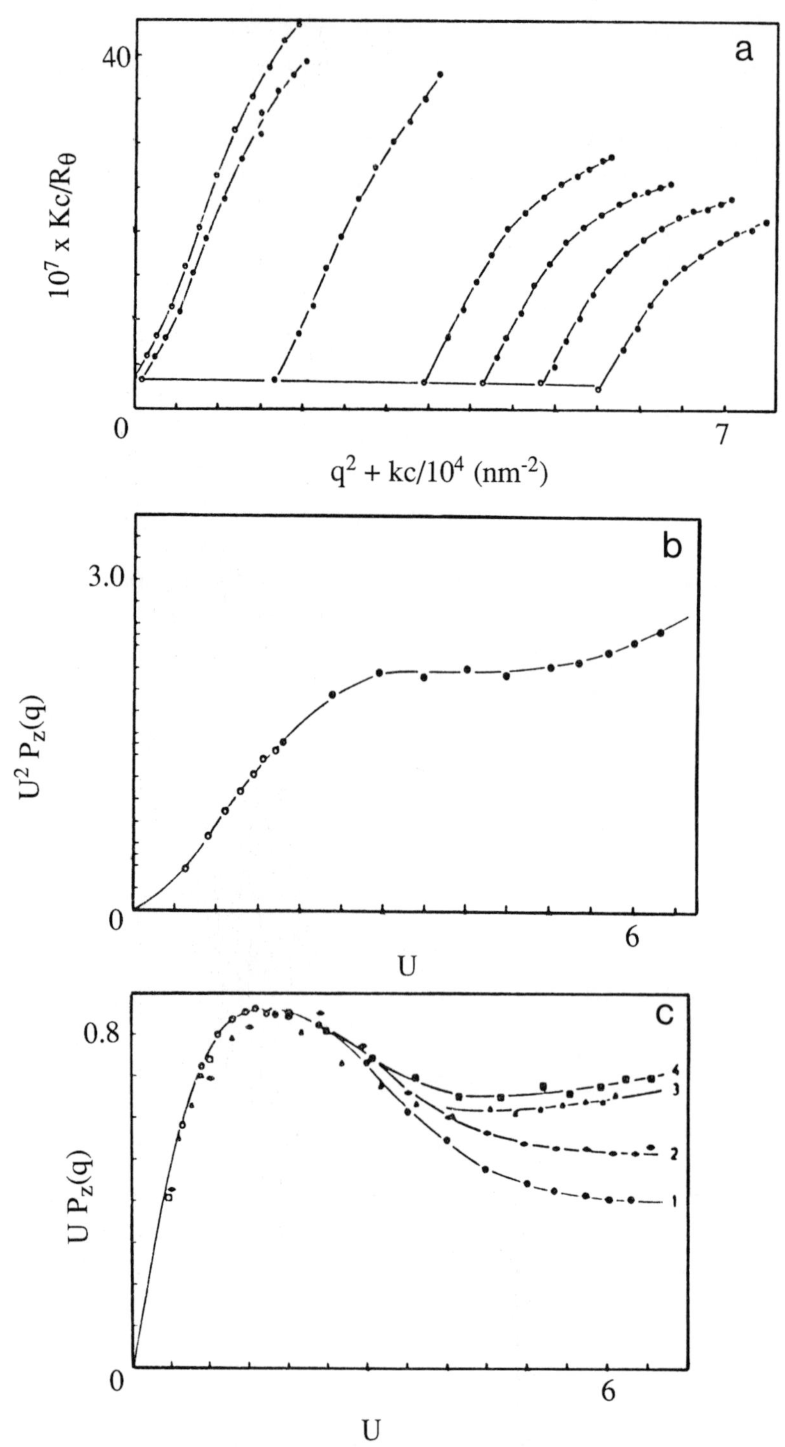

Figure 13. (a) Zimm, (b) Kratky and (c) Casassa-Holtzer plots of data obtained from α-casein[13].

There is a slight ambiguity in the results for the β-casein, which indicate a structure something between the two limits of a hard sphere and a Gaussian star-molecule. Surprisingly the α_S-casein, the major fraction in milk caseins, gave the much larger value of $\rho = 2.78$, and this would indicate stiff and long filaments.

6. STIFF CHAINS

Light scattering measurements confirm the view that the α_S casein structure is that of stiff, long filaments. Fig. 13 shows three different plots of the same data[13]: a Zimm plot, a Kratky plot, and what may be called a "Casassa - Holtzer" [14,15] or "bending rod" plot.

The Kratky plot reveals a change in behaviour from flexible Gaussian coils (plateau) to a rod like behaviour (asymptotic linear increase): to all familiar with the behaviour of semiflexible chains[16], this is indicative of stiff chains. The kink point, that would define the persistence length or the Kuhn segment length (l_K) cannot be well-localized in this plot, and another technique must be applied. Nevertheless, we can say that l_K must be very large, and of the order of a wavelength, because otherwise the Kuhn segment cannot be detected.

The procedure for analysis of such stiff chains can be demonstrated readily with the bacterial polysaccharide TA-1 the Zimm plot for which was shown in Fig. 2.

We noticed that a plot of qR_h/Kc *vs.* qR_g gives a much clearer picture for really stiff chains. Using the theory by Koyama we realized that the typical plateau of rods is approached from below if the rod is completely rigid, but the asymptote is reached from above when the chain has still some flexibility[18]. Figs. 14 and 15 show the theoretical and experimental results. The ratio of the maximum height to the plateau heights was found to be a measure of the number N_K of Kuhn segments; the plateau height gives the mass per unit length M_L, and (together with M_w) we can then find the contour length L_w and (with N_K) finally the Kuhn segment length, l_K: values for different polymeric chains are given in Table 4.

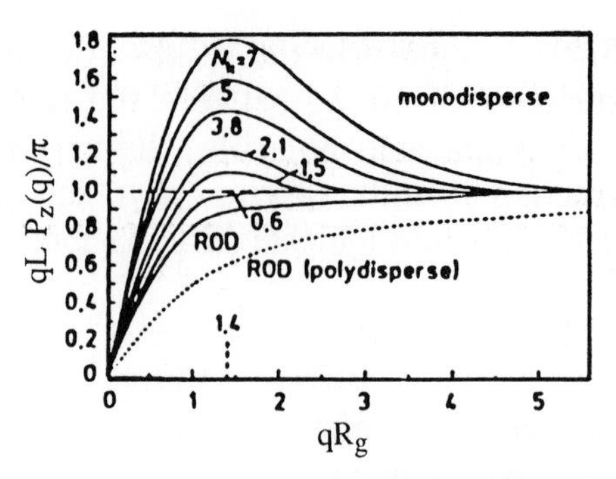

Figure 14. Plot of qP(q) against qR_g for chains of various chain stiffness according to Koyama's theory[17]. N_K is the number of Kuhn segments. All curves are normalized to the same asymptotic plateau.

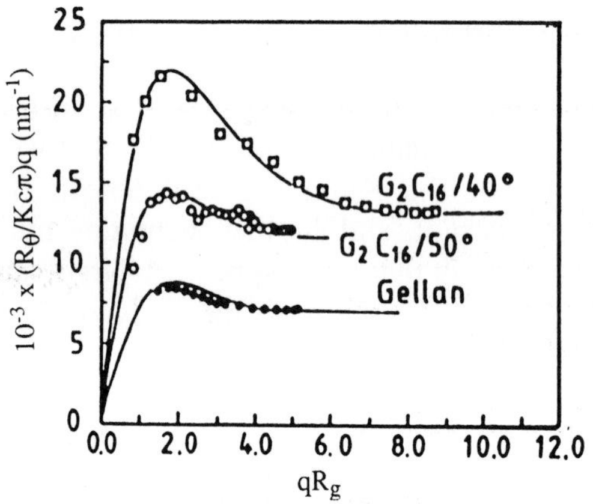

Figure 15. Plot of qR_Θ/Kc against qR_g for two micellar filaments[19] and for gellan[6].

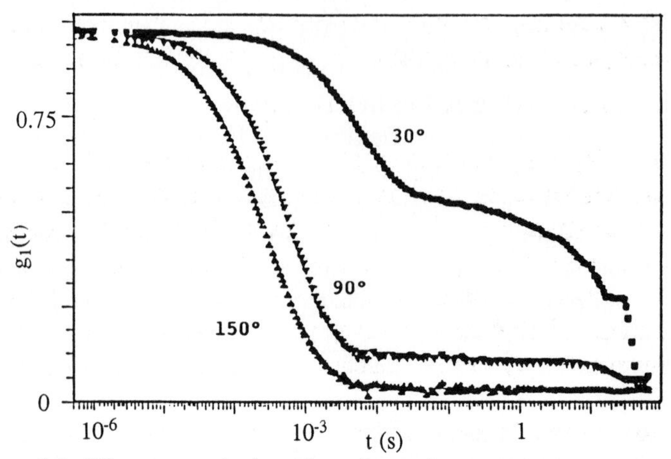

Figure 16. Time correlation function of an actin filament capped by Gelsolin. The curves correspond to dynamic LS measurements at 30°, 90° and 150°.

Table 4: Kuhn lengths (l_K), characteristic ratios (C_∞) and number of laterally aggregated strands (n_s) for different polymeric chains.

Polymer	l_k (nm)	C_∞	n_s
Schizophyllan	390	757	3
Gellan	322	625	2
Xanthan	255	494	2
TA1-EPS	152	314	3
Collagen	260	683	3
PBLG	190	499	1
DNA	80	235	2
Cellulose-tri-carbanilate	22	42.7	1
PS	2	9.8	1

The variation of the $\rho = R_g/R_h$ parameter and the coefficient C in eq. (10) as a function of the number of Kuhn segments have also been calculated by Schmidt and Stockmayer[4], but are not considered further in this Chapter.

7. ACTIN FIBRILS

Results of considerable interest were obtained for actin fibrils. The lengths of these biopolymers were controlled by the addition of Gelsolin to the monomeric G-actin before polymerization was initiated by a change in salt concentration. The contour length, L, was found by electron microscopy[22] to follow the relationship

$$L/l_0 = DP = [\text{ G-Actin }] / [\text{ Gelsolin }] \tag{21}$$

The time correlation function was measured[22] at various scattering angles, and for different proportions of G-actin to Gelsolin. One typical result is shown in Fig. 16. At low angles large clusters are seen, but at 90°-150° only one major process is seen which is diffusive, *i.e.* demonstrating $\sim q^2$ scaling. Inverse Laplace transformation of this time correlation function at Θ=90° gave the result shown in Fig. 16 for the diffusion coefficient distribution.

Knowing the length of the F-actin chains from Janmey's relationship[21] and having measured D_z by dynamic LS we can

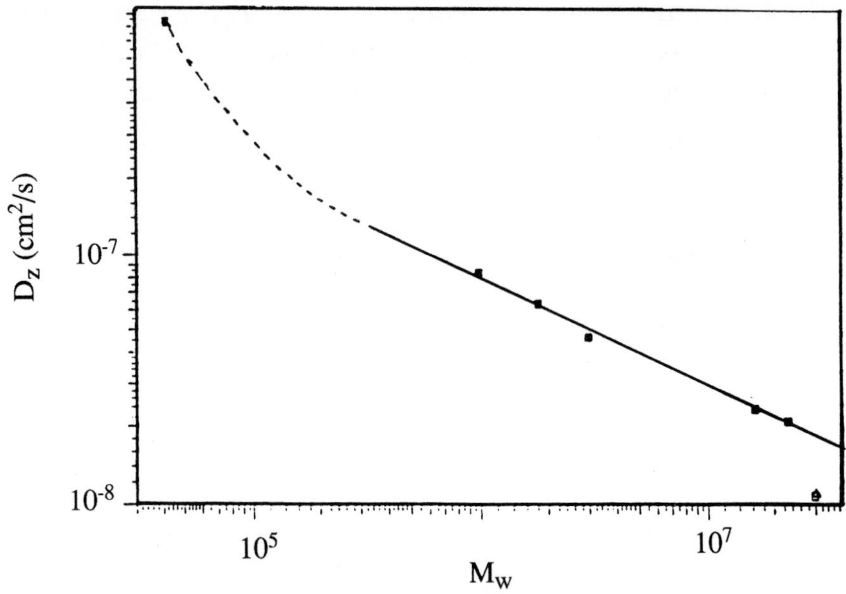

Figure 17. Calibration curve for D_z *vs.* M_w.

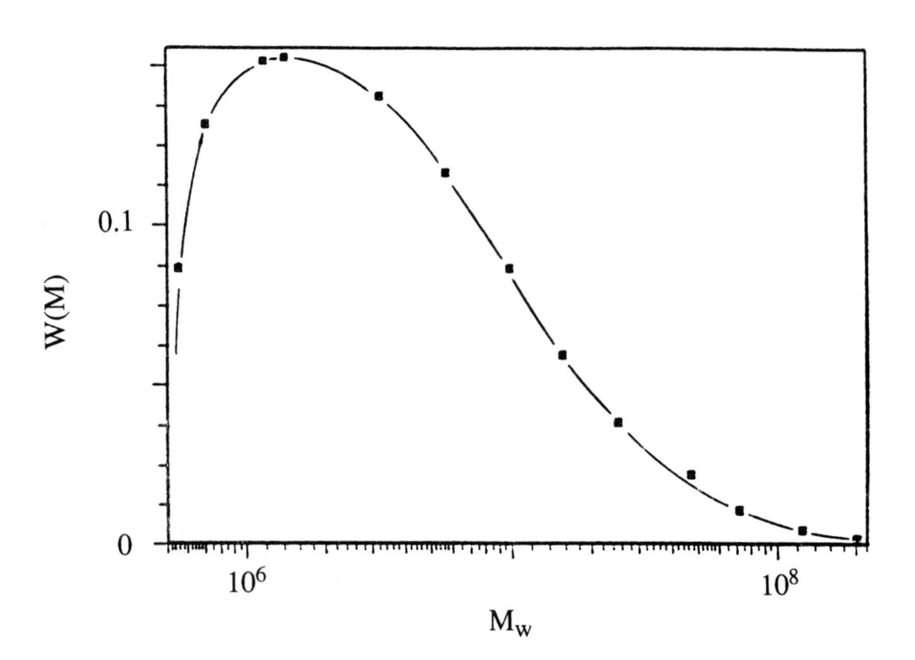

Figure 18. The molar mass distribution of the actin filaments obtained from the time correlation function in Fig. 16.

establish a calibration curve of D_z *vs.* M_w. (Fig. 17). With this calibration curve the weight distribution of Fig. 18 was finally obtained. These measurements on actin polymers reveal two important results:

(i) The exponent in

$$D \sim M^{-0.49} \tag{22}$$

is considerably smaller than -1, which is expected for rigid rods. An exponent of 0.45 - 0.50 is a clear indication of flexible chain behaviour. Thus the earlier results[23] from the Max Planck Institute in Munich are confirmed.

(ii) The molar mass distribution is apparently very broad. A value for $M_w/M_n \approx 2$ is expected for the most probable distribution. The larger polydispersity indicates other reactions than solely the capping reaction of the Gelsolin with F-actin filaments. Probably a non-specific association is superimposed on the specific filament formation, but more measurements should be performed to clarify this point.

ACKNOWLEDGEMENT

The preparations of the F-actin solutions were made together with Dr. Paul Janmey, from the Harvard Medical School at Boston. Dr. Janmey's generous help is gratefully acknowledged by the author who also wishes to thank his co-workers mentioned in the text who contributed so much to the results presented.

REFERENCES

1. H.B.J. Berne and R. Pecora, Dynamic Light Scattering', Wiley, New York, 1976.
2. P.N. Pusey, D.E. Koppel, D.W. Schaefer, R.D. Camerini-Otero and S.H. Koenig, *Biochemistry*, 1974, *13*, 952.
3. W. Burchard, M. Schmidt and W.H. Stockmayer, *Macromolecules*, 1980, *13*, 1265.
4. M. Schmidt and W.H. Stockmayer, *Macromolecules*, 1984, *17*, 509.
5. W. Burchard, *Chimia*, 1985, *39*, 10.
6. M. Dentini, T. Coviello, W. Burchard and V. Crescenzi, *Macromolecules*, 1988, *21*, 3312.
7. H. Yamakawa, "Modern Theory of Polymer Solutions", Harper and Row, Evanston, 1971.

8. M. Rinaudo and W. Burchard, in W. Burchard, *Adv. Polym. Sci.* 1983, *48*.
9. A. Reiner, Ph.D. Thesis Freiburg, 1981.
10. J. Selzer, *Polymer Preprints*, 1990, *31*.
11. (a) K.Huber, Ph.D. Thesis, Freiburg, 1987; (b) K. Huber and W. Burchard, *Polymer*, 1987, *28*, 1997.
12. A. Thurn, W. Burchard and R. Niki, *Coll. Polym. Sci.*, 1987, *265*, 653.
13. A. Thurn, W. Burchard and R. Niki, *Coll. Polym. Sci.*, 1987, *265*, 897.
14. E.F. Casassa, *J. Chem. Phys.*, 1955, *23*, 596.
15. A. Holtzer, *J. Polym. Sci.*, 1955, *17*, 432.
16. O. Kratky and G. Porod, *Rec. Trav. Chim., Pays-Bas*, 1949, *68*, 1106.
17. R. Koyama, *Phys. Soc. Japan*, 1973, *34*, 1029.
18. M. Schmidt, G. Paradossi and W. Burchard, *Makromol. Chem.; Rapid Commun.* 1985, *6*, 767.
19. P. Denkinger, W. Burchard and M. Kunz, *J. Phys. Chem.*, 1989, *93*, 1428.
20. P. Denkinger, M. Kunz and W. Burchard, *Coll. Polym. Sci.*, 1990, *26*, 513.
21. P. Janmey, *J. Biol. Chem.*, 1986, *261:18*, 8357.
22. M. Hauptmann, Diploma Thesis, Freiburg, 1990.
23. C.F. Schmidt, M. Bärmann, O. Tsenberg and E. Sackmann, *Macromolecules*, 1989, *22*, 3638.

2

The Use of Low Angle Laser Light Scattering with Gel Permeation Chromatography for the Molecular Weight Determination of Biomolecules

By K. Jumel, P. Browne[1] and J.F. Kennedy[2]

CHEMBIOTECH LTD., INSTITUTE OF RESEARCH AND DEVELOPMENT, UNIVERSITY OF BIRMINGHAM RESEARCH PARK, VINCENT DRIVE, EDGBASTON, BIRMINGHAM. B15 2SQ U.K.

[1]LDC ANALYTICAL, DIAMOND WAY, STONE BUSINESS PARK, STONE, STAFFORDSHIRE. ST15 0HH U.K.

[2]RESEARCH LABORATORY FOR THE CHEMISTRY OF BIOACTIVE CARBOHYDRATES, SCHOOL OF CHEMISTRY, UNIVERSITY OF BIRMINGHAM, PO BOX 363, BIRMINGHAM. B15 2TT U.K.

1. INTRODUCTION

Light scattering has been used as a method for determining polymer molecular weights and radii of gyration for a considerable time. However, data collection used to be rather tedious and time consuming, thereby largely restricting its use to mainly research purposes.

In the early 1970's the low angle laser light scattering (LALLS) photometer came onto the market[1]. The fundamentally different design of this instrument allowed light scattering measurements at very low angles (typically 3°-10°). Extrapolation to zero angle became unnecessary, which simplified data evaluation considerably. As a result the technique was adopted routinely for molecular weight determinations, although radius of gyration and associated information could no longer be obtained.

It was soon realized that this instrument could not only be used for static measurements, but also as a flow-through detector mainly for gel permeation chromatography (GPC)[2]. GPC/LALLS has been extensively used since that time and there are now a large number of applications reported in the literature. This review is intended to give a brief overview of GPC/LALLS as applied to the study of biomolecules, and focuses only on selected examples.

2. THEORY

The measurement of molecular weight by light scattering techniques makes use of the intensity of the light scattered at some angle θ as a function of the intensity impinging on the scattering volume, V. This can be put in the form

$$R_\theta = J_\theta / I_0 V \tag{1}$$

where R_θ is the Rayleigh factor, J_θ the intensity of scattered beam and I_0 the intensity of the incident beam.

In solution, light scattering data can be interpreted according to the fluctuation theory of light scattering which states that light is scattered as a result of changes of polarizability of the solution due to random fluctuations in the solute concentration in small volume elements[3]. It is therefore necessary to know the bulk concentration of the solute, and the solution refractive index as a function of the solute concentration. The relationship between the Rayleigh excess factor and the physical characteristics of the scattering molecule can thereby be given as

$$Kc/R_\theta = 1/M_w P(\theta) + 2A_2c + 3A_3c^2 + \ldots \tag{2}$$

where:

$K =$	the "polymer constant" $= (2\pi n^2/\lambda^4 N)\,(dn/dc)^2\,(1+\cos^2\theta)$;
$c =$	the concentration of solution;
$n =$	the refractive index of solution;
$\lambda =$	the wavelength *in vacuo*;
$N =$	Avogadro's number;
$dn/dc =$	the specific refractive index increment (*i.e.* the change in refractive index of the solution as a function of solute concentration;
$A_2, A_3 =$	virial coefficients;
$R_\theta =$	the Rayleigh excess factor $= R_\theta$(solution) - R_θ(solvent);

$P(\theta)$ = the form factor which depends on shape and size of the scattering molecules.

At limiting conditions where θ approaches zero angle, $P(\theta)$ approaches unity irrespective of size or shape of the scattering species. At low angles and concentrations, eq. (2) can therefore be rewritten as:

$$Kc/R_\theta = 1/M_w + 2A_2c \qquad (3)$$

3. DATA COLLECTION AND PROCESSING

Fig. 1 shows the design of the KMX-6 LALLS photometer and the block diagram in Fig. 2 shows the assembly of a GPC/LALLS system. The LALLS detector is connected before the concentration detector to prevent large backpressure on the concentration detector. The sample is injected as in normal GPC and two traces are obtained (see Fig. 3).

It is the Rayleigh excess factor R_θ which is of interest in light scattering and this is determined as follows

$$R_\theta = (G_\theta/G_0)\,(D/\sigma' l')$$

where G_θ and G_0 are the photomultiplier signals for the scattered and incident light respectively; D is the transmittance of the measuring attenuator; σ' is the solid angle through which the scattered light is collected; and l' is the length of the scattering volume parallel to the incident beam. The latter two parameters are geometrical constants and D can be optically determined, thus, R_θ is an absolute value. At low angles and solute concentration the relationship between R_θ and molecular weight is

$$Kc/R_\theta = 1/M_w + 2A_2c, \text{ or } M_w = 1/[(Kc/R_\theta) - 2A_2c]$$

Thus, if A_2 is known, M_w can be determined from a single measurement of R_θ which enables the continuous monitoring of GPC effluents. So from the GPC traces shown, the molecular weight for uniform intervals on the GPC elution curve is calculated from

$$Kc_i/R_{\theta,i} = 1/M_{w,i} + 2A_2c_i$$

The concentration c_i is most usually calculated from

$$c_i = (m\ x_i) / V_i\ x_i$$

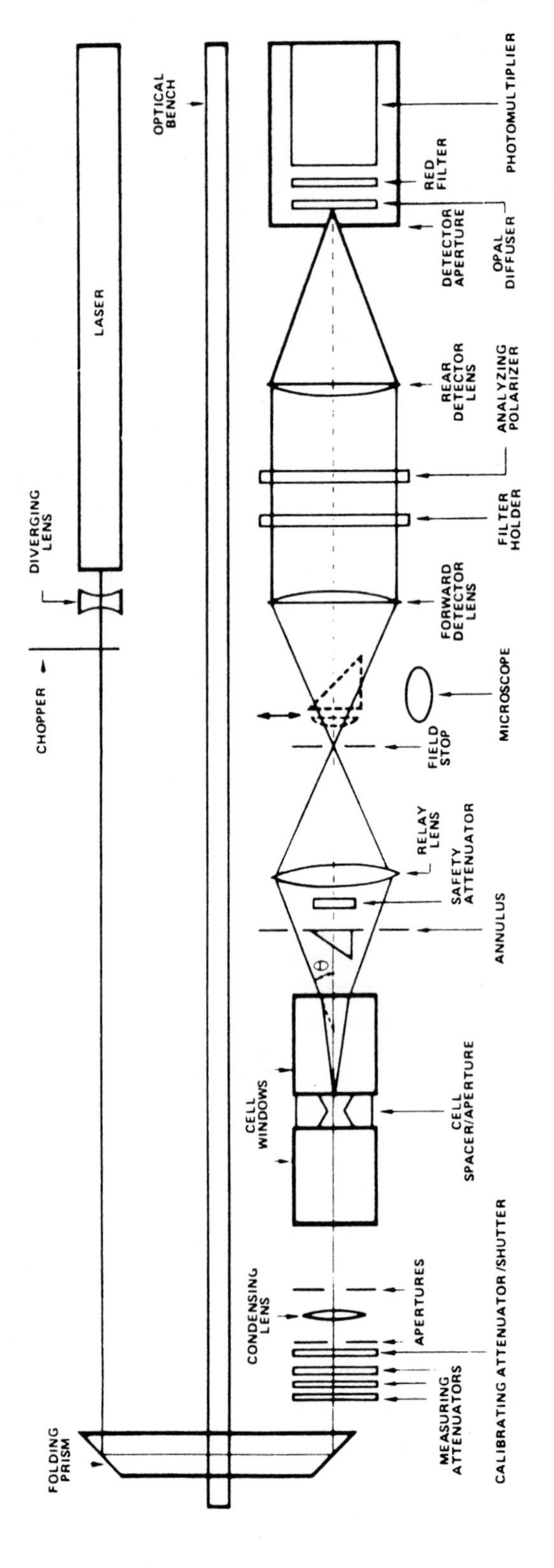

Figure 1. Optical diagram of a LALLS detector (KMX-6).

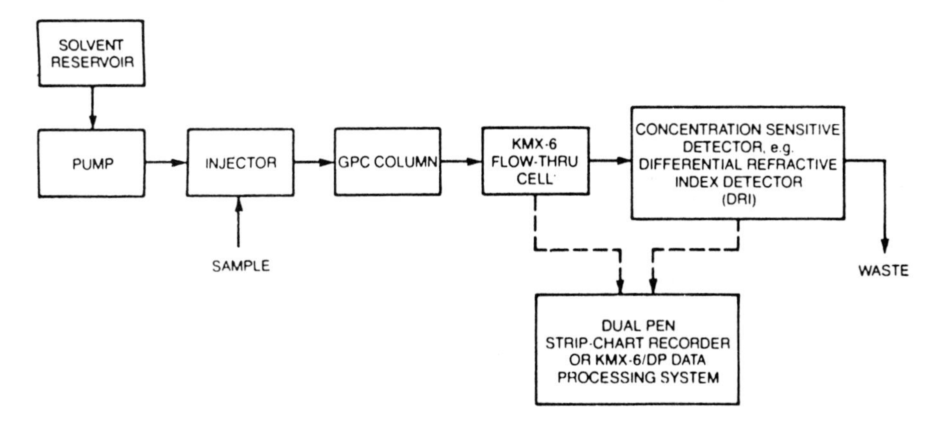

Figure 2. Block diagram of a LALLS (KMX-6)/GPC system.

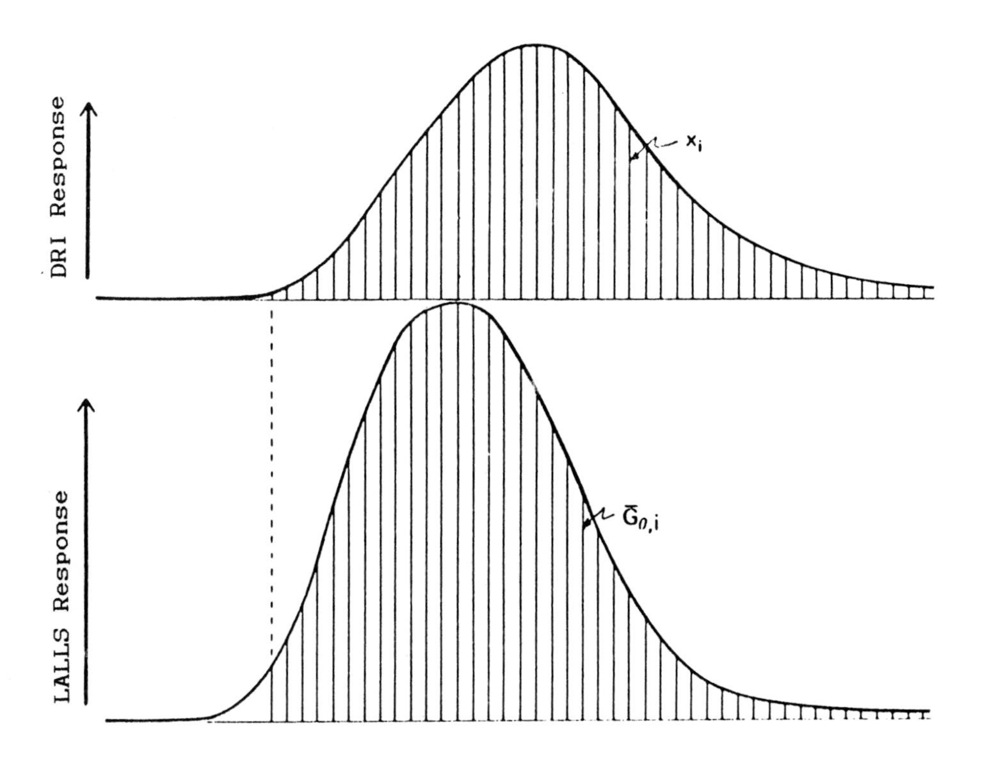

Figure 3. Elution curves generated by the differential refractive index detector and the LALLS detector.

Once M_i has been calculated at these intervals from the elution curve, the molecular weight averages are calculated as usual

$$M_n = \Sigma c_i / \Sigma (c_i/M_i)$$

$$M_w = \Sigma c_i M_i / \Sigma c_i$$

$$M_z = \Sigma c_i M_i^2 / \Sigma c_i M_i$$

These data can then be further manipulated to give cumulative and differential molecular weight distributions.

4. APPLICATIONS

Polysaccharides

Polysaccharides are a group of molecules which are often difficult to characterize due to their tendency to form aggregates and/or display a much larger than normal hydrodynamic volume in solution. It is, therefore, difficult to determine molar masses of these molecules accurately and calibrated methods are frequently insufficient.

Carrageenans Carrageenans are an important group of food additives due to their gelling properties. They are extracted from red seaweed and are composed of alternating α (1 - 3) and β (1 - 4) linked D-galactosyl residues. They are classified into various groups of which the three main ones are κ-, ι-, and λ-carrageenan (see Fig. 4), the difference is due to the occurrence of ester sulphate or anhydride formation in the 4-linked residue.

One of the first papers describing a GPC/LALLS investigation of absolute molecular weight distributions of carrageenans was published by Lecacheux *et al.*[4] These authors investigated the molecular weight distributions of carrageenans, because at that time the polymer fraction with molecular weight below 20,000 was thought to be responsible for a small toxic effect. Since carrageenans are extensively used in the food industry investigation of this problem was of considerable importance.

ι- and κ-carrageenans were dissolved in 0.1 M LiCl and the samples were run at 60°C. All food grade samples were found to have a low molecular weight tail and approximately 2% of the bulk molecular weight was found to be below 20,000. The method was found to be successful for the molecular weight monitoring of carrageenans.

A great deal of work has been carried out since then to investigate the behaviour and structural conformation of carrageenans and for some of this work LALLS and GPC/LALLS have been used. Rochas and Landry reported on the investigation into the presence of helical conformation in kappa carrageenan[5]. They used static LALLS for molecular weight determination and GPC/LALLS for the monitoring of the change in molecular weight due to the conformational change.

Rochas and Lahaye reported on the molecular weight distribution and weight average molecular weight of agarose and agarose -type polysaccharides by GPC/LALLS[6]. In conjunction with viscosity measurements they were able to relate molecular weight, chemical structures and rheological properties of agaroses.

GPC/LALLS has also been used in a similar manner for determining the molecular weight distribution of guar, xanthan, inulin and sinistrin[7,8,9].

Starch and its Components Starch is used extensively for chemical and feedstock applications and is an important source of high fructose syrups.

It is difficult to degrade starch completely because it gelatinizes on dilution with water which is thought to be due to retrogradation of one of its components (amylose). Amylose is a linear polymer consisting of α-D-glucopyranose units linked by (1 - 4) glucosidic linkages and has degrees of polymerization from hundreds to several thousand glucose residues. It makes up 20% - 30% of the carbohydrate content of most starches.

The other component, amylopectin, contains a backbone of α (1 - 4) linkages but also contains (1 - 6) linkages to form a highly branched structure and has degrees of polymerization between several thousand and several million. The structure and properties of these two components play an important part in the hydrolysis of starch. This is important from an industrial point of view, *i.e.* maximization of hydrolysis, and also from a purely research point of view because the true nature and structure of starch has so far eluded scientists.

In 1984 Hizukuri and Takagi reported on the use of GPC/LALLS for the molecular weight distributions of lily amylose and also compared degrees of polymerization of amyloses from different sources[10,11]. They did not discover any signs of aggregation but found that the molecular weight of their samples decreased with progressive determinations, a result attributed to amylase contamination. However, they found weight average degrees of polymerization between 3,200 and 6,700 units, Tapioca amylose giving the highest values and kuzu amylose the lowest. In order to verify the accuracy of the GPC/LALLS system, it was first calibrated with Pullulan standards of known molecular weight and gave very good results. Pullulan standards were used for calibration because of their similar structure to amylose (*i.e.* it is a linear α-D-glucan with (1 - 4) and (1 - 6) linkages) and is supposed to have the same dn/dc as amylose.

A study by Hong *et al.*[12] described the use of GPC/LALLS for monitoring amylase hydrolysis of starch as a function of time[12]. Chromatograms showed bimodal distributions of the hydrolysates, which was rather surprising since it is known that endoamylases act in a random-like manner. The authors attributed this effect to polymer-polymer interactions causing macromolecular associations. The second polysaccharide peak was due to glucose, maltose and oligomers. They detected precipitation material (*i.e.* solutions went cloudy) during all their hydrolyses. This material was found to be similar to the lower molecular weight fraction of the first peak from the bimodal distribution. They concluded that this material was similar to non-soluble amylose, but further investigations are still necessary.

In a further paper, Yu and Rollings used GPC/LALLS to determine molecular weight, molecular weight distributions and degree of branching of amylose, amylopectin, starch and glycogen in 0.5 M NaOH solution[13]. The method gave weight average molecular weights in fairly good agreement with manufacturer's data. In order to establish the universality of the method a calibration curve was constructed which was found to exhibit the required linear relationship. This also confirmed the applicability of GPC/LALLS for the indirect determination of Mark-Houwink coefficients, as

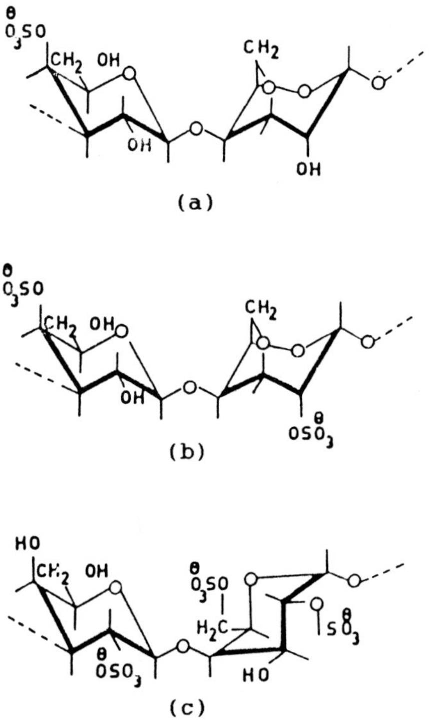

Figure 4. Primary structure of carrageenans: (a) κ-carrageenan, (b) ι-carrageenan, (c) λ-carrageenan.

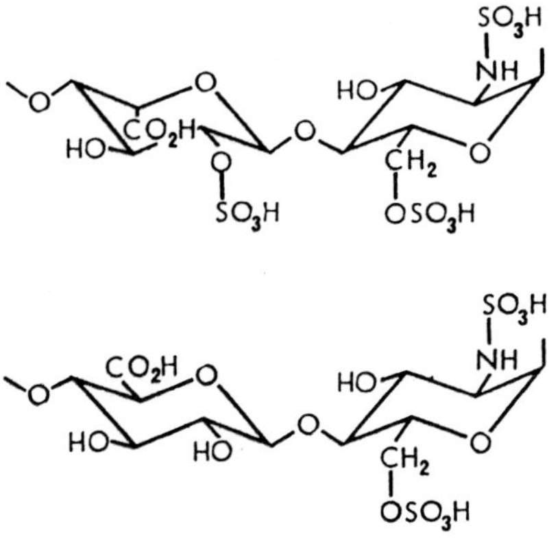

Figure 5. Primary structure of heparin.

previously reported by Cael *et al.*[14] Determination of branching parameters also confirmed expectations. Only in the case of starch, which is a mixed polymer, is the procedure inappropriate, due to the fact that the sample cannot be considered monodisperse. Without knowledge of the relationship between the sample's intrinsic viscosity and molecular parameters, the branching parameters of a mixed polymer cannot be determined directly.

Due to the difficulties associated with starch and its components regarding dissolution and retrogradation, the scientific community is still divided in its opinion on the validity of these results and research in this area is continuing.

Heparin Heparin is an acidic mucopolysaccharide consisting of repeating units of hexuronic acid and glucosamine (see Fig. 5). It acts as an anticoagulant and is, therefore, important in clinical chemistry.

Heparin molar masses cannot be determined accurately by the usual calibrated methods unless the system is standardized with heparins of identical animal and tissue source. Hennink *et al.* demonstrated that GPC/LALLS was a reliable and convenient method for the characterization of heparins provided that the polyelectrolyte effect is suppressed by a suitable method (in this case dialysis)[15]. Their study also confirmed that, neither calibration of the GPC system with standard dextrans, nor universal calibration could be applied to heparins. Results were checked by osmometry (to give M_n) and by analytical ultracentrifugation (to give M_w and M_z).

Proteins

Most proteins behave fairly similarly in solution and their molar masses can often be determined using stand-alone GPC. However, some proteins (*e.g.* glycoproteins) deviate from this standard behaviour and determination of their molar masses requires absolute methods. In addition, when molar masses of the individual components of protein mixtures are to be determined, a second absolute detector is often advantageous.

The molecular weight of amylase had been reported to vary between 49,000 and 54,000. The variations were attributed to possible changes in molecular species during preparation and measurement, for example proteolysis by proteases which could be present as contaminants. Takagi used GPC/LALLS to determine absolute M_W of this protein using a silica column to separate the material, and found the molar mass to be between 50,500 and 51,000 which agrees well with sedimentation studies[16].

The technique was also found reliable and accurate for the molecular weight determination of glycoproteins[17]. In 1983 Maezawa *et al.* reported on the molecular weight determination of membrane proteins in the presence of a non-ionic surfactant (octaethylene glycol n-dodecyl ether)[18]. They used porin, a γ-receptor protein from the *E.coli* outer membrane as test material and it was confirmed as trimeric under the conditions in which the analysis was performed. The GPC/LALLS technique was also successful in estimating the molecular weight of the oligomeric membrane protein. However, it was found that the micelle produced by the surfactant eluted very closely to the protein, so that correct molecular weights could not be given. For porin and the γ-receptor protein, molecular weights were found to be 113,900 and 143,000 respectively, which agreed well with values from other techniques.

In 1982, Bindels *et al.* published results achieved from bovine eye lens protein by GPC/LALLS. Cortical and nuclear calf lens extracts were separated into 13 crystalline fractions. Eight different β-crystalline fractions were found with almost all oligomeric structures from dimers to aggregates larger than dodecamers. They also managed to identify the structure by SDS gel electrophoresis and isoelectric focusing in the presence of urea which indicates that the technique also has a semi-preparative application.

5. CONCLUSIONS

Although only a limited number of examples have been discussed it has been shown that GPC/LALLS is a convenient and reliable method for the determination of molecular weight averages and distributions. It is far superior to calibrated methods, including universal calibration, and for materials which cannot be determined by calibrated methods it gives results of comparable accuracy to those obtained by other more time consuming techniques.

REFERENCES

1 W. Kaye, A. Havlik and J.B. McDaniel, *Polym. Lett.*, 1971, *9*, 695.
2 A.C. Ouano, *J. Chromatogr.*, 1976, *118*, 303.
3 K. Stacey, 'Light Scattering in Physical Chemistry', Academic Press, New York, 1956.
4 D. Lecacheux, R. Panaras, G. Brigand and G. Martin, *Carbohydr. Polym.*, 1985, *5*, 423.
5 C. Rochas and S. Landry, *Carbohydr. Polym.*, 1987, *7*, 435.
6 C. Rochas and M. Lahaye, *Carbohydr. Polym.*, 1989, *10*, 289.
7 B.R. Vijayendran and T. Bone, *Carbohydr. Polym.*, 1984, *4*, 299.
8 F. Lambert, M. Milas and M. Rinaudo, *Polym. Bull.*, 1982, *7*, 185.
9 W.-D. Eigner, P. Abuja, R.H.F. Beck and W. Praznik, *Carbohydr. Res.*, 1988, *180*, 87.
10 T. Takagi and S. Hizukuri, *J. Biochem.*, 1984, *95*, 1459.
11 S. Hizukuri and T. Takagi, *Carbohydr. Res.*, 1984, *134*, 1.
12 H.C. Hong, L.-P. Yu and J.E. Rollings, *Ann. N.Y. Acad.*,1987, *501*, 426.
13 L.-P. Yu and J.E. Rollings, *J. Appl. Polym. Sci.*, 1987, *33*, 1909.
14 J.J. Cael, D.J. Cietek and F.J. Kolpak, *J. Appl. Polym. Sci., Appl. Polym. Symp.*, 1983, *37*, 509.
15 W.E. Hennink, J.W.A. van den Berg and J. Feijen, *Thromb. Res.*, 1987, *45*, 463.
16 T. Takagi, *J. Biochem.*, 1981, *89*, 363.
17 S. Maezawa and T. Takagi, *J. Chromatogr.*, 1983, *280*, 124.
18 S. Maezawa, Y. Hayashi, T. Nakal, J. Ishii, K. Kameyama and T.Takagi, *Biochim. Biophys. Acta*, 1983, *747*, 291.
19 J.G. Bindels, B.M. deMan and H.J. Hoenders, *J. Chromatogr.*, 1982, *252*, 255.

3

Combined Differential Light Scattering with Various Liquid Chromatography Separation Techniques

By Philip J. Wyatt

WYATT TECHNOLOGY CORPORATION, SANTA BARBARA, CALIFORNIA 93130-3003, U.S.A.

1. INTRODUCTION

The combination of light scattering measurements with various particle/molecular separation techniques often permits an unparalleled characterization of the separated particles. In a sense, this is but an application of the so-called "inverse scattering" problem[1,2,3], *i.e.* from measurements of the light scattering properties of the particles, one can (may) determine some of their physical properties such as size and refractive index. Without separation, only average particle characteristics might be derivable while the distributions of the associated characteristics cannot be determined, in general. Yet, this inversion procedure is fraught with difficulties and the results are often inconclusive, particularly for particles of a refractive index significantly greater than that of the medium in which they are measured. In this latter case, only very special classes of homogeneous, spherical particles may be "inverted" successfully[4]. Fortunately, there is a very broad range of particles whose characterizations by classical light scattering techniques is relatively straightforward, even without separation. However, it is the purpose of this paper to concentrate on particles (molecules) that may be separated before the light scattering measurements are made.

The light scattering measurements to be discussed concern the detection of scattered light *intensity* which, in itself, implies the measurement of scattered light flux as a function of scattering angle with respect to the direction of an incident collimated beam. In keeping with the physics definition of the measurement of differential scattering cross sections, this intensity measurement is generally called differential light scattering[2] (DLS), total intensity light scattering (TILS - this Volume, or (rarely) multi-angle laser light scattering. Such incident light is generally chosen to be monochromatic and vertically polarized with respect to the plane in which the angular variation of the scattered flux is measured. It is also assumed that the frequency of the scattered light is essentially unchanged from that of the incident light. This does *not* preclude the possibility that some of the incident light energy may be absorbed by the scattering particles. Indeed, these absorptive properties of the scattering particles may be derived[5] under special conditions. Measurements concerned *only* with scattered light undergoing slight frequency shifts due to the relative *motion* of the scattering particles are not considered here, but have been discussed in other papers in this Volume.

In the next section, some important elements of the theory are summarized with particular emphasis on the Rayleigh-Gans-Debye (RGD) approximation which forms the basis by which measurements of the scattering properties of dissolved molecules are characterized. Section 3 of this paper presents several examples of particle/molecular separations and the associated light scattering measurements by which such particles are analyzed. The final section of this paper contains a brief discussion of the future promise for such separation methods combined with light scattering.

2. THEORY

In its most general form, the scattering of light by a *single* particle of arbitrary shape and dielectric structure may be described by a suitable application of Maxwell's equations. That is, given the shape, structure, and orientation of the particle with respect to the incident plane wave illumination, one can, in principle, *predict* the scattered light intensity. For all but the simplest of shapes and structures, this becomes a formidable task since it requires a detailed evaluation of so-called "boundary conditions" at every part of the particle surface, as well as at any interior regions of dielective discontinuity. Suffice to remark, that for spherical[6] or spherically symmetric[7] particles, this procedure is straightforward. Only with the advent of digital computers did such calculations of scattered light intensity from more complex particles become practicable.

The ability to predict how a single particle scatters light (and, by superposition, how an ensemble of particles will scatter light) is a far cry from being able to *invert* the process. In general, these inversions require the ultimate of separation, *viz* measurement of single particles, one at a time. If some *a priori* characteristics of the particles are known, then one can reduce the inversion process to a "least squares" fit to a parameterized "model." For example, certain types of bacterial spores[8] have a spherically symmetric structure (determined from electron microscopy) on which basis they may be characterized by four parameters: radius, coat thickness, and refractive index of core and coat, respectively. The "best fit" parameters may then be derived, but the process is difficult[9] and time consuming and the resultant parameters may not be unique. Were the spores measured in a suspension, then the deduction of the average values of these four parameters would be even more formidable since a *distribution* of each parameter also must be derived. There is a need, therefore, to be able to obtain suspensions of monodisperse particles if such inversion procedures are to be successful. There are currently under development separation techniques to achieve these goals. They will be described in the next section of this paper.

A significant simplification of the formal electromagnetic theory was developed by Rayleigh and others that permits an inversion of the measured light scattering intensities directly to yield particle characteristics for a broad and important class of *particles in suspension*. This Rayleigh-Gans-Debye (RGD) theory involves an approximation more familiarly known to physicists as the "Born approximation" and has broad application, as we shall see. There are two fundamental requirements for application of the RGD theory:

$$|m-1| \ll 1 \quad (1)$$

and

$$2ka|m-1| \ll 1 \quad (2)$$

where $m = n/n_0$, the ratio of the particle refractive index n to the refractive index n_0 of the suspending medium (solvent). The propagation constant $k = 2\pi n_0/\lambda_0$, a is the mean particle radius, and λ_0 is the wavelength of the incident light in vacuum. Eq. (1) states that the amount of light that is scattered by the object (relative to the amount incident) is negligible, while eq. (2) states that the total phase change of a wave passing through the particle is negligible relative to the wave passing only through the pure solvent.

On the basis of eqs. (1) and (2), the scattering characteristics of suspended particles are easily predicted. Each small volume element of the particle will scatter independently of, and unaffected by, any other volume element of the same particle. The excess scattering, $R(\theta)$, into the direction between θ and $\theta + d\theta$ for the total particle may be calculated by adding up the individual contributions from each volume (mass) element and including the difference in their phase based on their spatial differences in the time varying field of the incident light. The excess scattering $R(\theta)$ (often called the Rayleigh *excess* scattering ratio, or just Rayleigh ratio) is defined by

$$R(\theta) = f \frac{(I_\theta - I_{s\theta})}{I_0} \tag{3}$$

where I_θ is the measured scattered intensity from the particle (or suspension of particles) at the angle θ into a solid angle $d\Omega_\theta$ (= $\sin\theta \, d\theta \, d\phi$) subtended by a detector at the angle θ. The corresponding scattering from the solvent alone is $I_{s\theta}$ and I_0 is the incident light flux per square unit of illuminated area. The term f corresponds to the absolute calibration coefficient which must be corrected for the geometry of the particular scattering cell and detector configuration. It is also a function of solvent and cell refractive indices.

The RGD approximation has been applied to a variety of particles on which basis their physical characteristics often may be determined. Although we shall not discuss them here, application of the approximation to even particles as large as bacterial cells has been very successful for a variety of bacterial shapes and sizes[10,11]. The particular field on which we shall focus our attention for most of this paper is the application of the RGD approximation for the characterization of macromolecules in suspension.

The mean refractive index of solvated molecules is generally almost equal to the solvent itself and such particles are perhaps the best examples for application of the RGD approximation. A variety of molecular structures have been studied by application of the RGD approximation. At very low molecular concentrations, the excess Rayleigh scattering ratio may be shown[12,13] given by the expression

$$\frac{R(\theta)}{K^*c} = M_w P(\theta) [1 - 2A_2 M_w P(\theta) c] \tag{4}$$

to order c^2. The weight average molecular weight for the suspension is M_w, c is the concentration, A_2 is the second virial coefficient and $P(\theta)$ is the scattering function which depends on the molecular configuration. For vertically polarised incident light, the term $K^* = 4\pi^2 n_0^2 (dn/dc)^2/(N_A \lambda_0^4)$ where N_A is Avogadro's number, and dn/dc the refractive index increment associated with the dissolved molecules. With respect to a single molecule, Debye[14] and others[15] have shown in general that $P(\theta)$ may be represented by the series

$$P(\theta) = 1 - (h^2/3) \langle r_g^2 \rangle + a_4 h^4 - a_6 h^6 + \ldots \qquad (5)$$

where

$$\langle r_g^2 \rangle = \frac{1}{M} \int r^2 \, dM \qquad (6)$$

and $h^2 = (2k \sin \theta/2)^2$; the integration being taken with respect to the molecule's centre of *gravity*. The higher order terms depend on the molecular shape. Defining

$$\langle r^{2m} \rangle = \int_0^\infty r^{2m} W(r) \, 4\pi r^2 \, dr \qquad (7)$$

where W(r) is the distribution function[15] of the distance r between any two scattering centres, the higher order coefficients of eq. (5) become

$$a_{2m} = \frac{1}{(2m+1)!} \langle r^{2m} \rangle \qquad (8)$$

Note that the second term of eq. (5) is *independent* of the molecular configuration and involves only the z-average square radius (or its misnomer "square radius of gyration"). This is often referred to as the Guinier relation[16]. In the limit of very small h^2 (or small scattering angle) where eq. (5) may be represented by the first two terms and at sufficiently low concentrations such that terms proportional to $2A_2M_wc$ may be neglected, the *slope* of $R(\theta)$ near $\theta = 0$ yields an immediate value for this "radius of gyration."

Returning to eq. (4), we see that from measurements of $R(\theta)$ at several angles and for varying concentrations, one can, in principle, obtain a best fit to M_w, $\langle r_g^2 \rangle$ and A_2. In a chromatographic separation, however, only a *single* concentration value is available at each eluting fraction and A_2 cannot be determined from a single injection. In eq. (4), the term $2A_2M_wP(\theta)c$

is generally very small compared to 1. Thus A_2 may be measured either off-line as an average over the entire eluting sample, or on-line at each eluting fraction by measuring the separated solution at two different injection concentrations. Details of the means by which the molecular parameters are derived from the experimental data may be found in reference 13.

3. MEASUREMENTS

Fig. 1 presents an exploded view of a refraction flow cell[17] by which means the scattered light intensity from an eluting sample is measured in the DAWN®-F detector. A separation system such as a GPC column is attached at the inlet manifold, a secondary concentration sensitive detector connected to the outlet manifold, and the glass cell surrounded by an array of collimated detectors. Fig. 2 shows the refraction geometry with respect to a typical detector located at the angle θ' with respect to the bore of the cell. The laser beam illumination passes through the bore in a direction parallel to the flow by means of windows in the manifolds. Typically, the bore diameter is 1.25mm, the detectors are hybrid transimpedence photodiodes, and the glass is selected to have a higher refractive index than the eluting mobile phase. The total flow cell volume between connectors is about 60μl.

A typical chromatographic configuration is shown in Fig. 3 for use with gel permeation columns. The light scattering photometer/detector is placed between the columns and a refractive index detector used to monitor molecular concentration. Other standard chromatography components include pump, injector, pulse dampers, and an on-line degasser. The entire system is computer-controlled to provide for auto-injection, storage of collected data (typically, 50 to 100 measurements at each slice for each of 15 detectors), and data processing. Obviously, such a configuration has many permutations. For example, the RI detector may be replaced by a UV detector or even viscometer. The GPC columns may be replaced by reverse phase columns, TREF (temperature rising elution fractionation) columns, field flow fractionation channels, or even eliminated. For the remainder of this section, we shall present a broad variety of light scattering measurements based on different separation techniques. Except where stated otherwise, all measurements reported here are made with vertically polarised incident light at the He-Ne laser wavelength of 632.8 nm. Significant

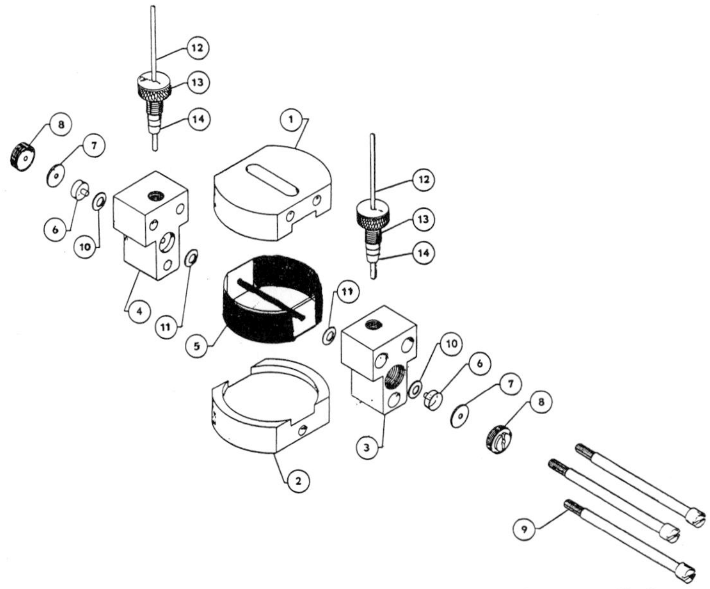

Figure 1. Exploded view of the refraction flow cell for making light scattering measurements from a chromatographic elution.

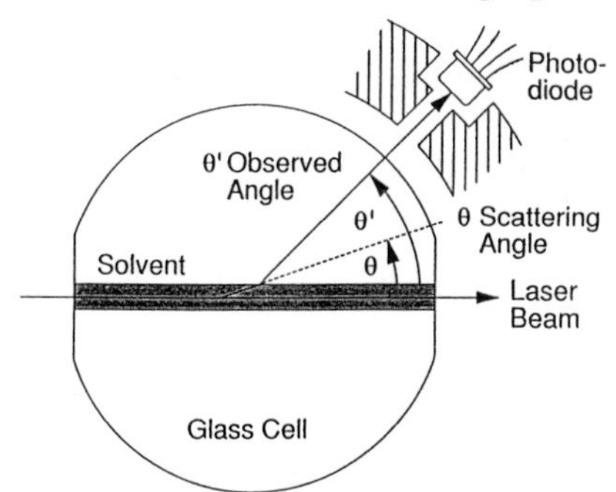

Figure 2. Refraction geometry with respect to a fixed detector. The scattering angle θ and detector angle θ' are related by $n_g\cos\theta' = n_s\cos\theta$, where n_g and n_s are the refractive indices of the glass and solvent, respectively.

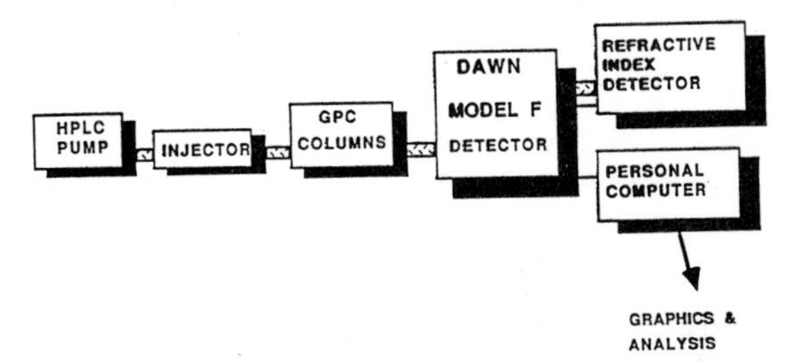

Figure 3. Typical configuration in a GPC mode of a chromatographic system used with a light scattering detector.

signal-to-noise ratio improvements[18] have been achieved using an Argon Ion laser operating at 488nm.

Once the data have been collected, they may be processed to yield many important parameters of the separated particles. For the case of molecules, this includes number-, weight-, and z-average molecular weights and the corresponding moments of the mean square radius. The differential and cumulative distributions of weight average molecular weights and corresponding distributions of root mean square radii may be derived also for molecules of sufficient size. By measuring the slope of log M *versus* log r_g plots, molecular conformation may be derived as well. For *unseparated* molecules, one may derive only the weight average molecular weight, z-average square radius, and the second virial coefficient. For individual particles of spherical shape measured one at a time, one may derive the particle refractive index and radius, as well as distributions thereof[19]. For non-spherical particles, extraction of physical parameters is more difficult, but includes size and structure (from depolarization measurements).

For homopolymers, some off-line measurements are required, as is evident from the parameters of eq. (4). These include the refractive index n_0 of the solvent and the refractive index increment, dn/dc, at the measurement wavelength, λ_0. For copolymers, a more complex strategy is required since dn/dc may vary at each elution fraction. If the copolymer is homogeneous in the distributions of its component homopolymer fractions, then dn/dc will be constant over the entire size distribution and the measurement of an average dn/dc off-line will suffice. Another simple case corresponds to copolymers made from homopolymers whose dn/dc values are nearly equal. In this latter case, again, an average dn/dc value may be used. For the examples in this paper, we shall restrict ourselves to examples whose dn/dc values are generally constant over the band. Note, that the use of an RI detector to monitor *concentration* requires a constant dn/dc since the detector response (output signal, ΔV) is directly proportional to dn/dc, *i.e.*

$$\alpha\,\Delta V = \Delta n = \frac{\Delta n}{\Delta c}\Delta c = \frac{dn}{dc}\Delta c \tag{9}$$

where α is a calibration constant. Once the molecular weight of a polymer falls below about 10,000, dn/dc may change significantly.

Fig. 4 presents the chromatograms of a separated pullulan sample of nominal molecular weight 800K daltons. Shown in the figure is the RI signal and the 90° scattered intensity as a function of elution volume. Here, the aqueous mobile phase separation was achieved using GPC columns (ultrahydrogel columns, Waters). Fig. 5 shows a similar result for a mixture of three polystyrene standards in tetrahydrofuran (THF). Their nominal molecular weights were 550K, 207K, and 47K.

A three-dimensional plot of the light scattering chromatograms from the three polystyrene standards mixture as a function of elution volume and scattering angle is shown in Fig. 6. Each different scattering angle is shown as a separate chromatogram with the scattering angle increasing towards the back of the figure. At the rear of the figure is shown the corresponding RI signal aligned in volume to compensate for the volume displacement between the light scattering detector and the RI detector (see Fig. 3).

A gradient phase measurement is shown in Fig. 7 of the Rayleigh excess factors at 90° scattering for a mixture of BSA (200μg), PK (80μg), and GMP(4μg). Here, a UV detector (280nm) was used and its output is superimposed on the 90° light scattering curve. The mobile phase was a linear NaCl gradient from 0 to 0.5M.

Fig. 8 presents 90° light scattering and UV chromatograms for a polyacrylamide sample separated at the University of Utah (M. Benincasa) by means of a cross flow field flow fractionation[20] channel. The flow rate was only 0.17ml/min while the total cross flow was 0.38ml/min. The corresponding smoothed data are shown in Fig. 9.

Fig. 10 shows the scattering at 90° from a mixture of three polystyrene latex particle samples as a function of retention volume. Their nominal sizes were 150 nm, 251 nm and 500 nm. They were separated with a du Pont Sedimentation Field Flow Fractionation (SFFF) device. Note the relative heights of the peaks from these three equal injected mass samples. Similar data, obtained at 21°, are illustrated in Fig. 11. The separations are significant, since between samples the baseline is nearly reached even between the higher masses.

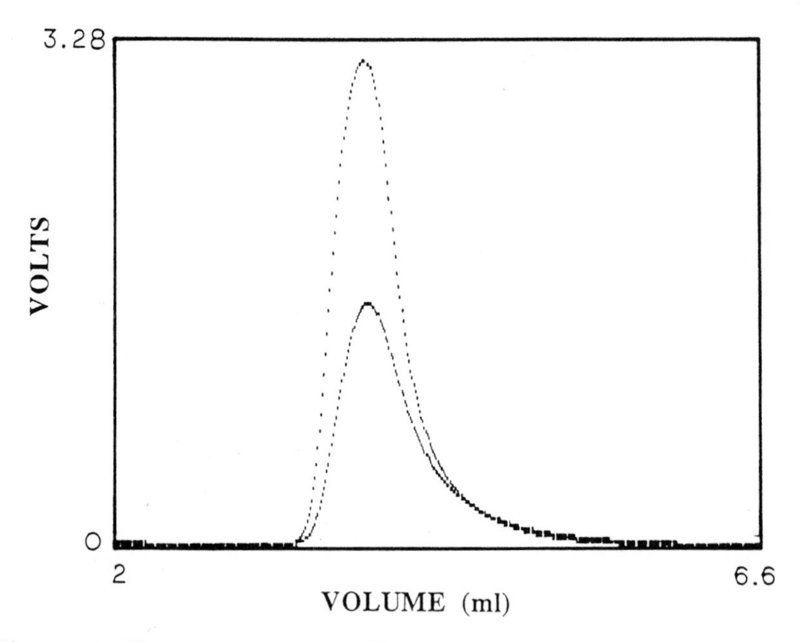

Figure 4. Chromatogram of pullulan standard (Shodex Standard P-82 Grade P-800). The figure shows the output of the DAWN®-F 90° detector and the differential refractometer signal (Waters 410) against elution volume. The sample was run through two Waters μ-bondagel E-linear columns at a flow rate of 0.3ml/min using 0.15M NaCl as the mobile phase. The higher intensity signal is from the DAWN®-F. Data are unsmoothed.

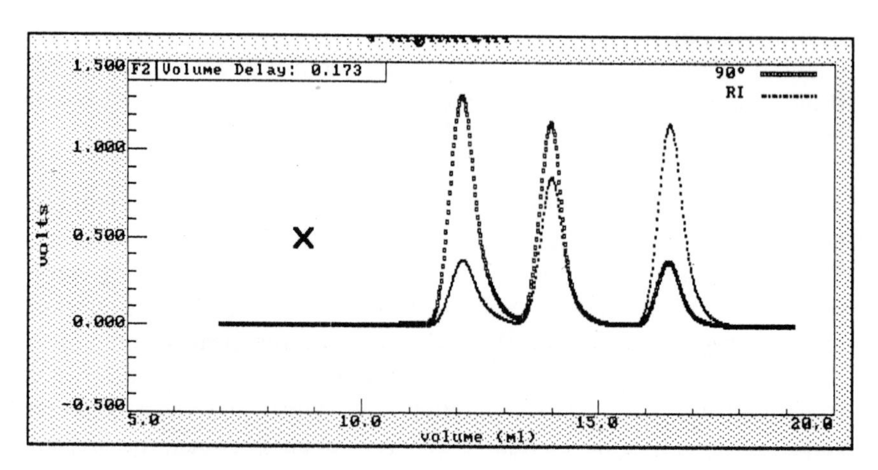

Figure 5. Chromatogram of three polystyrene latex standards (560K, 207K and 47K) in THF. Data are unsmoothed (PSS columns).

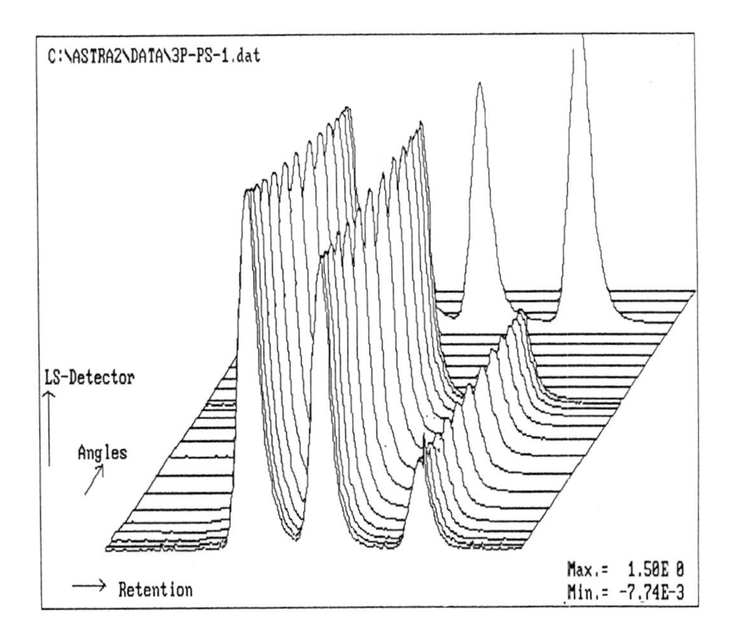

Figure 6. Three-dimensional plot of light scattering signals at different scattering angles from three polystyrene standards with corresponding RI signal at rear of figure.

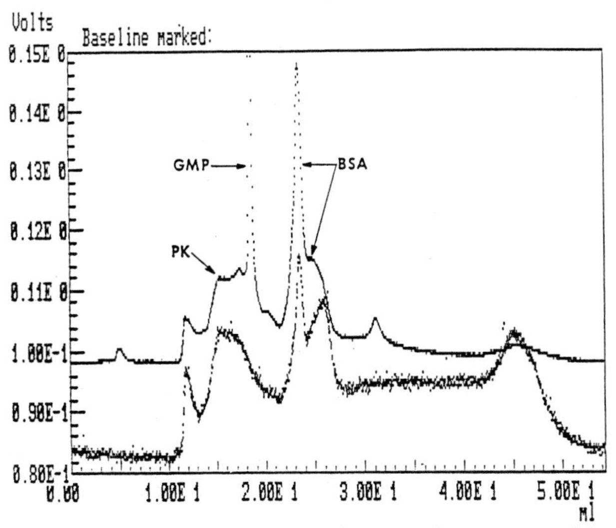

Figure 7. Gradient phase measurement of pyruvate kinase (PK), bovine serum albumin (BSA), and guanosine monophosphate (GMP) at injection masses of 80μg, 100μg, and 4μg, respectively. The top curve is the UV (280nm) trace, the lower corresponds to the 90° light scattering signal. NaCl gradient from 0 to 0.5M; tris buffer 0.05M, pH 8.5; TSK Spherogel DEAE-5 PW column.

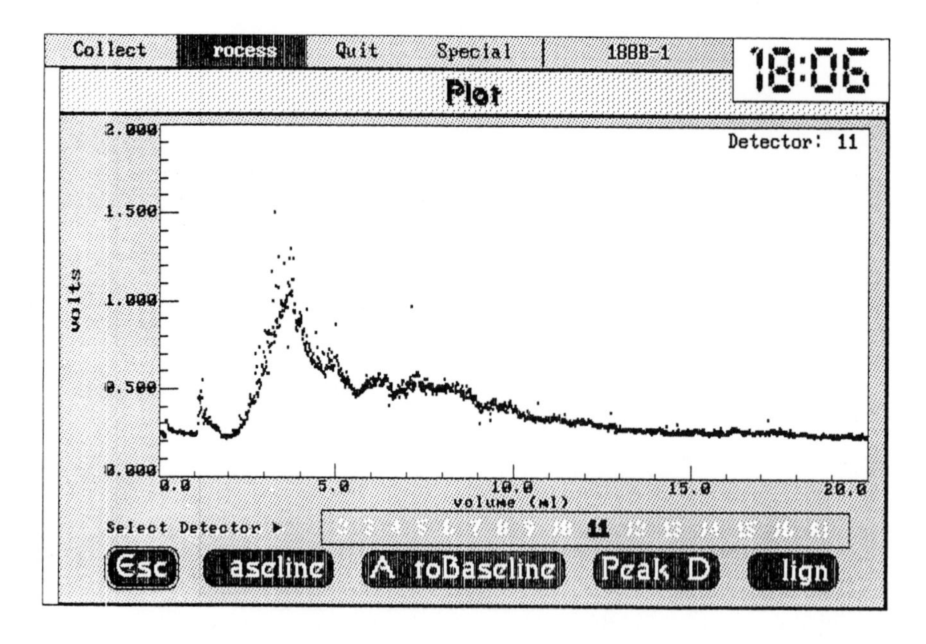

Figure 8. Polyacrylamide sample scattering at 90°, from separation by cross flow field flow fractionation. Unsmoothed data.

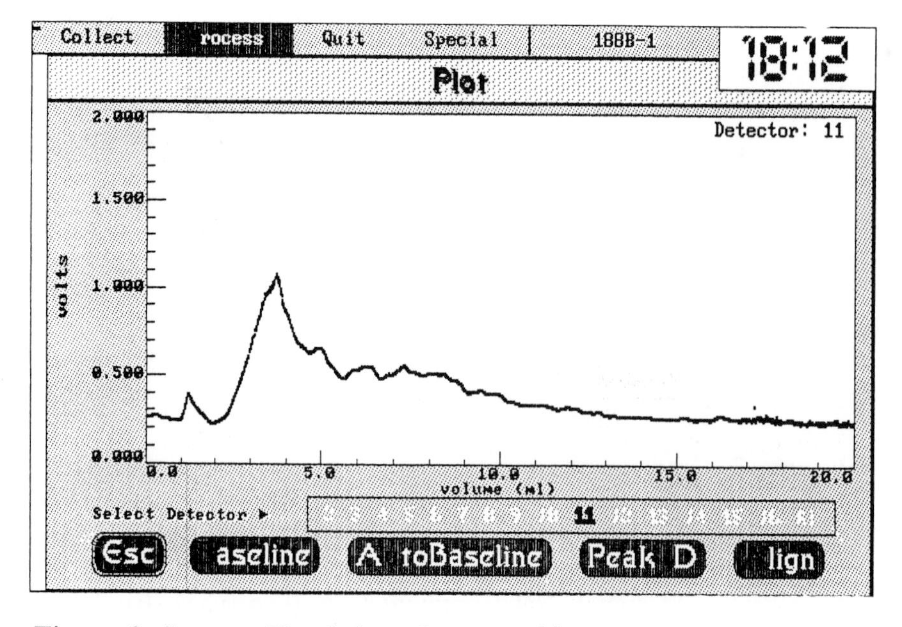

Figure 9. Same as Fig. 8, but after smoothing.

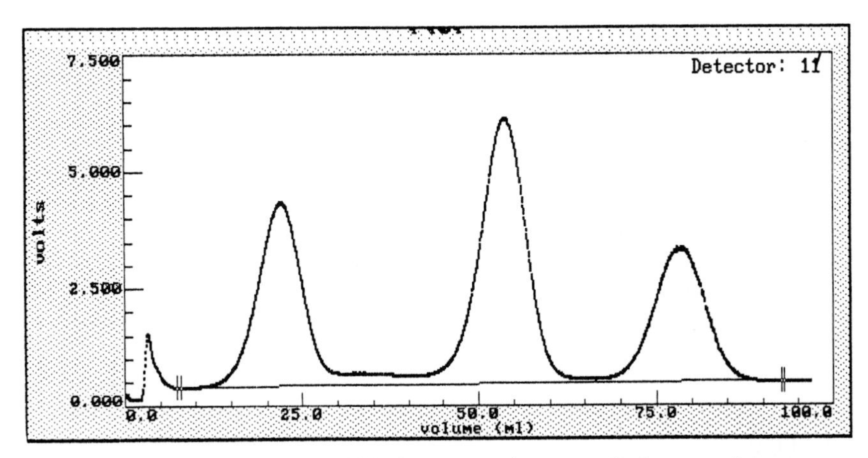

Figure 10. Scattering at 90° from a mixture of three polystyrene latex samples of nominal diameters 150nm, 261nm, and 500nm, respectively. Du Pont SF³ system.

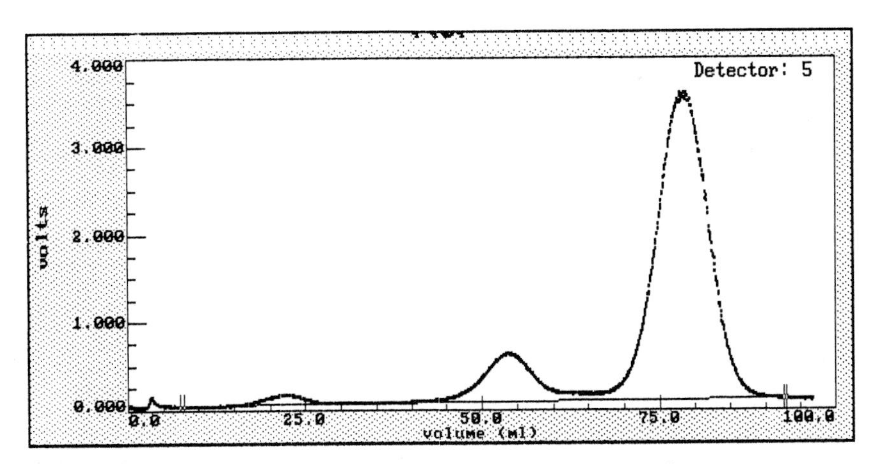

Figure 11. Same as Fig. 10, but at 21° scattering angle.

Fig. 12 presents a three-dimensional plot of a similar set of latex particles (232 nm, 330 nm, and 426 nm) fractionated using a somewhat smaller force field of an FF Fractionation Inc. SFFF device. Note, the dip in the 426 nm peaks midway in the scattering angles. This type of behaviour is characteristic of the governing Lorenz-Mie scattering theory[6,7]. The spikes in the background correspond to noise from air bubbles in the UV detector. The data were collected by M. Moon at the University of Utah.

An unsmoothed chromatogram of the 90° scattering from a mixture of polystyrene in THF separated by a thermal field flow fractionation column is shown (Fig. 13). As with most FFF separations, the smaller molecules or particles elute first. The peaks correspond to molecular weights from 40K through over 4M. The maximum temperature gradient across the channel was about 600°C/cm. The data were produced by T. Walter at the University of Utah.

Figs. 14 and 15 represent two final examples by GPC separation of biomolecules. Fig. 14 presents the 90° scattering from a pectin sample. The RI signal shows a double peak which changes in time with the lower molecular weight fractions polymerizing to increase the higher molecular weight fractions. Even higher molecular weight components are shown by the light scattering peak at the far left which is barely detected by the RI detector. The 90° scattering data from heparin are shown in Fig. 15. Here, the light scattering shows an even more extreme high molecular weight peak not even detected by the RI detector. In addition, the light scattering shows the appearance of a double peak where the RI shows only a single peak. The heparin samples were measured at 488 nm using an argon ion laser source. We shall return to these unusual data presently.

The data collected at the many scattering angles may now be combined together with the RI or UV concentration data to calculate the molecular weights and sizes at each elution fraction. From these may be generated the cumulative and differential distributions, the molecular weight and root mean square radius *versus* elution volume, log M_W *versus* log r_g, *etc.* Figs. 16 to 19 present examples of such analyses from the collected data in the GPC separation mode. Fig. 16 presents the molecular weight *versus* elution volume for the pullulan sample shown earlier in Fig. 4. Fig. 17 shows another type

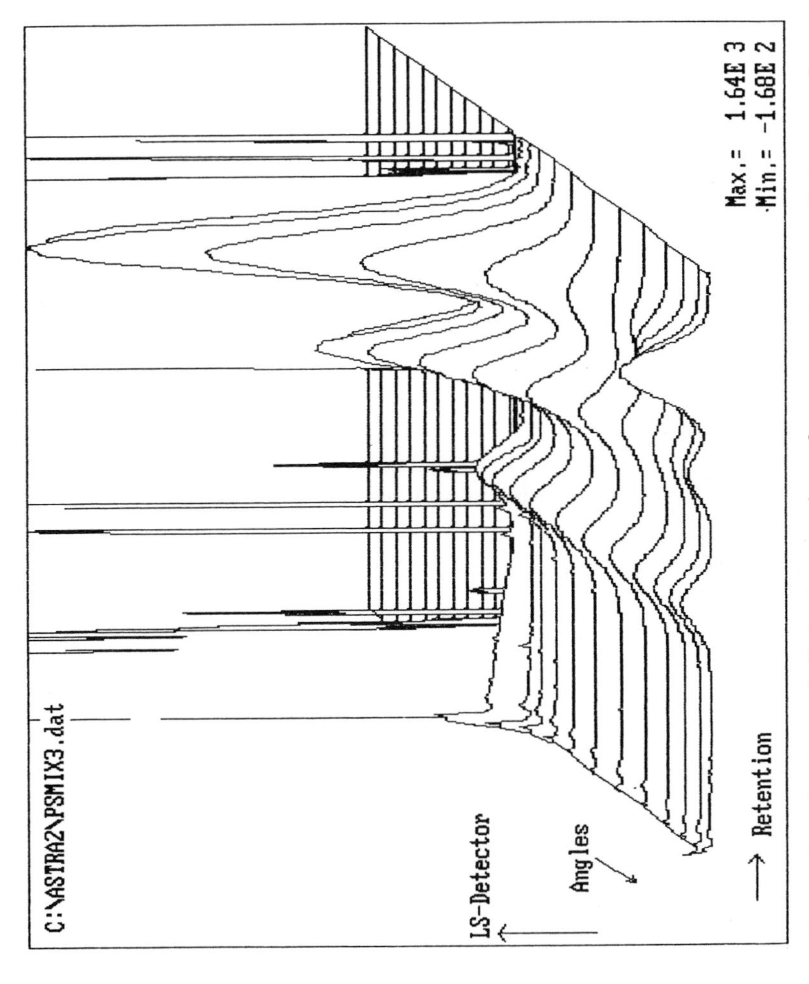

Figure 12. Three-dimensional plot of light scattering from a mixture of three polystyrene latex particles of nominal diameter 232nm, 330nm and 426nm.

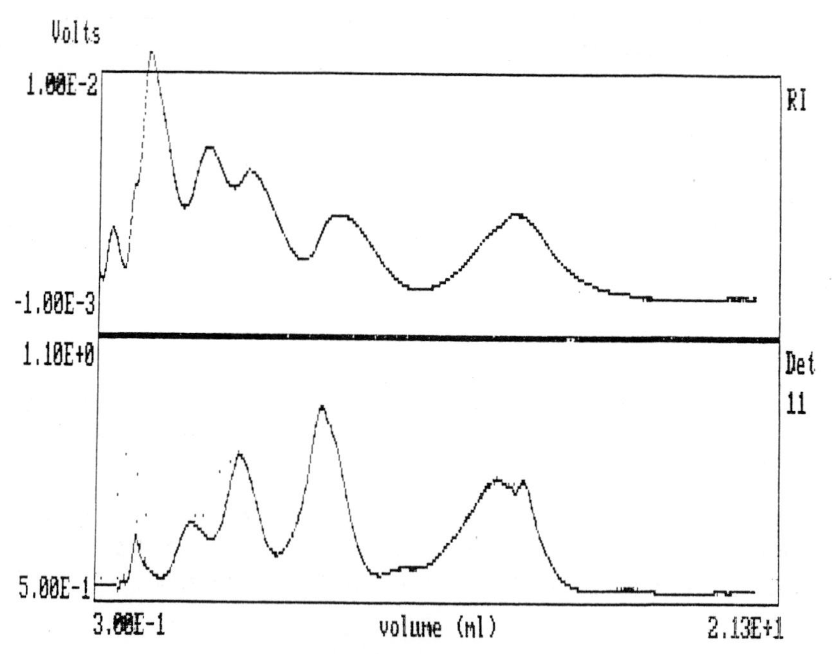

Figure 13. Light scattering at 90° after separation by thermal field flow fractionation. Polystyrene mixture in THF of molecular weights from about 40K through 4M. The top curve is the corresponding UV signal.

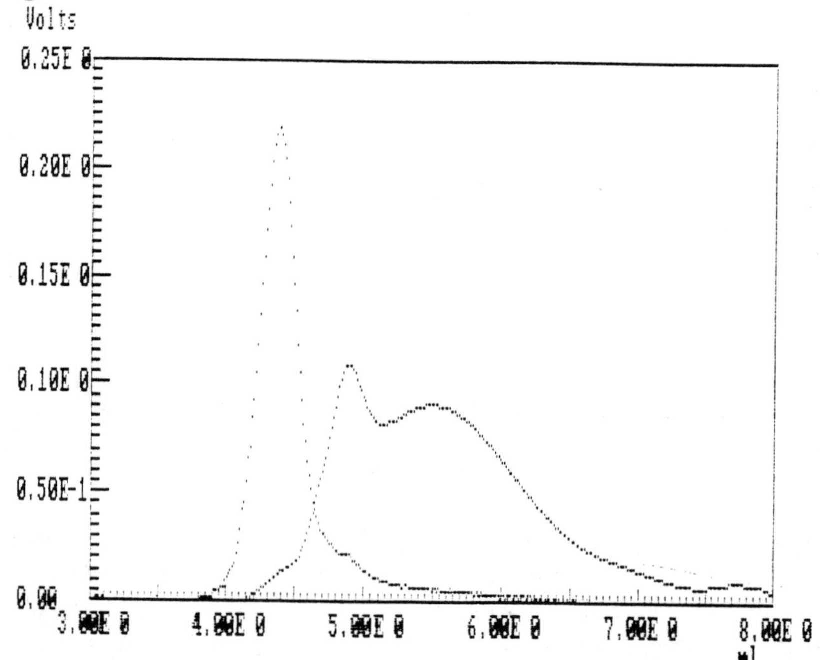

Figure 14. Light scattering chromatogram at 90° for a pectin sample contrasted to RI signal.

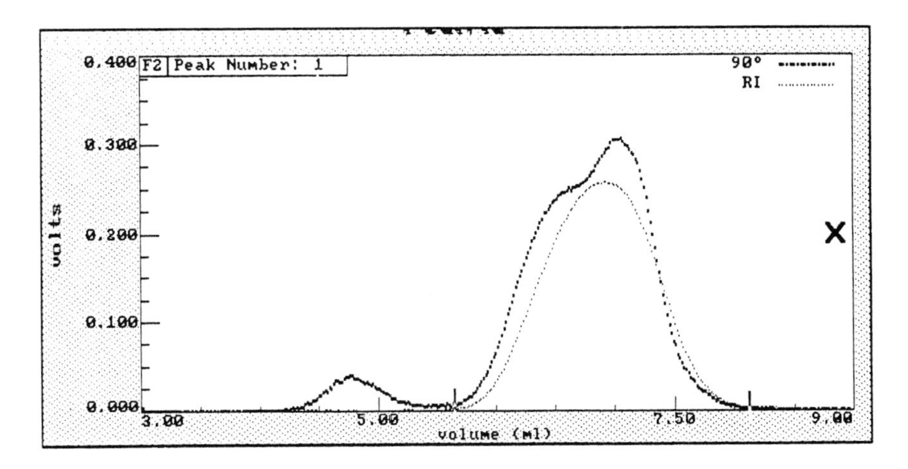

Figure 15. Light scattering chromatogram at 90° from a heparin sample contrasted with RI signal. Note the indication of a double light scattering peak and the small excluded peak not seen by the RI.

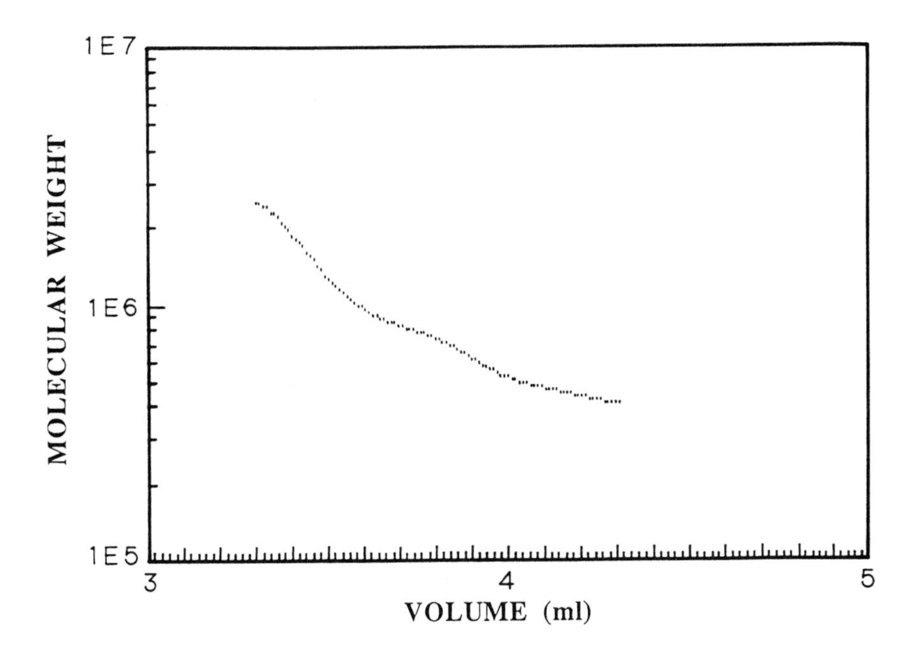

Figure 16. Molecular weight *versus* elution volume for the pullulan standard of Fig. 4. It shows the absolute weight average molecular weight for each elution slice. The average values for the sample were: M_n = 692,300; M_w = 774,500; M_z = 883,400; M_w/M_n = 1.12. The number average radius of gyration = 32 nm.

of analysis possible from the recorded light scattering data. It presents the differential molecular weight distribution of thyroglobulin. Fig. 18 shows the unexpected variation of molecular weight with elution volume for the heparin sample of Fig. 15. The high mass, small size fraction responsible for the reversal of the curve could only be seen *via* light scattering measurements. This fraction probably consists of branched or microgel-like constituents of the heparin sample. Heparin samples from various sources show this variation. Similar curves have been obtained from microgels of polystyrene[21].

Fig. 19 presents a log M *versus* log r_g plot for the broad polystyrene standard NIST-SRM 706. The slope of this curve, 0.567, corresponds to the expected variation for a random coil structure in THF. Measurements were made using an argon ion laser operating at 488nm. The fluctuations correspond to lack of adequate size resolution at small radii. Since only positive values of $\langle r_g^2 \rangle$ can be plotted on this logarithmic scale, the actual statistical fluctuations cannot be shown in such plots.

The final illustration (Fig. 20), contrasts the mass variation with retention volumes for the three polystyrene samples of Fig. 5. At the left are the results for good quality PSS columns, while the curve at the right shows the effect of column deterioration.

The selection of data presented above confirms the versatility of the multi-angle light scattering detection for various chromatographic separation techniques. At the molecular level, measurement of the slope of $R(\theta)/(K^*c)$ or its reciprocal near $\theta = 0$ yields the square mean radius at each slice for molecules whose dimensions are of the order of $\lambda/20$ or greater, where λ is the wavelength in the medium. If the conformation of the molecules are known, then such data may be used directly at *all* angles to extract $\langle r_g^2 \rangle$. Alternatively, the method of Mijnlieff and Coumou[15] may be employed so that all angular data can be utilized without prior knowledge of the molecular conformation. Including all angular data, especially those at higher angles, tends to improve the size resolution and permit measurement somewhat below the $\lambda/20$ range. Naturally, by making more measurements over longer periods of time, the precision of $\langle r_g^2 \rangle$ values may be improved by $\sqrt{N}$, where N is the number of independent measurements.

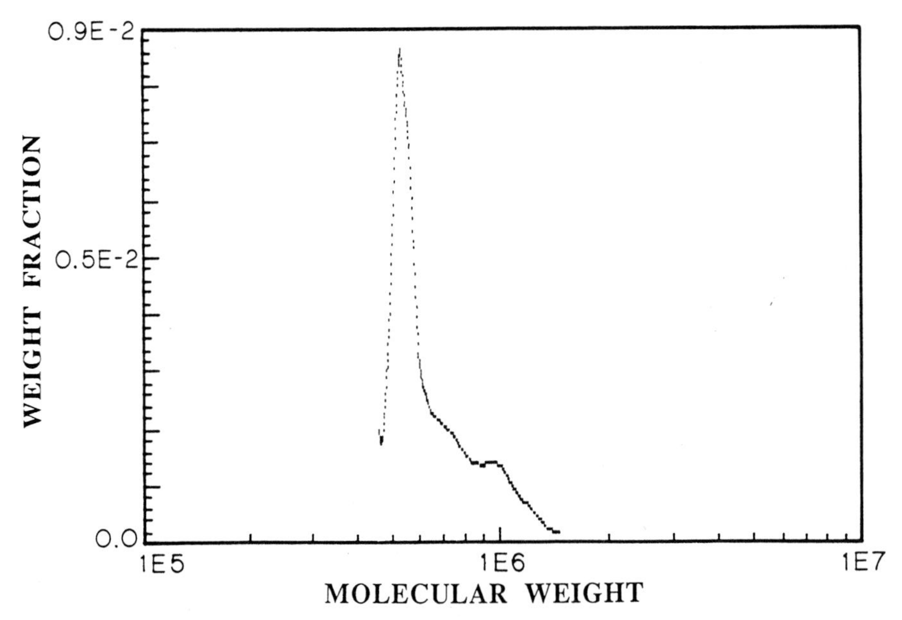

Figure 17. Differential molecular weight distribution for a sample of thyroglobulin. The average values for the sample were: M_n = 621,000; M_w = 670,000; M_z = 736,600; M_w/M_n = 1.08. The number average radius of gyration = 18 nm.

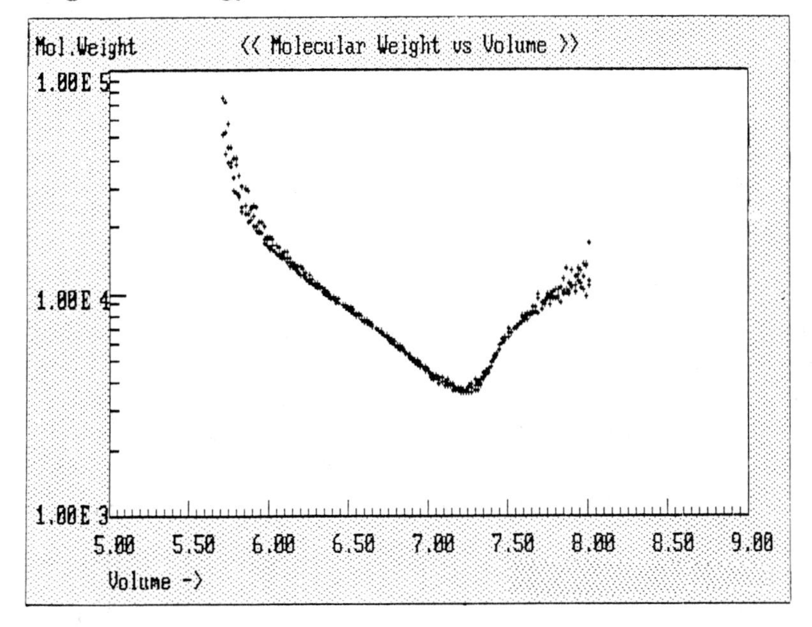

Figure 18. Log of M_w *versus* elution volume for a heparin sample eluting from a linear PSS column. Measurements made at 488nm with Argon Ion laser.

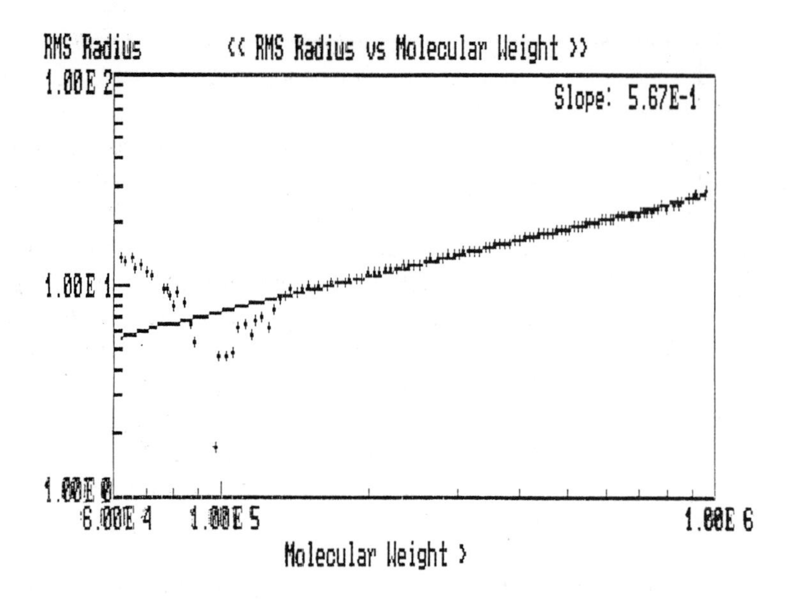

Figure 19. Log M_w *versus* log r_g for SRM 706. Measurements made at 488nm.

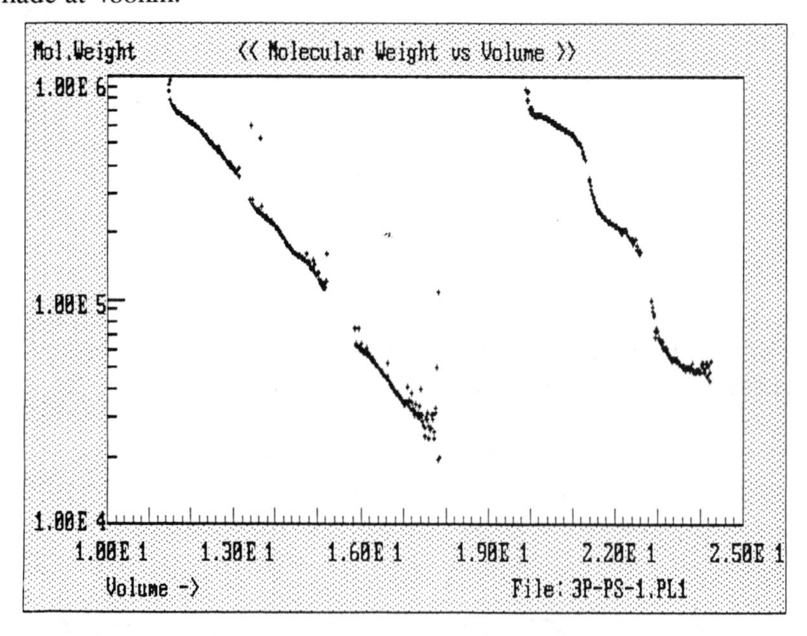

Figure 20. Comparison of elution curves for high quality and deteriorating columns.

For sufficiently large linear molecules, a plot of log M *versus* log $<r_g^2>^{1/2}$ should yield a straight line whose slope yields directly the conformation: 1 for a rod, 0.5 to 0.6 for a random coil, and 0.33 for a sphere. If the separated molecules include branched fractions, microgels, or networks, the resultant log - log plot will depart significantly from such straight line behaviour. This indicates a mixture of molecular conformations is present. The heparin samples of Fig. 15 would yield such departures were the molecules larger and the measurements able, thereby, to produce meaningful size information. Since the *masses* calculated appear to be very accurate, it would be possible to extract the hydrodynamic radii from sequential intrinsic viscosity measurements. The reversal of mass in the light scattering elution curve seems characteristic of the presence of such complex structures. C. Johann has found similar results for polystyrene gels fractions[21].

Now, it might be argued that the light scattering measurements of Fig. 15 are not due to changes in molecular structure, but rather a change in dn/dc. Thus, the "apparent" increase of molecular weight with increasing elution volume is just due to a significant departure of this refractive index increment. Even a cursory examination of Fig. 15 (and the more detailed log - log plots of Johann[21]) show that this is not possible. Indeed, A. Matsumoto[22] has explained these anomalies by considering the covalent binding of large macromolecular fractions to form microgels with greater density and correspondingly smaller size than might be expected from simple aggregates.

The improvement expected in the precision of mass determined by direct application of eq. (4) can be predicted heuristically from consideration of small molecules whose effective mean square radii are very much smaller than $\lambda/40$. In this event, $P(\theta) = 1$ and, for the low concentrations associated with chromatographics separations, eq. (4) reduces to

$$\frac{R(\theta)}{K^* c} = M \qquad (10)$$

independent of angle. Thus, *each* angle, θ_i, yields a generally independent value of M, whose *precision* would improve with the number of angles, N, measured.

Thus, the average molecular weight for a given slice would be given *by*

$$M = \frac{1}{N}\Sigma M_i \pm \delta M \qquad (11)$$

where

$$\delta M = \sqrt{\frac{\sum_i \delta M_i^2}{N}} \qquad (12)$$

$\delta M_i = \delta R(\theta_i)/(K^*c)$ and $\delta R(\theta_i)$ is the standard deviation of the Rayleigh excess ratio measured at θ_i. For larger molecules, the inclusion of different angles would be expected to improve the accuracy of the extrapolation of eq. (4) to $\theta = 0$.

4. CONCLUSIONS

The ability to make classical light scattering measurements over a broad range of scattering angles on an effluent following chromatographic separation represents a powerful tool for the analyses of such separated samples. Because of the absolute nature of the light scattering measurement itself, the absolute size and mass of particles satisfying the Rayleigh-Gans-Debye criteria may be determined. In this latter case, the "size" determined is actually the mean square radius of the particle/molecule measured with respect to its centre of gravity. Extraction of the weight averaged molecular weight from each eluting fraction requires an independent determination of both the molecular concentration at each fraction and dn/dc, the refractive index increment. If either of these quantities is not available, the mean square radius can *still* be measured.

Extraction of the mean square radius, or z-average square radius, generally requires that the molecules measured be of an overall size greater than about $\lambda/20$, where λ is the wavelength of the light in the solvent medium. For very small molecules, therefore, there will be no discernable variation of the measured Rayleigh excess ratio, with the measurement at each angle being about the same. Both for such small molecules and larger ones, the multiple measurements invariably result in a more precise determination of molecular weight than are obtainable at a single, low angle.

If the sample being analyzed contains a distribution of molecular sizes, then a log - log plot of mass *versus* root mean square radius will yield the molecular configuration. Again, for such a result, meaningful size information must be extractable from the data.

Newer separation techniques, such as the field flow fractionation methods being developed by Giddings' group at the University of Utah, show tremendous promise for the separation (and subsequent analyses by differential light scattering methods) of high molecular weight samples and particles.

For the case of particles separated by sedimentation field flow fractionation, differential light scattering represents the only means for their rapid characterization. If the particles are homogeneous spheres, the inversion to extract radius and refractive index is straightforward. For more complex structures, light scattering measurements promise a host of possible characterizations including measures of particle structure by means of ancillary depolarization measurements.

Finally, the application of differential light scattering to gradient phase separations, temperature rising elution fractionation, and many other separation techniques under current development and to be developed in the future may well represent the most important, if not the only, technique to characterize absolutely the separated species. With the advent of better software and an improved theoretical understanding of the underlying chemicophysical processes, by which the studied particles/molecules have evolved, the future growth of the light scattering interpretations discussed briefly in this paper is assured. The importance of classical light scattering measurements as an analytical tool will continue to grow in the years to come.

ACKNOWLEDGEMENTS

The author would like to acknowledge the remarkable talents of the young scientists who made so many of the measurements reported here. They include Christian Jackson, Lena Nilsson, and Janet Howie from our laboratories; Torsten Walter, M.H. Moon, and Marie Anna Benincasa from Professor Giddings' University of Utah laboratory; and Christoph Johann of Polymer Standards Service (PSS) for his having supplied us with both exceptional GPC columns and continually challenging comments.

REFERENCES

1. B.A. Fikhman, *Biophys.*, 1963, *8*, 441.
2. P.J. Wyatt, *Appl. Optics*, 1968, *7*, 1879; P.J. Wyatt, *Nature*, 1969, *221*, 1257.
3. *Cf.* H.P. Baltes, 'Inverse Source Problems in Optics', Springer Verlag, Berlin ,1978; A.J. Devaney, *Opt. Letters*, 1982, *7*, 111; K. Shimizu and A. Ishimaru, *Appl. Optics*, 1990, *29*, 3428.
4. J.V. Dave, *Appl. Optics*, 1971, *10*, 2035.
5. P.J. Wyatt, *Appl. Optics*, 1980, *19*, 975.
6. L.V. Lorenz, *Videnski. Selsk. Skrifter*, 1890, *6*, 1. Translated into French in 'Oeuvres Scientifiques de L. Lorenz', Libraire Lehmann, Paris, 1898.
7. P.J. Wyatt, *Phys. Rev.*, 1962, *127*, 1837; *Ibid.*, 1964, *134*, ABI.
8. P.J. Wyatt, *J. Appl. Bact.*, 1975, *38*, 47.
9. P.J. Wyatt and D.T. Phillips, *J. Theor. Biol.*, 1972, *37*, 493.
10. P.J. Wyatt, *Nature*, 1960, *226*, 277.
11. P.J. Wyatt in 'Methods in Microbiology', Vol. *8*, J.R. Norris and D.W. Ribbons, eds., Academic Press, New York, 1973.
12. B.H. Zimm, *J. Chem Phys.*, 1948, *16*, 1099.
13. P.J. Wyatt (in press).
14. P. Debye, *J. Phys. Colloid Chem.*, 1947, *51*, 18.
15. P.F. Mijnlieff and D.J. Coumou, *J. Colloid and Interface Sci.*, 1968, 42nd National Colloid Symposium.
16. A. Guinier, *Ann Phys.*, 1939, *12*, 161.
17. *Cf.* S.T. Phillips, J.M. Reece, and P.J. Wyatt, U.S. Patent No. 4, 616, 927 (1986).
18. *Argon Ion Laser*, Wyatt Technology Application Note #8 (1990).
19. V.R. Stull, *J. Bact.*, 1972, *109*, 1301.
20. J.C. Giddings, *J. Chromatogr.*, 1976 *125*, 3; *Ibid.*, *Chem and Eng. News*, Oct. 10 (1988).
21. C. Johann, *Labor Praxis*, pp. 1106-1109 (Dec. 1989).
22. A. Matsumoto, Poster presentation, ACS National Meeting, Washington, D.C., Sept. 1990.

4

New Tools for Biochemists: Combined Laser Doppler Microelectrophoresis and Photon Correlation Spectroscopy

By F.K. McNeil-Watson and A. Parker

MALVERN INSTRUMENTS LIMITED, SPRING LANE SOUTH, MALVERN, WORCESTERSHIRE. WR14 1AQ U.K.

1. INTRODUCTION

Laser Doppler micro-electrophoresis has until recently been confined to laboratories with some expertise in physics and instrumentation. However, advances in electronics and microcomputers have now made these measurements accessible to all research workers. The use of a combined laser Doppler micro-electrophoresis / photon correlation spectroscopy instrument for biochemical applications will now be described. The ability to measure both size and zeta potential of bio-particles and colloids provides a powerful means to characterise such systems. Particular advantages of the instrument we will discuss include precise control of the applied electric field, measurements of very low electrophoretic mobilities, and fast routine measurements.

2. THE 'ZETASIZER 3'

An overview of the instrument is shown in Fig. 1. (A fuller description has been given in ref. 1.)

Capillary Cell

The capillary cell has a four-electrode format with separate driving and sensing electrodes so that the field can be measured precisely. The capillary is mounted in a thermostatted optical vat so that the sample is held at a constant known temperature and flare from the walls of the capillary where the laser beams pass is minimised by the

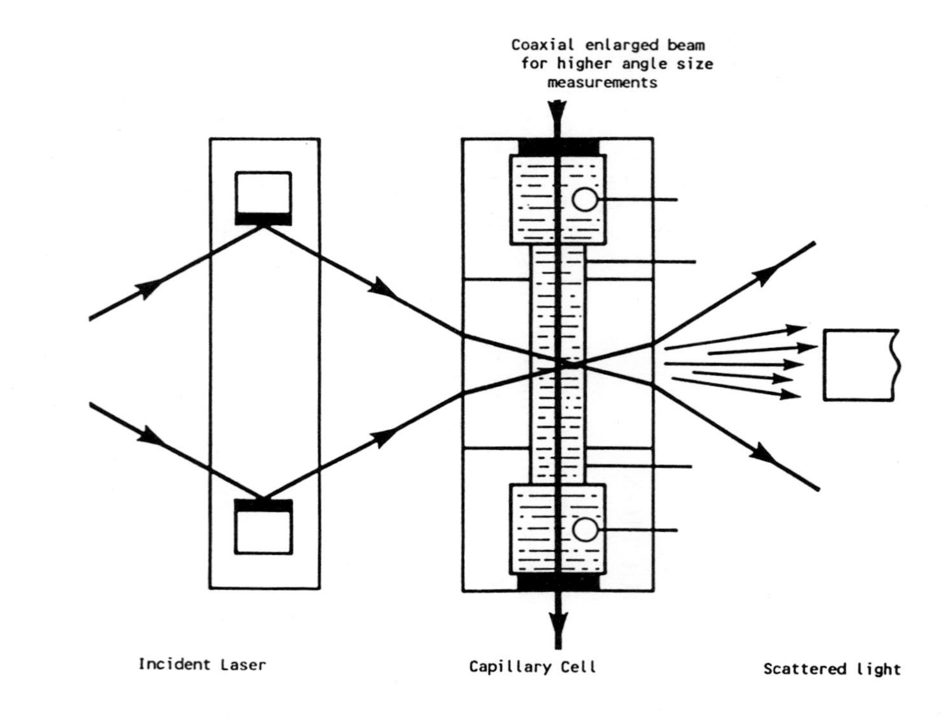

Figure 1. Schematic Layout of Zetasizer 3.

refractive-index matching fluid. The cell can be positioned automatically so that the 'probe-volume' can be scanned across the cell: this enables measurement of the electro-osmotic flow due the 'wall charge' to be made using a Van Gils plot. In routine measurements the cell is positioned so that the probe volume corresponds to one of the 'stationary levels' (Fig. 2) where electro-osmotic flow cancels.

The Detector

The detector is a photon-counting photomultiplier on a goniometer whose axis coincides with the beam crossover point, so that measurements can be made at different scattering vectors. An auxiliary expanded laser beam can be brought along the axis of the cell so extending the range of scattering angles for particle size analysis up to 140°. The detector can also be positioned coaxially with one of the crossed laser beams, which is then attenuated, so that a portion of unscattered light is mixed with scattered light from the other beam. This 'heterodyne' detection enables measurements of the highest resolution on multicomponent systems such as the mixture of two latex populations with an intralipid emulsion shown below: for more routine studies the normal or 'homodyne' detection scheme in which light scattered from particles in each laser beam is detected simultaneously.

Scattering Angle

The ability to vary the scattering angle and hence the scattering vector is particularly useful in size measurements by photon correlation spectroscopy where the strong variation of size with angle for particles above the Rayleigh limit can be exploited in improving the resolution of the method[2]. It can also be used to separate the Brownian diffusion from other modes available to the sample, such as rotation of non spherical particles, or flexing of non-rigid ones. For electrophoretic mobility measurements Uzgiris[3] has discussed the various contributions of Brownian motion, field inhomogeneity, analyser resolution, Joule heating as well as genuine distribution of surface charge density. Essentially the peak that we see in the Doppler spectrum of an electrophoresis measurement is broadened by a number of independent factors, of which for small particles (<100 nm) Brownian motion is the most important. Choosing a small scattering angle reduces this broadening, but it has

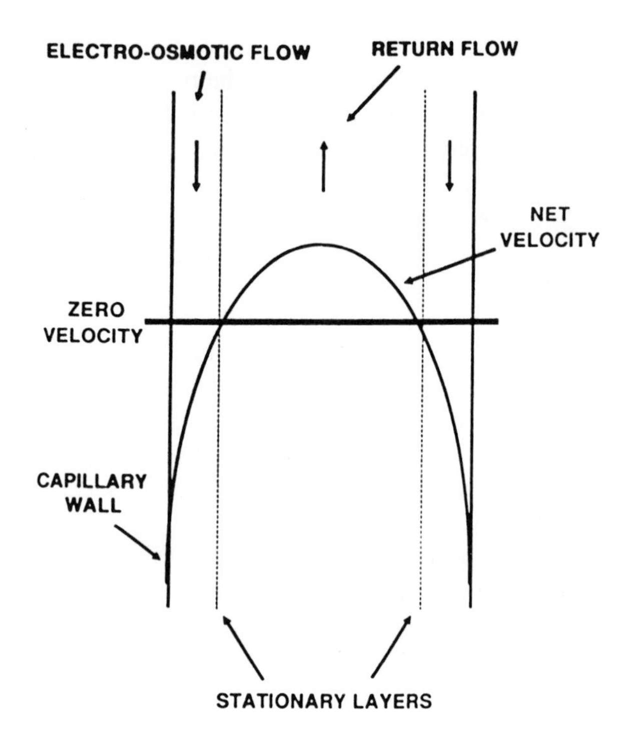

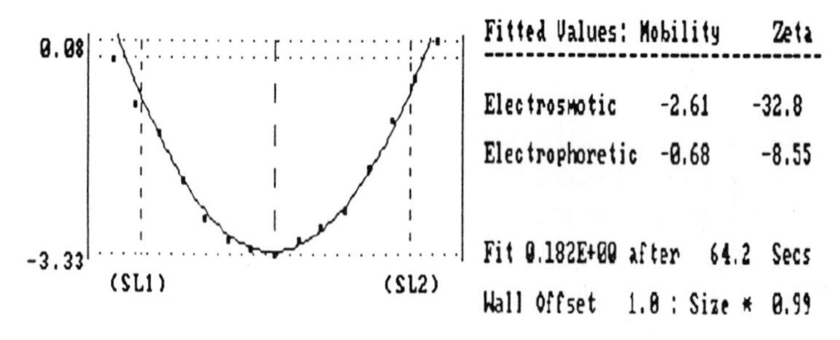

Figure 2. Electro-osmotic flow profile in a capillary and a 'Van Gils' plot.

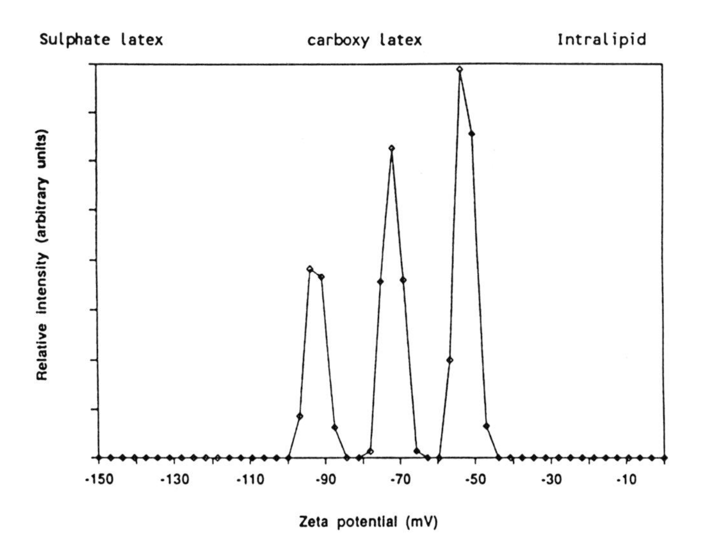

Figure 3. Mixture of three populations resolved in 'heterodyne' mode.

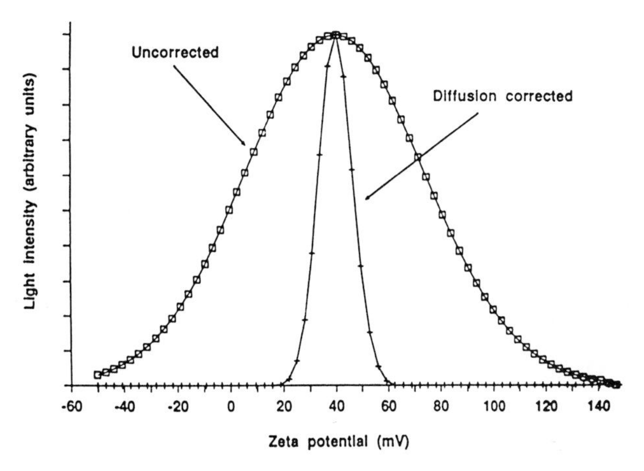

Figure 4. Removal of diffusion broadening of 15 nm particles.

other deleterious side effects: the Doppler shift of the electrophoretic motion is also reduced although the shift is proportional to Q (the scattering vector), whereas the Brownian width is proportional to Q^2. However the lower frequency region is more prone to disturbance by background vibration, longer accumulation times are needed to achieve the same accuracy, and scattering from dust particles is more serious due to the strongly forward peaked scatter of larger particles. In the present instrument we compromise on a scattering angle (for homodyne detection) of 32° in the sample, and apply a novel signal processing method to remove the effects of Brownian diffusion broadening.

This has been described in more detail in refs 4 and 5. We use the ability to 'duty-cycle' the cell (*i.e.* control the time during which the AC field is applied, and also during which the correlator acquires data) to measure both the spectrum with 'field-on' and 'field-off'. The latter contains only the effects of diffusion and fundamental instrumental resolution. This therefore contains the essential information about the broadening that is also present during the 'field-on' cycle and is deconvoluted from the spectrum to give the spectrum with only electrophoretic other sources of broadening remaining. The effect is demonstrated in Fig. 4 on data taken on very small alumina particles (mean size about 15 nm).

Minimizing Joule Heating

Joule heating of the cell due to the high currents that may be passed in solutions such as 0.145 M NaCl may be minimised by running at a low 'duty-cycle'. It may also be reduced by using a smaller diameter capillary tube, possibly with the aid of a methyl-cellulose coating to reduce the wall charge since it is then more difficult to optimally select the stationary level of the cell.

Phase Modulator

The measurement of very small mobilities, and indeed the 'field-off' measurements described above are made possible by the use of a phase modulator that frequency shifts one of the laser beams so inducing an effective 'apparent Doppler shift' from stationary particles in the probe volume. Comparison of the detected frequency with the known modulator frequency then enables the system to deduce even a very small shift, as well as discriminate the sign of the zeta potential. This ability is particularly useful in the measurement of the 'isoelectric point' (iep).

3. SUMMARY

In summary the benefits of this type of instrument to research in the biochemical field are the extension of a previously slow and tedious manual measurement, restricted to particles of 200 nm size or more, of at least a moderate mobility, to particles of 10 nm or so with mobilities essentially zero. The speed and sensitivity of the technique means that kinetic processes can be followed. The ability to measure size distribution on the same sample can also be a powerful advantage since the effects of polydispersity on heterogeneous samples will generally be different if the sample has to be prepared at different concentrations for a differing measurement technique.

The automation of processes like a Van Gils plot: vital in checking that a microelectrophoresis cell is clean and well aligned is also a powerful incentive for the experimenter to make use of the best methods in performing measurements previously the domain of the expert.

REFERENCES

1. F.K. McNeil-Watson, 'A New Instrument for Particle Size Analysis and Electrophoretic Mobility Measurements' ed. P.J. Lloyd, Wiley, 1988.
2. P.G. Cummins and E.J. Staples, *Langmuir*, 1987, *3*, 1109.
3. E.E. Uzgiris in 'Cell Electrophoresis' ed A.W.Preece and D Sabolovic, Elsevier, 1979, p. 369.
4. F.K. McNeil-Watson and A. Parker, to be published in the proceedings of the International Symposium on Colloid and Surface Engineering, San Diego, 1990.
5. Application report 'High Resolution Electrophoresis measurements', Malvern Instruments Ltd., 1990.

5

An On-line Dynamic Light Scattering Instrument for Macromolecular Characterisation

By Paul Claes, Martin Dunford, Andrew Kenney and Penny Vardy

*OROS INSTRUMENTS LIMITED, 715, BANBURY AVENUE, SLOUGH, BERKSHIRE. SL1 4LJ U.K.

1. INTRODUCTION

The use of light scattering techniques in biochemistry has been limited by the expense and complexity of available instrumentation[1-3]. The use of the dynamic light scattering (DLS) method in particular has been hampered both by the size and cost of the optical equipment and by the need for expensive "hard wired" autocorrelators. In order to overcome these limitations we have developed a compact and economical instrument by using miniature solid state optical components and modern digital signal processing technology in place of a dedicated autocorrelator.

In order to make the equipment suitable for as wide a range of applications as possible, two versions have been produced for use either as an on-line liquid chromatography detector or for molecular characterisation by manual introduction of a small sample.

Use of Dynamic Light Scattering to Determine Molecular Size

The Model 801 Molecular Size Detector uses DLS to determine the translation diffusion coefficient (D_T) of particles including macromolecules in solution. The technique is based on the

---

* Now: Biotage (U.K.) Limited, Harforde Court, Foxholes Business Park, John Tate Road, Hertford. SG13 7NW U.K.

measurement of fluctuations in scattered light intensity caused by the relative movements of macromolecules in solution[4]. Light from a near infra-red semiconductor laser (780 nm) is passed through a measurement cell and is scattered by molecules moving under "Brownian" motion. Photons which are scattered at 90° to the incident laser beam are collected by a lens and conducted to an Avalanche Photo Diode (APD) *via* an optical fibre. The APD produces a single, electrical pulse for each photon detected and these pulses are stored by an integral computer. The miniature optical components are mounted on an optical bench that measures only 25 x 5 x 5 cm.

Light scattered by macromolecules interferes either constructively or destructively and the manner and extent of interference changes as molecules move, causing a fluctuation in scattered light intensity between zero and twice the average intensity. The time scale of these fluctuations depends on the speed of movement of molecules, hence the translational diffusion coefficient (D_T) can be obtained from measurements of these time scales. Comparatively small, fast moving particles such as macromolecules scatter very little light and cause rapid intensity fluctuations, and it is necessary to count individual scattered photons to obtain measurements of D_T. Photons are counted into time windows or channels and the time constant of intensity fluctuation obtained by autocorrelation of this data. Once D_T has been determined, the hydrodynamic radius, R_H, can be calculated from D_T using the Stokes-Einstein equation:

$$D_T = \frac{kT}{6\pi\eta R_H} \tag{1}$$

where k is Boltzmann's constant, T is the absolute temperature and η is viscosity. The molecular weights of molecules detected are estimated using the relationship between molecular weight and R_H determined for a series of standard proteins. This estimation of molecular weight is based on two assumptions. First, it is assumed that proteins are approximately spherical in shape and, secondly, it is assumed that all proteins have a constant density relation to their size in order to calculate mass from the molecular volume. These assumptions break down in certain cases, for example non-globular proteins have R_H values greater than those predicted from the molecular weight, and literature values for molecular weights are often based on DNA or amino acid sequence data and do not take into account the contributions from glycosylation. Heavily glycosylated proteins will thus exhibit R_H values which are larger than expected. Fig. 1 shows a plot of $\log_{10} R_H$ against $\log_{10}$ molecular weight superimposed on the line generated by a plot of the relationship used by the Model 801 to estimate molecular weight from R_H. The data obtained for most proteins examined to date lies on or close to the line, indicating that the assumptions used provide a good estimate of molecular weight in most cases.

Operation of the Model 801 Molecular Size Detector

After sample introduction, the instrument determines the whether there are enough molecules in the sample to produce a result. Photon correlation then continues until sufficient data have been gathered to provide 99% accuracy in R_H. Autocorrelation of data proceeds as the scattered photons are counted. After autocorrelation is complete the autocorrelation function obtained analysed to provide D_T, R_H, molecular weight, polydispersity and statistical parameters indicating the reliability of the measurement. This data can be printed, transferred to a personal computer *via* a serial link and/or output to a chart recorder.

2. CHROMATOGRAPHIC APPLICATIONS

Cation Exchange of Egg White Proteins

In the process of developing chromatographic separation methods for proteins, a significant amount of time is spent on the analysis of the resulting eluate fractions in order to determine the molecular composition and hence the elution position of components of interest. This off-line analysis is often accomplished by means of electrophoretic techniques, measurement of biological activity or immunoassay. On-line DLS can be used to facilitate the identification of eluted peaks and is illustrated in Figure 2 which depicts the chromatogram obtained during the separation of egg white proteins by cation exchange. It was known that the sample contained three major protein components, ovalbumin (molecular weight 43,000), lysozyme (molecular weight 14,000) and ovotransferrin (molecular weight 76,000), but their elution positions in the gradient could not be predicted in the absence of knowledge of their "pI"s. Different components were identified by the Model 801 Molecular Size Detector in each of the 3 peaks. The first peak (unbound material)

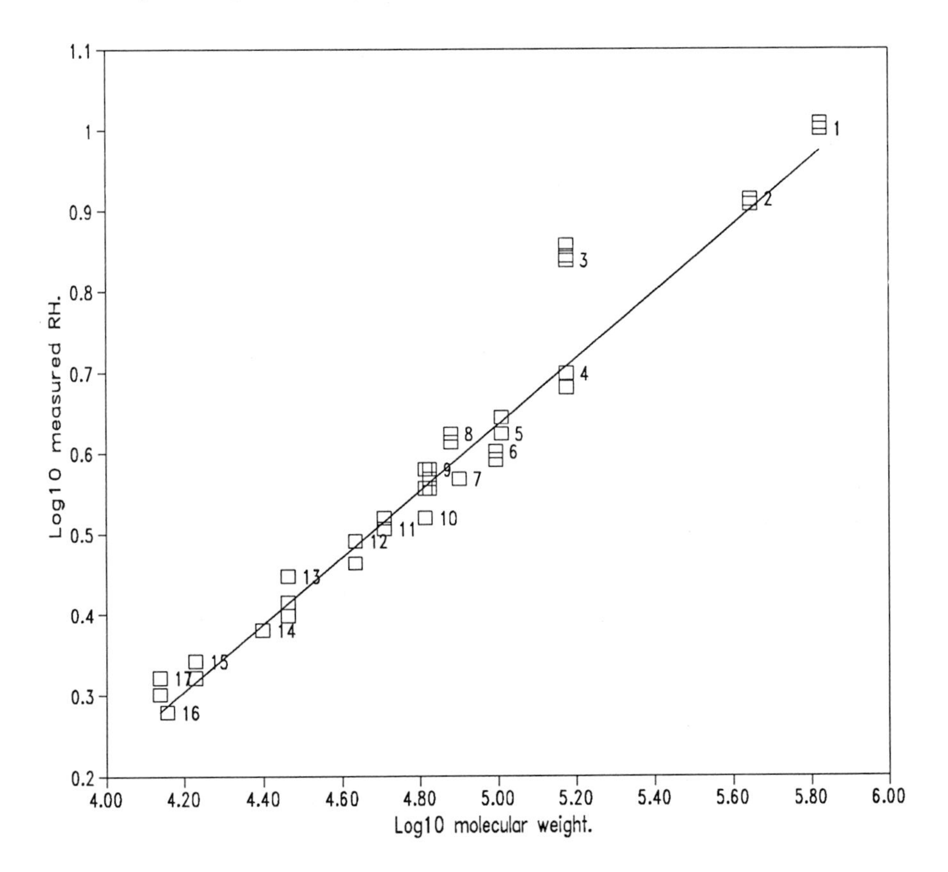

Figure 1. Plot of $\log_{10} R_H$ against $\log_{10}$ molecular weight. Proteins measured were: (1) thyroglobulin; (2) apoferritin; (3) IgG; (4) yeast alcohol dehydrogenase; (5) hexokinase; (6) amyloglucosidase; (7) horse alcohol dehydrogenase; (8) transferrin; (9) bovine serum albumin; (10) haemoglobin; (11) hexokinase sub-unit; (12) ovalbumin; (13) carbonic anhydrase; (14) chymotrypsinogen; (15) myoglobin; (16) lysozyme; (17) ribonuclease A. The relationship between $\log_{10} R_H$ and $\log_{10}$ molecular weight used in the estimation of molecular weight from measured R_H is also shown (——).

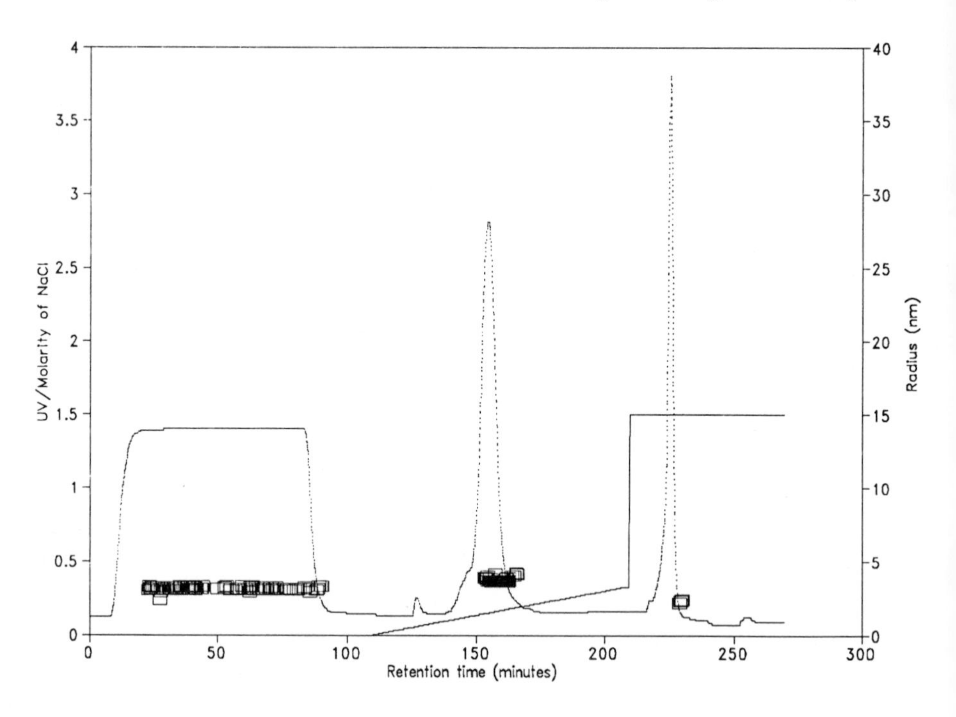

Figure 2. Molecular size analysis during cation exchange of egg white proteins. Chromatography was carried out on an Oros Instruments Model 100 Touch Screen Laboratory Scale Chromatograph. Prepared egg white sample (diluted 6-fold in Buffer A, 5 mM NaH_2PO_4; pH 6.0 and gel filtered to remove aggregated material; 60 ml volume) was loaded at a flow rate of 2 ml/min on to a pre-equilibrated 1.6 cm x 32 cm column of S-Sepharose Fast Flow. After sample loading the column was washed with 70 ml 5 mM NaH_2PO_4; pH 6.0. Bound proteins were eluted with a gradient of 100% Buffer A to 68% Buffer A, 32% Buffer B (1 M NaCl, 5 mM NaH_2PO_4; pH 6.0) over 96 min followed by a step to 1.5 M NaCl, 5 mM NaH_2PO_4; pH 6.0. The measured R_H in nm (□), molarity of NaCl in the elution buffer (----) and the measured count rate per second (.......) are shown plotted against the retention time in minutes.

had an average R_H of 3.1 nm, estimated molecular weight 45,000, peak 2 had an average R_H of 3.8 nm, estimated molecular weight 74,000 and peak 3 had an average R_H of 2.3 nm, estimated molecular weight ~22,000. This suggested that ovalbumin passed through the column without binding and that the first bound and eluted peak contained ovotransferrin, and the second contained mainly lysozyme. Subsequent gel electrophoresis confirmed this. The rapid identification of peaks can greatly accelerate the development of chromatographic methods for proteins since the time consuming off-line analyses can often be eliminated.

Gel Permeation of Hepatitis B Surface Antigen Preparation

As part of the routine production of recombinant Hepatitis B surface antigen particles for use as a vaccine it is important to characterise their size and polydispersity (size distribution). This can be achieved by gel permeation chromatography in conjunction with on-line molecular size measurement as depicted in Fig. 3. A single UV peak was observed and superimposed on this are plotted the hydrodynamic radius measurements in nm. The average radius across the peak apex was 15.4 nm. Values obtained either side of the apex are less reliable because of the diminished particle concentration. Such information can be used to study batch-to-batch variations to ensure product consistency but the technique can also be used to study the effects of solvent composition and pH, preparation method, and host strain selection on the size of these particles in solution.

3. MOLECULAR CHARACTERISATION

Polysaccharides

The measurement of hydrodynamic size by DLS can provide a simple and rapid means to estimate the average molecular weight of polysaccharides in solution. The Model 801 Molecular Size Detector has been used to measure the hydrodynamic size of dextran and pullulan fractions of varying molecular weight. The correlation of hydrodynamic size with molecular weight and intrinsic viscosity was also investigated. Commercially available dextran fractions are usually characterised by their weight average molecular weight. This parameter is often obtained by means of gel permeation chromatography (which separates molecules according to their hydrodynamic size) in conjunction with viscometry or classical light

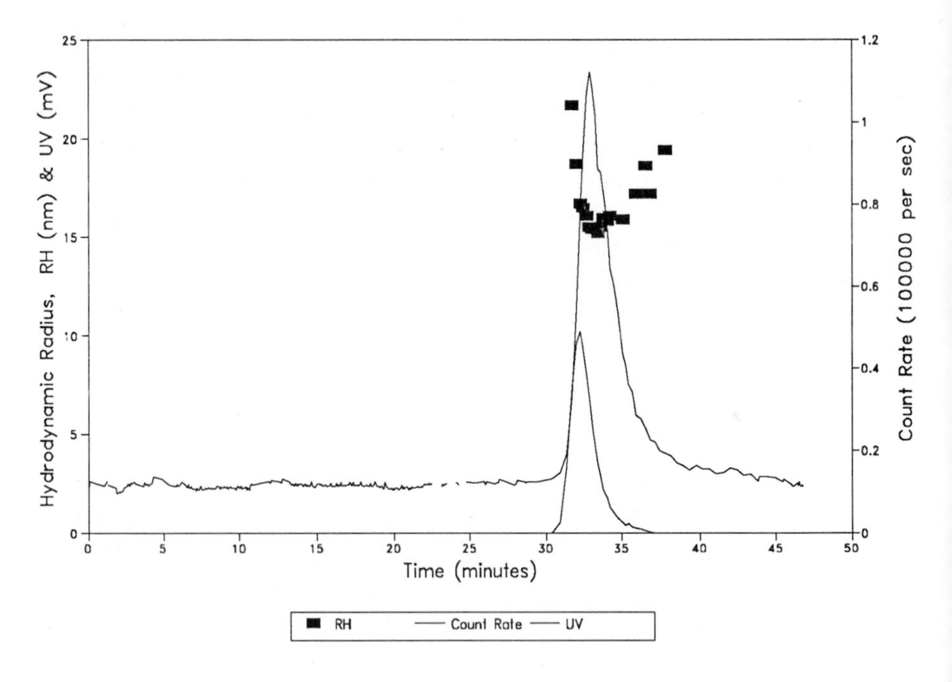

Figure 3. Gel permeation HPLC of hepatitis B surface antigen. The sample supplied was clear and colourless. No pretreatment of the sample was performed. A 0.2 ml loop was injected onto a DuPont Zorbax GF250XL column. The mobile phase was phosphate buffered saline solution (PBS) pH 7.6, which had been filtered to 0.2 μm and partially degassed under vacuum before use. The Model 801 Molecular Size Detector was installed immediately after the UV (280nm) detector. A 0.1 μm ceramic membrane filter, 25 mm diameter, was installed in the Model 801. The filter was changed before every injection onto the HPLC. The flow rate was 1 ml/min.

scattering equipment. The hydrodynamic size of molecules in solution can be obtained by means of DLS which directly measures the translational diffusion coefficient directly. The diffusion coefficient may be used to obtain the hydrodynamic radius from the Stokes-Einstein equation. Fig. 4 shows the relationship between the experimentally determined hydrodynamic radius and the weight-average molecular weight, M, for a series of dextran standards of M between 40,000 and 2 x 10^6. The dextran standards were obtained from Sigma Chemical Co. and were dissolved to a concentration of 0.5 mg/ml in 0.1 M NaOH. This calibration curve could now be used to estimate the weight average molecular weight of other, similar dextran fractions.

Intrinsic viscosity values also correlate well with the measured hydrodynamic radius (Fig. 5). The possibility exists of combining the calibrations of hydrodynamic radius against the product of weight average molecular weight and intrinsic viscosity to provide a "universal calibration" that could be used with a range of water soluble carbohydrates of differing sizes and shapes.

This principle can also be applied to other polysaccharides, for example, pullulan. Fig. 6 depicts the calibration curve obtained with some pullulan fractions from Showa Denko with molecular weights between 5,800 and 853,000. The slopes of the dextran and pullulan curves are 0.42 and 0.51 respectively: this difference may reflect the difference in the degree of branching between the two materials.

Macromolecular Interactions

Dynamic light scattering is an ideal technique for the study of macromolecular interactions in solution. It can provide real time information for the study of interactions with time scales measured in seconds and only small samples are required which may be recovered after measurement. Applications include the study of protein-protein interactions for instance, immune complex formation; protein refolding and denaturation including the recovery of insoluble recombinant proteins from microorganisms; protein aggregation or fragmentation and the study of the physico-chemical basis of protein crystallisation[5,6]. All macromolecules can be studied by DLS and the technique is not limited to proteins or molecules with distinct chromophores. Hence the binding of proteins to other macromolecules of interest may be studied, including nucleic acids,

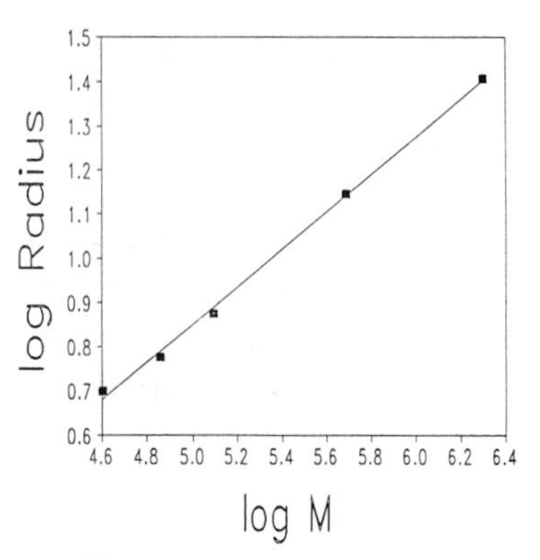

Figure 4. Dextran calibration curve. log R_H *vs.* log M. The dextran standards of molecular weight between 40,000 and 2 x 10^6 were obtained from Sigma Chemical Co. and were dissolved to a concentration of 0.5 mg/ml in 0.1 M NaOH.

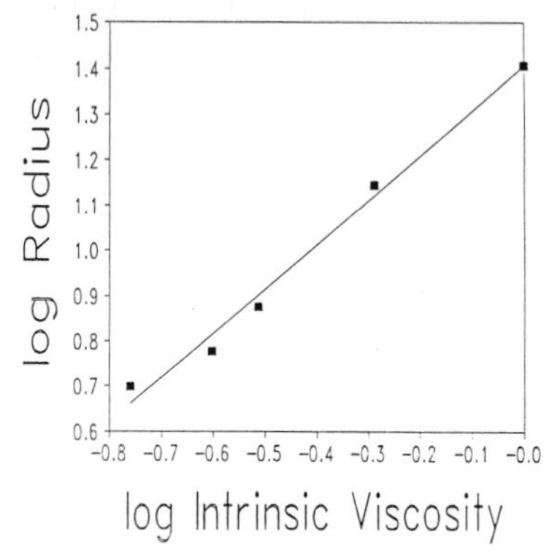

Figure 5. Dextran calibration curve. log R_H *vs.* log [η].

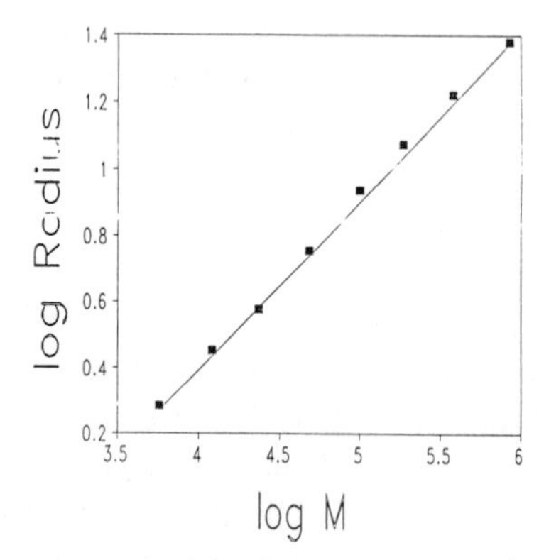

Figure 6. Pullulan calibration curve - log R_H *vs.* log M. The pullulan fractions were obtained from Showa Denko with molecular weights between 5,800 and 853,000 and were dissolved in water for measurement.

carbohydrates, detergent micelles and lipids.

4. CONCLUSION

The development of miniature optical components has facilitated the design of a small, low cost, easy-to-use dynamic light scattering instrument. These features will enable the DLS technique to be used more widely in the rapid, non-invasive characterisation of macromolecules in solution.

The construction of a fast, software-based autocorrelator has made it possible to obtain DLS measurements on-line during the separation of macromolecules by liquid chromatography. On-line DLS has been shown to be effective in providing peak composition data which can remove the need for further off-line analysis. This technique will be particularly useful in the development of protein separation methods and in the routine quality control of purified protein products.

ACKNOWLEDGEMENTS

The authors wish to thank the following individuals for making samples, results and helpful comments available: D. Pepper, Scottish National Blood Transfusion Service; F. Van Wijnendaele, SmithKline Biologicals; J. Park, Alpha-Beta Technology, Inc.; P. Dubin, Indiana-Purdue University.

The Model 801 Molecular Size Detector includes technology developed by the UK Ministry of Defence Royal Signals and Radar Establishment and licensed to Oros Instruments Limited by Defence Technology Enterprises Limited (DTE).

REFERENCES

1. R.J.G. Carr, R.G.W. Brown, J.G. Rarity and D.J. Clarke, Laser Light Scattering and Related Techniques. In 'Biosensors. Fundamentals and Applications' A.F.P. Turner, I. Karube and G.S. Wilson (Eds) OUP, Oxford, 1987, p. 679.
2. R.G.W. Brown. On-line Analysis of Proteins using Optical Techniques. In 'Separations For Biotechnology' M.S. Verrall and M.J. Hudson, (eds) Ellis Horwood Ltd., Chichester, 1987, p.430.

3. R.G.W. Brown, *Trends Biotechnol.*, 1985, *3*, 200.
4. B.J. Berne and R. Pecora, 'Dynamic Light Scattering' Wiley, New York, 1976.
5. J.G. Rarity, R.N. Seabrook and R.J.G. Carr, *Proc. R. Soc. Lond.*, 1989, *A423*, 89.
6. Z. Kam, H.B. Shore and G. Feher. *J. Mol. Biol.*, 1978, *123*, 539.

6

An Overview of Current Methods of Analysing QLS Data

By Robert M. Johnsen and Wyn Brown

INSTITUTE OF PHYSICAL CHEMISTRY, UNIVERSITY OF UPPSALA, BOX 532, S-751 21 UPPSALA, SWEDEN

1 INTRODUCTION

A number of research workers in the field of quasi-elastic light scattering accepted an invitation to participate in a comparison of methods currently applied by them to the analysis of QLS experiments. The methods submitted are briefly described.

Broad band autocorrelators (lag-times in the range 1 μs to 1 min) are now commonly used. Such instruments permit determination of relaxation times over a broad range (*e.g.* 6-8 decades) and consequently stimulate speculation as to optimum ways of treating such data. This is illustrated with the analysis of an extremely broad distribution of relaxations (a semi-concentrated Θ-system of high molecular weight polystyrene in cyclohexane).

A method of analysis of closely spaced relaxations is suggested for application to dilute solutions of high polymers exhibiting intramolecular motions ('internal modes').

2. OVERVIEW OF METHODS

A few years ago (1985) we attempted a comparison of methods in current use. The methods applied by the participants are listed below, with some comment as to their use. Detailed information will be left to the original references, (see the Appendix for a list of participants and the methods applied by them).

Of the four sets of test data which were analyzed in this comparison, only one will be shown here. These data represented the sum of two exponentials, as seen in Fig. 1. The position of the vertical bars in the figure indicate the line-width of the relaxation and the breadth of the bars, their relative amounts. In those cases where interpretation of analysis varies with prior knowledge, the results have been presented in favour of prior knowledge of the presence of discrete modes. The Profiled Singular Value, for example, produced a monomodal solution with a broad profile, and a bimodal solution with a narrow profile. The bimodal solution is shown in this diagram.

Brief Description of Numerical Methods

Cumulants[1] A polynomial fit to $\log(g_2\text{-}1)$ *versus* lag-time, yielding an estimate of mean line-width and relative variance in line-width distribution. This is a rapid and long popular method of estimating polydispersity in a distribution.

Marquardt[2] A Marquardt algorithm for a nonlinear least squares fit of the sum of exponential functions to the square root of g_2-1 data. Requires initial estimate of amounts and line-widths.

S-exponential sums[3] The method fits a positive exponential sum to a given data set providing a best weighted least squares fit. No initial setting of any of the parameters is required and the number of exponential coefficients is determined by the number of components apparent above the noise level. Requires data at lag-times an integer multiple of a constant sample time.

Lambda depression Essentially a stripping-off method. The slowest relaxation rate is estimated and g_1 is multiplied by $\exp(\Gamma t)$ yielding, in the case of a bimodal distribution, a single exponential decay to a baseline equal (in height) to the amount of the stripped-off exponential.

Linear programming with sequence statistics[4] The research group engaged in dynamic light scattering at the Institute of Macromolecular Chemistry, Czechoslovak Academy of Science in Prague has been looking at alternative means of achieving a good fit, and alternative means of judging goodness of fit. In the methods described below, Zimmermann used sequence statistics to judge the

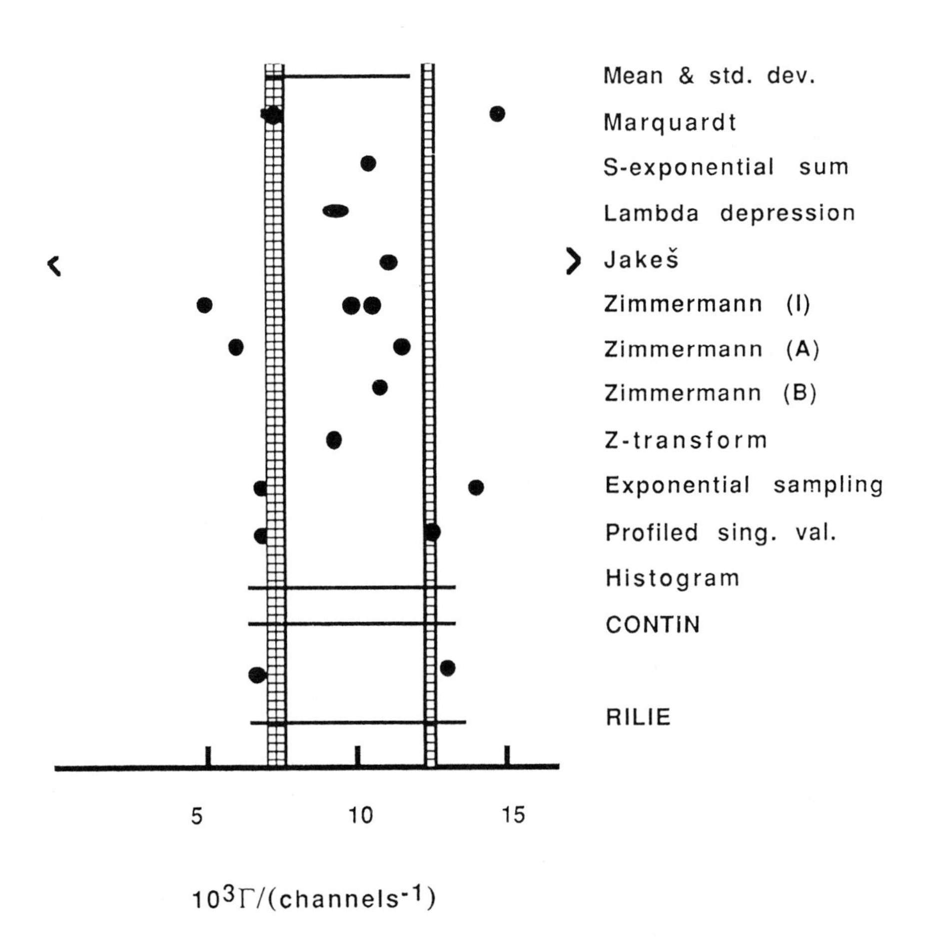

Figure 1. Methods of analysis applied in the 1985 comparison.

optimum number of exponential terms to be included in the model. He thus calculated the distribution of runs in residuals of the same sign and compared the result with the distribution expected for that number of residuals. Too many short runs thus indicated over-fitting (too many exponential terms in the model), and too many long runs indicating under-fitting.

Zimmermann I. Working with the square root of g_2-1 data, the distribution was modeled with a sum of delta-spikes. For a given number of spikes, the sum of absolute value of residuals was minimized.

Zimmermann A. Working with g_2-1 data directly the distribution was modeled with a histogram, and the sum of absolute value of residuals was minimized.

Zimmermann B. The model was again the delta-spike model, and fitting was done to g_2-1. Here, the sum of squared residuals was minimized.

Jakeš.[5] The delta spike representation was again used for the distribution of line-widths. Fitting was done to g_2-1 directly, and optimization was done with respect to absolute values of residuals.

Z-transform with spike recovery[6] Requires data equally spaced in lag-time. Uses the Z-transform to estimate an increasing number of exponential terms. At each stage of the procedure, a delta spike (slower than any relaxation in the distribution, with amplitude equal to mean amount of all exponential terms) is added to the data. The procedure is allowed to continue so long as the added spike is recovered among the estimated exponentials.

Exponential sampling[7] Lays out a grid of suggested line-widths and estimates (linear regression) the amount of each component. The grid is laid out with equal spacing in log(line-width), and the spacing should reflect the signal/noise ratio.

Profiled singular value[8,9] The profiling function is not intended to describe the profile of the distribution (of line-widths). It is the profile of a smoothing function applied to the line-width distribution. The breadth of the profile function plays an important role in the resolution of the method. Without additional knowledge of the system under observation, a narrow profile is applied to samples with a greater relative variance (causing a bimodal split-up) and a broader profile, to data sets with a small relative variance.

Histogram[10] The distribution of line-widths is represented by a series of histogram bins, equally broad on a log (Γ) scale, and used to compute the g_1 values most closely resembling (by least squares criterion) the measured values. The number of bins, and their breadth, may be chosen according to the estimated precision of the experiment.

CONTIN[11,12] As applied to QLS, a grid of line-widths is laid out with equal spacing in log(Γ) and a preliminary unsmoothed solution is sought. Thereafter, a penalizing function (regularizer) is added to the model and additional solutions are sought with increasing weight of the regularizer. For each solution, a statistic, the 'probability to reject', is calculated *via* a comparison of the sum of squared residuals for this degree of regularization and that obtained in the preliminary calculation. The regularizer used is based on the sum of second derivatives of G(Γ) with respect to Γ. The solution for which the 'probability to reject' is closest to 0.5 is recommended as the 'chosen solution'. All fitting is done to the g_1 values, which are calculated from the g_2-1 values obtained from a homodyne experiment.

RILIE[13] An iterative regularization procedure adapted for application on a desktop computer. It generally yields results somewhat smoother than CONTIN's chosen solution.

REPES A constrained regularization procedure similar to CONTIN's PSC package, using a second order regularizer. Unlike CONTIN, it fits directly to g_2-1 and, after performing its preliminary (unsmoothed) calculation, it will find directly the solution at the requested 'probability to reject'.

MAXENT[14,15] A maximum entropy procedure implemented by Livesey and co-workers. A grid of line-widths is laid out with equal spacing in log(Γ), and the distribution in G(Γ) is sought such that the sum of squared weighted residuals (in g_2-1) has reached a predetermined level, subject to the constraint that the entropy decrease caused by deviation of the distribution from a rectangular distribution is a minimum.

3. USING CONTIN

During the past two decades there has been collaboration in the field of polymer chemistry between the Institute of Physical Chemistry in Uppsala and the Institute of Macromolecular Chemistry, Czechoslovak Academy of Sciences in Prague. In recent years we have used QLS to investigate polymer-solvent systems in the concentration region from dilute solution to very concentrated solution, with a variety of solvents. Of very special interest to the polymer chemist is the so called theta (Θ), or Flory, condition. For a given polymer-solvent pair there can often be found a temperature at which solutions behave as ideal solutions, just as a gas will behave as an ideal gas at the Boyle temperature. This temperature, at which the solution has ideal behaviour, is called the theta temperature. Some rather unexpected results were found for such systems, as shown below (in Fig. 8, for example). This excessive amount of detail is very disturbing and led Jakeš in Prague to take a closer look into the workings of CONTIN.

Jakeš[16] simulated a number of experiments, each containing about 1000 points distributed semi-logarithmically over a lag-time span from 2 μs to 30 s. The simulations were based on various relaxation time distributions, as seen in the following. Analyses are presented in terms of a distribution of relaxation times (τ), rather than line-widths (Γ). The field correlation is described as:

$$g_1(t) = \int_0^{\infty} G(\Gamma) \exp(-\Gamma t) \, d\Gamma = \int_{-\infty}^{\infty} \Gamma \, G(\Gamma) \exp(-\Gamma t) \, d\ln(\Gamma)$$

$$g_1(t) = \int_0^{\infty} A(\tau) \exp(-t/\tau) \, d\tau = \int_{-\infty}^{\infty} \tau \, A(\tau) \exp(-t/\tau) \, d\ln(\tau)$$

The simulated homodyne intensity correlation measurement will then be the square of $g_1(t)$ plus added noise.

Integration Jakeš' first discovery was that, contrary to Stock and Ray's conclusion[17] that CONTIN can oversmooth, CONTIN would undersmooth at short relaxation times and oversmooth at long

relaxation times. This is demonstrated clearly in Fig. 2, where the fit to the rectangular bin at long relaxation time is quite adequate, but to that at short relaxation time, quite unsmoothed. This he attributed to regularizing in G(Γ) rather than ΓG(Γ) when working with a grid equally spaced in log (Γ). Simply turning off integration (setting the quadrature flag, IQUAD, to one) produced the desired effect. This is equivalent to switching from Simpson to trapezoidal integration.

Baseline Application of standard CONTIN to distributions which are negatively skewed (in a G(Γ) description) produces artifacts in the slow relaxation region, as seen for the Pearson distribution (Fig. 3) and more clearly for the Gaussian distribution (in Γ) in Fig. 4.

For these distributions, the majority of relaxation is in the millisecond region and the g_2-1 values for lag-times beyond 100 ms. will be negligible. Since CONTIN fits to g_1, it must take the square root of g_2-1. For $g_2<1$ CONTIN uses -square root of $|g_2-1|$, which can introduce a bias if a considerable portion of the data fall into this category. Reconstructing the above distributions with ~ 7% added internal baseline removes the hump from the Pearson distribution, and suppresses, to a great degree, the artifacts in the Gaussian distribution, as seen in Fig. 5. Jakeš achieved the same improvement with his own *Re*gularized *P*ositive *E*xponential *S*ums (REPES) program, which operates directly on g_2-1.

The importance of this baseline effect should not be exaggerated, however. There is little to be gained by including a large number of data at long lag-times if they contain only random noise.

Edge effects One should also be aware that a sharp edge in a distribution can produce 'ripples' in the the distribution spectrum. The presence of a spike in a distribution can produce waviness across an adjacent smooth distribution, as in Figs. 6 and 7.

4. APPLICATIONS

Combined diffusive and viscoelastic relaxations

The system which started Jakeš on his search for numerical abberations in the CONTIN analysis was polystyrene in cyclohexane

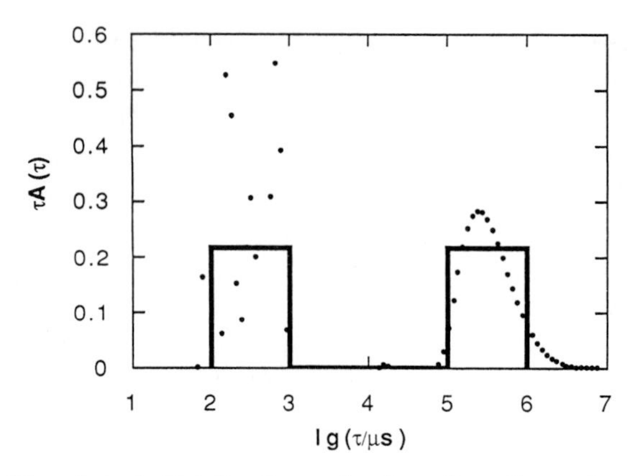

Figure 2. Simulated distribution of relaxation times in two equally large rectangular bins. Fitting with standard CONTIN. (Jakeš).

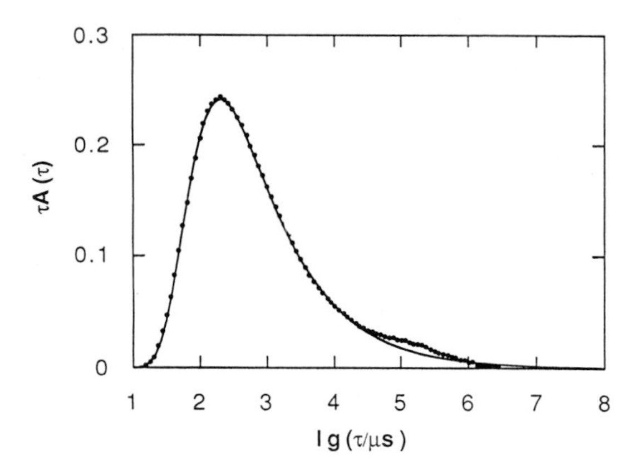

Figure 3. Simulated Pearson distribution (solid line) and results from CONTIN with IQUAD = 1. (Jakeš).

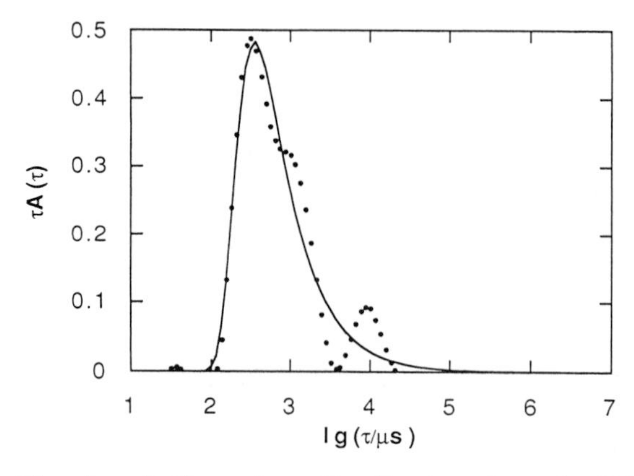

Figure 4. Simulated Gaussian distribution in t (solid line) and results from CONTIN with IQUAD = 1. (Jakeš).

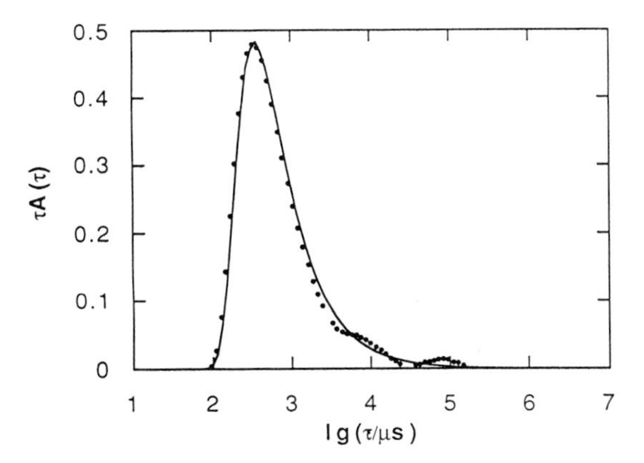

Figure 5. Simulated Gaussian distribution, as in previous figure, but with 7% baseline added to g_1 (solid line) and results from CONTIN with IQUAD = 1. (Jakeš).

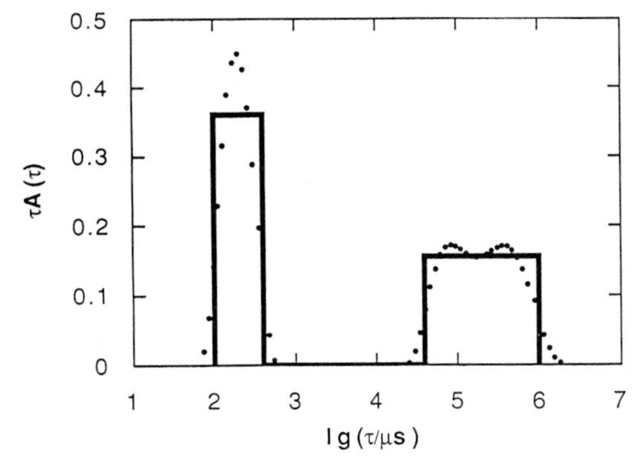

Figure 6. The influence of bin width on the edge effect in a CONTIN analysis. (Jakeš).

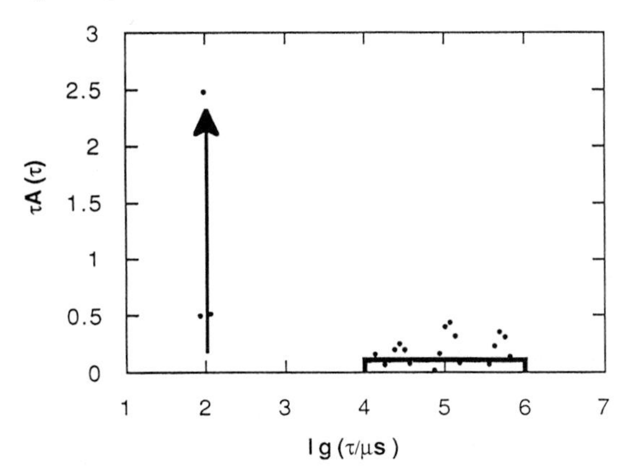

Figure 7. A delta spike producing an edge effect on a rectangular bin. (Jakeš).

at the *theta* temperature (≈35°C), as seen in Fig. 8. The extraordinarily detailed structure does not disappear upon turning integration off (IQUAD=1), nor is it in any way related to a baseline effect. It is present in analyses performed with REPES (Jakeš regularizing routine mentioned above) and with MAXENT (Figs. 9 and 10). These two routines operate on g_2-1 directly, in which case the baseline effect need not be considered.

The most rapid relaxation, at about 70 μs relaxation time, accounts for nearly half of the scattering intensity. Could this sharp edge produce the detail in the long relaxation time region? Subtracting this fast relaxation from the experimental data and re-analyzing with REPES, virtually no change in the long relaxation region (Fig. 11) was found.

Changing the scattering angle yielded similar detail in the resulting spectra of relaxation times, with a shift in position of only the fastest relaxation (Fig. 12). This line-width is directly proportional to the square of the scattering vector, q, (Fig. 13) indicating that it is a true diffusive mode.

Analysis of a composite data set, in which measurements at varying angles are added together (the lag-times being the same in all experiments), yielded, in Fig. 14, a smeared diffusive mode and sharp q-independent modes. This would indicate that we are observing a complex pattern of viscoelastic relaxations in concentrated high polymer solutions at the theta temperature.[18,19]

Internal Modes Dilute solutions of monodisperse linear polymers exhibit, in the field intensity autocorrelation spectrum,[20] relaxations due to translational diffusion and, if $qR_g>1$, intramolecular motions (where q is the scattering vector $((4\pi n/\lambda)\sin(\theta/2))$ and R_g is the radius of gyration of the polymer coils). In the Rouse-Zimm model, these intramolecular motions can be described in terms of a series of internal modes. Writing only the first internal mode explicitly, the field autocorrelation function is given by[21]

$$g_1(t) = \exp(-\Gamma_1 t)\ (A_0 + A_1\exp(-\Gamma_2 t) + \ldots)$$

Γ_1 is q^2 times D, the translational diffusion coefficient and $\Gamma_2 = 2/\tau_1$, the relaxation time of the first internal mode.

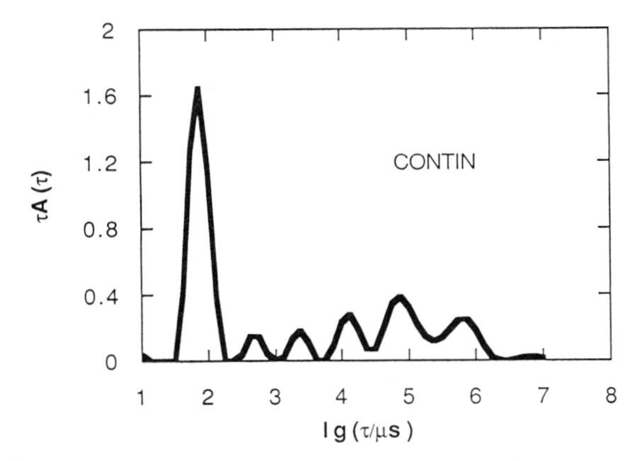

Figure 8. CONTIN (IQUAD = 1) analysis of a concentrated Θ-solution of polystyrene (M = 3.8 x 10^6, cyclohexane, 35°C, concentration 12.5 g/dl).

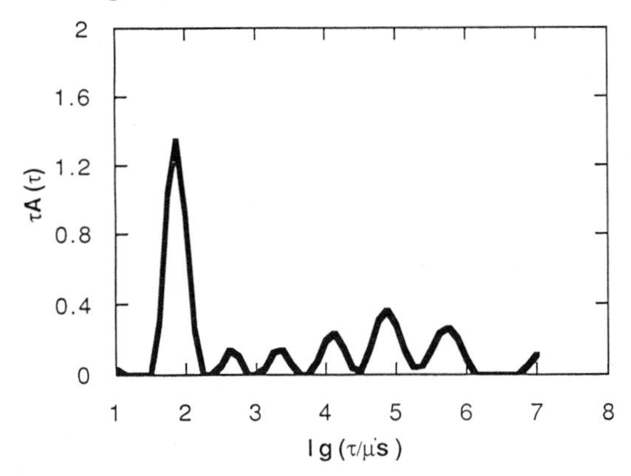

Figure 9. As in previous figure, but analyzed with REPES.

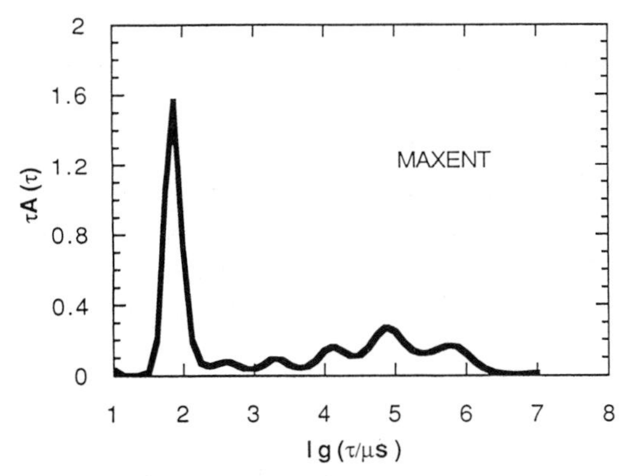

Figure 10. As in previous figure, but analyzed with MAXENT.

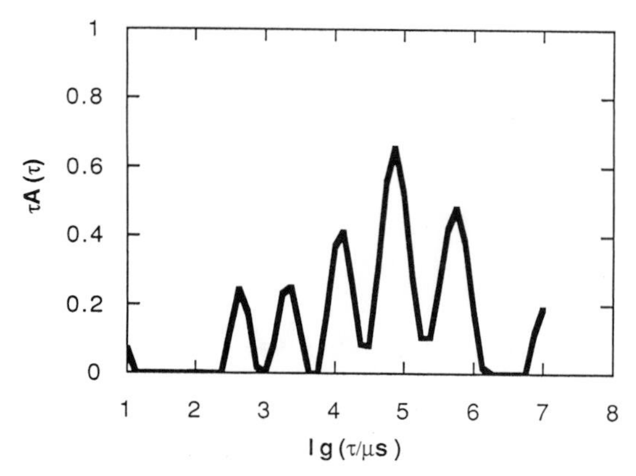

Figure 11. REPES analysis of PS Θ-solution after subtraction of the diffusion relaxation from the g_2-1 values.

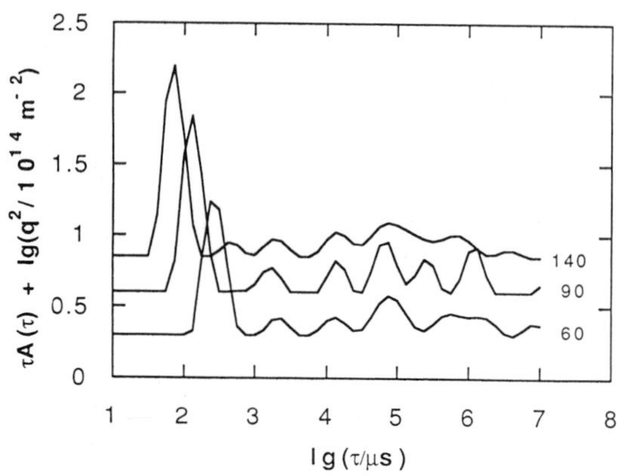

Figure 12. REPES analysis of three experiments at different detection angles for the PS-cyclohexane system. Same solution as in previous figures, but wavelength 632 nm rather than 488 nm.

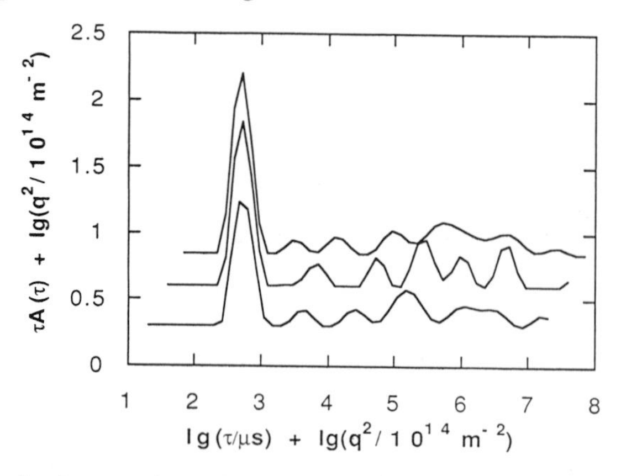

Figure 13. As in previous figure, but shifted along the abscissa by lg $(q^2/10^{14}\ m^{-2})$.

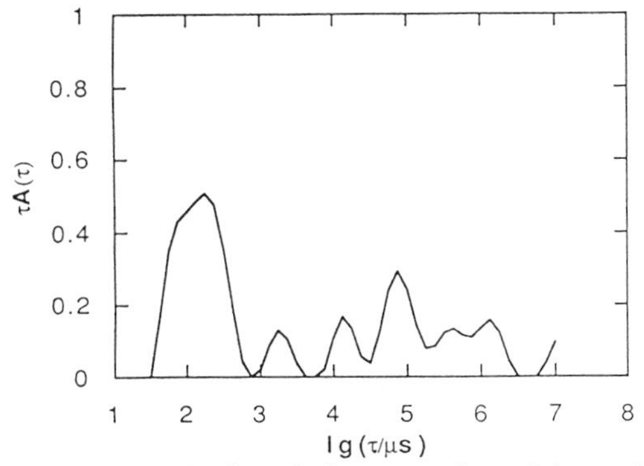

Figure 14. REPES analysis of data produced by adding together the g_2-1 data for the three different angles. The q-independent relaxation modes should remain stationary (reinforce each other) and the q-dependent modes should smear.

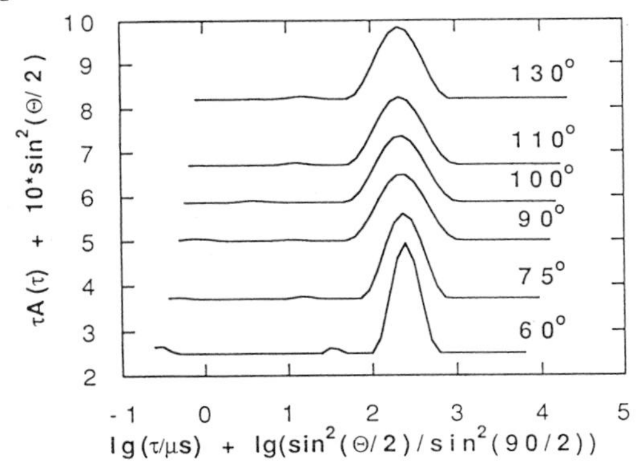

Figure 15. REPES analysis of a very dilute (0.0336 g/dl) solution of PS in 2-butanone. The abscissa values are shifted to compensate for the effect of angle on the relaxation time of a diffusive mode.

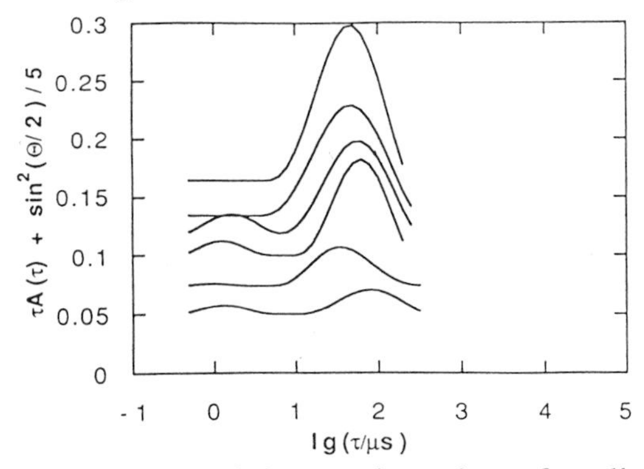

Figure 16. Re-analysis of the previous data after dividing the g_2-1 values by exp $(-2Dq^2t)$, revealing the relaxation due to internal modes which are of such low amplitude that they are not visible in Fig. 15.

In Fig. 15 the REPES analysis of measurements at several angles of the system polystyrene (M=4.9 x 10^6) in 2-butanone at 25°C. the translational diffusion is well represented but the internal mode, at slightly smaller relaxation time, is indistinguishable. Using a translational diffusion coefficient determined from measurements at lower angles, where the contribution from internal modes is negligible, the g_2-1 values for these experiments were divided by $\exp(-2Dq^2t)$ and reanalyzed with REPES, Fig. 16. The internal mode is clearly revealed at higher angles. The value used for the translational diffusion coefficient will necessarily strongly influence the final analysis, and should therefore be determined with the highest accuracy possible.

ACKNOWLEDGEMENTS

We are grateful to Jaromír Jakeš for the use of his REPES program. We are also indebted to Dr. Taco Nicolai for the long duration angle-dependence measurements on the PS-cyclohexane system, and to Johan Fundin for the PS-2-butanone measurements.

APPENDIX - PARTICIPANTS IN THE PROJECT: "COMPARISON OF METHODS FOR ANALYSIS OF DATA FROM DYNAMIC LIGHT SCATTERING"

In the case of research groups, the contact person is listed first.

1. A. Cantor and R. Pecora, Stanford University, U.S.A., using CONTIN and DISCRETE.
2. B. Chu, H. Dhadwal, and Y. Georgalis, Stony Brook, N.Y., using cumulants, RILIE, CONTIN.
3. J. Jakeš, Institute of Macromolecular Chemistry, Czechoslovak Academy of Sciences, using Positive Exponential Sum.
4. T.A. King and I. Butler, University of Manchester, using S-exponential sum.
5. K.H. Langley and Ki Min Eum, University of Massachusetts, Amherst, using cumulants, multiple exponentials, CONTIN.
6. S.W. Provencher, EMBL, Heidelberg, using CONTIN.
7. J.G. Rarity, P.N. Pusey, E.R. Pike and F.K. McNeil-Watson, Malvern, using cumulants, exponential-sampling, profiled singular value.
8. Z.A. Schelly, University of Texas at Arlington, using Z-transform.

9. K.S. Schmitz, University of Missouri, using Lambda depression, histogram.
10. K. Zimmermann, Institut Jacques Monod, Paris.

REFERENCES

1. D.E. Koppel, *J. Chem. Phys.*, 1972, *57*, 4814.
2. D.W. Marquardt, *J. Soc. Ind. Appl. Math.*, 1963, *11*, 431.
3. P.J. Nash and T.A. King, *J. Chem. Soc. Faraday Trans. 2,* 1985, *81*, 881.
4. K. Zimmermann, M. Delaye and P. Licinio, *J. Chem. Phys.*, 1985, *82*, 2228.
5. J. Jakeš, *Appl. Mat. (Czechoslovakia)*, 1988, *33*, 161.
6. J. Szamosi and Z.A. Schelly, *J. Phys. Chem.*, 1984, *88*, 3197.
7. N. Ostrowsky, D. Sornette, P. Parker and E.R. Pike, *Optica Acta,* 1981, *28*, 1059.
8. M. Bertero, P. Brianzi and E.R. Pike, *Inverse Problems*, 1985, *1*, 1.
9. M. Bertero, P.Brianzi, E.R. Pike, G. de Villiers, K.H. Lan and N.Ostrowsky, *J. Chem. Phys. Lett.*, 1985.
10. G.C. Fletcher and D.J. Ramsay, *Optica Acta*, 1983, *30*, 1183.
11. S.W. Provencher, *Biophys. J.*, 1976, *16*, 27.
12. S.W. Provencher, *Comput. Phys. Commun.*, 1982, *27*, 213.
13. B. Chu, 'Methods of Enzymology', Academic Press, NY, 1985, Vol. 101.
14. A.K. Livesey, M. Delaye, P. Licinio and J.E. Brochon, *Faraday Discuss. Chem. Soc.*, 1987, *83*, 247.
15. A.K. Livesey, M. Delaye and P. Licinio, *J. Chem. Phys.*, 1986, *84*, 5102.
16. J. Jakeš, *Czech. J. Phys. B*, 1988, *B38*, 1305.
17. R.S. Stock and W.H. Ray, *J. Polym. Sci. Polym. Phys. Ed.*, 1985, *23*, 1393.
18. T. Nicolai, W. Brown, R.M. Johnsen and P. Štěpánek, *Macromolecules,* 1990, *23*, 1165.
19. W. Brown, T. Nicolai, S. Hvidt and P. Štěpánek, *Macromolecules,* 1990, *23*, 357.
20. T. Nicolai, W. Brown and R. Johnsen, *Macromolecules.,* 1989, *22*, 12795.
21. R. Pecora, *J. Chem. Phys.*, 1965, *43*, 1562.

7

Low-bias Macroscopic Analysis of Polydispersity

By Stephen W. Provencher

MAX-PLANCK-INSTITUT FÜR BIOPHYSIKALISCHE CHEMIE,
POSTFACH 2841, D-3400 GÖTTINGEN, GERMANY.

1. INTRODUCTION

In light scattering, as well as many other areas, the data, y_j, are often modelled by a linear superposition of apparently independent contributions,

$$y_j = \int_a^b s(z)\, K(z,t_j)\, dz + \varepsilon_j, \qquad j = 1, 2, ..., N_y \qquad (1)$$

where s(z), which characterizes the "polydispersity," is to be estimated. In addition to actual size polydispersity, the distribution s(z) may describe additional processes, such as non-translational motions. At most, the covariance matrix of the noise components, ε_j, is approximately known. The kernel, K(z,t), is assumed to be known and typically contains a scattering form factor, a weighting factor for the particular type of distribution function in the variable z, and an exponentially decaying factor, *e.g.*, exp(-zt), as in quasi-elastic light scattering (QLS), where the y_j are measurements of the electric-field autocorrelation function magnitude at delay times t_j. However, the methods described here can handle other K(z,t).

This is now widely recognized as an *ill-posed problem*, where there is an infinite set, Ω, of possible solutions, all of which fit the data to within experimental error. Even worse, the errors are unbounded. Thus, for most physically reasonable K(z,t) and true s(z), and an arbitrarily *large* error amplitude, A, there always exists a large enough ω such that[1]

$$s(z) + A \sin(\omega z) \in \Omega \tag{2}$$

for arbitrarily *small* but non-zero $|\varepsilon_j|$. There are infinitely many non-negative members of Ω. For *each* of these non-negative members there are infinitely many wildly oscillating members of Ω, such as those in eq. (2).

The traditional way of dealing with this apparently hopeless problem was to simply restrict s(z) to a fixed form (*e.g.* a Schulz-Zimm distribution or a sum of one or two δ-functions) with a few parameters that are fit to the data. This can yield excellent results if this fixed form happens to be correct, but otherwise there is great danger of misleading results. This highly biased approach can also prevent unexpected discoveries, which are the most informative ones. This approach can be useful if the fixed form is based on prior knowledge or on a single model with few parameters that is to be tested on a series of data sets obtained under a wide range of conditions. Increasing the number of adjustable parameters decreases the reliability of their estimates and the credibility and significance of the model.

Over the past twelve years, (non-negatively) constrained regularization methods, such as implemented in the program package CONTIN[1-5] and recently in maximum-entropy (MaxEnt[6-8]) routines, have been the most successful in avoiding the bias of fixed-form models and still providing reliable information on s(z). Here a more model-independent bias toward parsimony is purposely introduced to select the member of Ω that has the maximum entropy or "smoothness," as defined in CONTIN. Parsimony tends to protect against artifacts; the selected member of Ω may very well not have all the detail of the true solution, but the detail that it does have should be "*demanded* by the data"[5,6] and therefore less likely artifact.

We shall see that this is not always true. When the true s(z) strongly disagrees with the bias imposed by the regularizor of either of these methods, then it can disastrously backfire and cause artifact peaks in the estimated s(z). The problem is that we should have taken the term "ill-posed problem" literally. We are asking for too much when we attempt to estimate s(z) on the microscopic scale (*i.e.* at each grid point). Regularization methods are optimal for this type of problem, but the uncertainties in the estimates at the individual grid points are still so great that their isolated values are not of use to us. CONTIN and most other methods immediately take this microscopic solution and compute macroscopic statistics of the peaks in the solution, such as their amplitudes, means, standard deviations and skewnesses. These are the useful quantities because they can be accurately and reproducibly estimated, although they are biased by the regularizors. However, if we are only interested in well-determined macroscopic statistics, we do not need regularization. This paper presents two methods that avoid these regularizors and their biases completely.

The occurrence of an artifact peak can be disastrous because peaks are usually the most important features of a distribution function. They are often interpreted as evidence for a separate species, size class or type of process. Peaks together with their cumulants or other macroscopic statistics are in fact the main information obtained from an experiment, since the inverse problems occurring in light scattering are generally so ill-posed that shoulders or higher-order critical points cannot be reliably determined. One very direct way of avoiding artifact peaks is to simply select from Ω the non-negative member with the best fit to the data and the fewest peaks. A limited regularized version of this has already been crudely implemented[2,1] in CONTIN as "peak-constrained solutions." The first method (MinPeak) uses an improved unregularized version. Macroscopic statistics such as cumulants are computed for each peak. Strictly speaking, MinPeak is still slightly biased because it will select a member of Ω that replaces several unresolvable peaks in the true solution by a single peak.

The other method, MinClass, does not have this bias. It works with more general classes, which are not constrained to be single peaks. A class is the density, s(z), in a closed interval of the z-grid. All possible partitions of the z-grid are exhaustively searched for the partition with the minimum mean-square predicted relative error in the total density in each class. MinClass has always yielded results similar to MinPeak, but it has the advantages of simplicity, of a guaranteed global optimum through the exhaustive search, and of the complete lack of bias (except that inherent in the non-negativity constraint, which is usually based on prior knowledge). At the present early stage of development, the two methods are probably best used in combination, since MinPeak provides a good stopping criterion with the minimum number of peaks necessary to fit the data[2,1], and MinClass provides nearly unbiased estimates of the corresponding peak statistics. They yield an automatic hierarchical analysis with a series of possible solutions, starting with the best one-peak or one-class solution and increasing in complexity to the best n-peak solution, where n is the smallest number of peaks in Ω.

We shall also see that the usual error estimates for the unregularized microscopic solution and the macroscopic statistics are grossly pessimistic, because the non-negativity is neglected. We use a Monte Carlo procedure to get error estimates, which agree well with the actual errors. The macroscopic peak statistics can be remarkably accurate. In addition to avoiding artifacts, MinPeak and MinClass can yield significantly higher resolution, particularly of sharp or overlapping peaks that are broadened by the biases of CONTIN and MaxEnt.

2. METHODS

Simulation and Regularization

The noisy second-order (intensity) QLS autocorrelation function

$$Y_j = B\,(1 + \gamma^2\, y_j^2) \qquad (3)$$

was simulated assuming that it followed Poisson statistics with background $B=10^7$ and coherence factor $\gamma^2=0.3$. The IWT=5 option[2,4] of CONTIN was used to estimate the y_j and least-squares weights, which were used for all the other methods as well. There were 15 groups of delay times with a total of 128 channels; the first group had 16 channels from 0.2 μs to 3.2 μs; the second had eight channels from 3.6 μs to 6.4 μs; each succeeding group had eight channels with twice the delay time of the corresponding channel in the preceding group; the longest delay time was 52.4288 ms. In practice[2], γ is unknown, and the γy_j are used as data in eq. (1). The integral over the unnormalized solution, $\gamma\ s(z)$, provides an estimate for γ.

In eq. (1), s(z) was a decay-constant distribution in ln Γ; *i.e.*

$$K(z,t) = \exp(-\Gamma t), \qquad z = \ln \Gamma \tag{4}$$

The z-grid had 85 points in equal intervals of z from $\Gamma=1\ s^{-1}$ to $\Gamma=10^7\ s^{-1}$. The true s(z) was simulated as a sum of δ-functions at the grid points, and eq. (1) was analyzed as a sum of exponentials. Thus there were no discretization or numerical integration errors that could have been suspected of causing the artifact peaks. Similarly, error-free values of B were used. [Artifact peaks were also obtained with more realistic simulations, as well with the simple linear case, where eq. (3) was replaced by $Y_j = y_j$.]

The regularized solutions, $s_\alpha(z)$, are obtained by solving

$$VAR(\alpha) + \alpha^2 R[s_\alpha(z)] = \text{minimum}, \qquad s_\alpha(z) \geq 0 \tag{5}$$

$$VAR(\alpha) \equiv \sum_{j=1}^{N_y} w_j(\hat{y}_j - y_j)^2 \tag{6}$$

on the z-grid, where w_j is the least-squares weight, $\hat{y}_j$ is the fit value obtained by substituting $s_\alpha(z)$ into eq. (1), and the regularization parameter, α, determines how strongly the regularizor, $R[s_\alpha(z)]$, will bias the solution. With CONTIN, $R[s_\alpha(z)]$ is the sum of the squares of the second differences of $s_\alpha(z)$ at the grid points[2,1]; with MaxEnt, it is the negative entropy[9,8]. For a given α, both have unique solutions with no secondary minima.

The choice of α is critical. The bias and VAR(α), the variance in the fit to the data, increase monotonically with α. First CONTIN computes a nearly unbiased "Reference Solution,"[2,1] $s_0(z)$, and VAR(0), which effectively correspond to $\alpha = 0$. For a series of increasing α values, CONTIN then computes $s_\alpha(z)$ and PROB1(α), which is a rough probability level that VAR(α)/VAR(0) is so large that it is due to bias[2,1]. A detailed search was made until an α was found with

$$0.45 \leq PROB1(\alpha) \leq 0.55 \tag{7}$$

For a given Reference Solution, PROB1(α) depends only on VAR(α). Therefore with MaxEnt, α was adjusted until the narrow range in eq. (7) was also satisfied by its VAR(α). Thus the solutions from the two methods had practically the same variance of fit to the data and could be directly compared. Also for comparison with MaxEnt, CONTIN imposed no boundary conditions on s(z) at the ends of the grid.

The "Classic MaxEnt"[9,8] form,

$$R[s_\alpha(z)] = \sum_i [f_i \ln (f_i/m_i) + m_i - f_i] \quad (8)$$

is used. The sum extends over the grid points, and f_i and m_i are respectively the amplitudes and the prior expected amplitudes of $s_\alpha(z)$ in the discretized problem. The f_i are biased toward the m_i. Since the z-grid is in equal intervals of ln Γ, a constant $m_i = m$ is used[6]. Equation (8) accounts for the fact that the normalization constant for s(z) is unknown and therefore that m is also to be estimated. For a fixed α and with the parameters m and $x_i \equiv \ln f_i$, eq. (5) is an unconstrained optimization problem. This was solved with the full Newton method using the standard Marquardt modification[10]. Each iteration is computationally expensive, but reliable, with Bryan's convergence test[8] $t < 10^{-20}$, indicating very accurate convergence. A simple *Regula falsi* search of VAR(α) *versus* α yielded a series of solutions with different fixed α, converging to a solution that satisfied eq. (7). It can be important to include a "Dust Term"[3] with Γ = 0 to account for processes with decays too slow to be measured. To avoid questions of how to assign a prior to such a term, MaxEnt was given the advantage of prior knowledge that the simulations in this paper had no Dust Term, and MaxEnt had none. The other methods had a Dust Term.

MinPeak

MinPeak searches for the non-negative member of Ω with the fewest peaks that best fits the data. In principle, MinPeak with up to two peaks can be implemented with the peak-constrained analyses of CONTIN using a small α that still avoids error messages due to numerical instabilities. However, MinPeak has no regularizor and no equality constraints. In this case, some of the operations in CONTIN can be eliminated or performed exactly, and the inequality constraints can be converted to simple non-negativity constraints.[2] A small zero-order "regularizor," essentially a Marquardt parameter,[10] can be used if numerical instabilities occur. This simplifies, stabilizes, and speeds up the computations. Large gradients of the objective function at the binding constraints can provide useful indications of where to shift the peak or valley positions to get faster to the optimum, somewhat analogous to gradient methods for nonlinear optimization.

MinPeak starts with a one-peak analysis and increases the number of peaks until PROB1 drops down below 0.5. Usually the transition is from greater than 0.99 for n-1 peaks to less than 0.01 for n peaks, and the choice of n peaks is clear. Otherwise, both solutions must be seriously considered[2,1]. We shall see that these low-bias solutions can yield remarkably accurate macroscopic peak statistics.

MinClass

MinClass substitutes $s_0(z)$ into eq. (1) to obtain the fit to the data, $\hat{y}_j$. Then N_r replicate noisy data sets are produced by adding pseudo-random noise (with variances inversely proportional to the w_j) to the $\hat{y}_j$. The $s_0(z)$ for each of these replicate data sets are computed from eq. (5). In this paper, $N_r = 50$ was used. For each pair of grid points with $z_i < z_k$,

$$I_v(z_i,z_k) = \int_{z_i}^{z_k} s_0(z)\, dz \tag{9}$$

is numerically evaluated, where $v = 0$ denotes using the original $s_0(z)$ in eq. (9), and $v > 0$ a replicate $s_0(z)$. A useful measure of the relative uncertainty in $I_0(z_i,z_k)$ is

$$\sigma_{ik}^2 \equiv \frac{1}{N_r} \sum_{v=1}^{N_r} \left\{ \frac{I_v(z_i,z_k)}{I_0(z_i,z_k)} - 1 \right\}^2 \tag{10}$$

where σ_{ik}^2 is set to a very large number if the original $s_0(z)$ has no density on the interval and $I_0(z_i,z_k) = 0$. These error estimates account for the non-negativity; the usual linear estimates of the covariance matrix from least squares implicitly assume that the s(z) can go negative, and greatly overestimate the errors.

Given n, the number of classes, MinClass searches all possible partitions of the z-grid into n contiguous closed intervals and chooses the partition with the smallest sum over the σ_{ik}^2. In the sense of the above criterion, this partition makes the most accurate statements about the distribution. The macroscopic statistics of the classes, such as cumulants, are nearly unbiased, and their uncertainties are estimated from σ_{ik} in eq. (10) with $I_v(z_i,z_k)$ then denoting the corresponding cumulant. The sample correlation coefficients between the I_0 estimates for neighbouring classes are computed from the I_v of eq. (9).

Strictly speaking, statistics such as moments are indeterminate if the z-grid extends arbitrarily far. A very slow process has an infinite possible range in ln Γ until $\Gamma = 0$, and the fast end of the z-axis is completely indeterminate. MinClass therefore has the option of adding an "End Class" at each end of the z-grid. The σ_{ik}^2 of the End Classes are not included in the sum to be minimized. This can be very useful in automatically splitting off nearly indeterminate density at the ends of the grid. The statements about all the classes remain correct; *i.e.* the error estimates of the statistics in these End Classes are usually very large, and those in the inner classes are correspondingly smaller. We allow End Classes at both ends, although the slow End Class is usually empty, because the Dust Term serves a similar purpose.

MinClass computes a series of solutions with increasing n, starting with n = 1. Typically, an n is reached where going to n+1 classes causes a large increase in two of the σ_{ik} and a large negative correlation coefficient between the two classes; then the choice of n classes is fairly clear. However, the stopping criterion of MinPeak is much more reliable and unambiguous. MinClass is faster than CONTIN, because part of the computations for the 50 replicate solutions need only be done once and because many combinations in the exhaustive search can be skipped as soon as the partial sum of the σ_{ik}^2 exceeds the minimum sum found so far.

3. RESULTS

Failure of MaxEnt

A broad unimodal s(z) was simulated with four overlapping Gaussian profiles in z, with standard deviations all equal to $\ln(\sqrt{10})$ and means at $\Gamma = 10^2, 10^{8/3}, 10^{10/3}, 10^4\ s^{-1}$. The amplitudes were all $\gamma/4$, with $\gamma^2 = 0.3$ in eq. (3). Ten analyses were performed with this s(z) and ten sets of pseudo-random noise (see Section 2). Figure 1 shows a typical analysis. In all ten cases, MaxEnt produced such artifact peaks.

In one case CONTIN also yielded artifact peaks when eq. (7) was satisfied, but yielded an excellent solution when PROB1 = 0.76. In the other nine cases excellent results as in Fig. 1 were obtained with eq. (7). Similarly, in this one borderline case, the MinPeak one-peak solution had PROB1 = 0.59; in the other nine cases, PROB1 ranged from 10^{-4} to 0.16, indicating a completely adequate fit with one peak. It is therefore important to consider at least all solutions in the range $0.1 \leq$ PROB1 ≤ 0.9 as significant possibilities[2,1]. However, for Table 1, the solutions obeying eq. (7) were used.

From each solution, γ, the integral of the unnormalized solution, and κ_j, the cumulants[11,12], are computed for the entire solution and for each peak (of CONTIN, MaxEnt, and MinPeak) and class (of MinClass). For sharp peaks, the standard deviation in z, $\sqrt{\kappa_2}$, is to first approximation the standard deviation over the mean in Γ. The useful dimensionless shape statistics, $R_j \equiv \kappa_j / \kappa_2^{j/2}$, are computed from j=3 to j=6, where R_3 is the skewness and R_4 the excess, also called kurtosis[11]. A normal distribution has zero R_j for all $j \geq 3$. A distribution with positive R_4 tends to be more sharply peaked than a normal; *e.g.* double exponentials, exp $(-c_1|z-c_2|)$, all have $R_4 = 3$; conversely, rectangular distributions all have $R_4 = -1.2$.

Table 1 shows the errors in these statistics (up to R_4) from the ten one-class solutions of MinClass and from the entire solutions of the other methods. Considering the uncertainties in the solutions on the microscopic scale, these statistics are remarkably accurate with CONTIN, but also with MinPeak and MinClass, which are not dependent on the smoothness of the true solution. The error estimates for the peak statistics from MinClass using analogs to eq. (10) are in good agreement with the actual errors in Table 1. The s(z) simulated here happens to have no density at the ends of the grid and no Dust Term. Therefore, provision for these in MinClass and, to a lesser extent, MinPeak, leads to somewhat less accuracy in Table 1. The biases in CONTIN and MinPeak tend to cluster s(z) together, away from the ends of the grid, and CONTIN always correctly found a zero Dust Term. These three methods also provide useful estimates of R_j up to j = 6, and possibly higher. However, this is an unusually broad well-

Table 1. Average absolute values of the errors in the analyses of ten simulated data sets from a broad unimodal distribution. The true values of the peak statistics and Dust Term of the unnormalized solution are given in parentheses. Percent errors are shown, except for absolute errors when the true value is 0.

	(True value)	CONTIN	MaxEnt	MinPeak	MinClass
Amplitude, γ	($\sqrt{0.3}$)	0.01%	0.41%	0.27%	0.68%
Mean, κ_1	(6.908)	0.09%	0.21%	0.32%	0.54%
Standard deviation, $\sqrt{\kappa_2}$	(2.07)	0.6%	6.5%	1.5%	2.7%
Skewness, $\kappa_3 / \kappa_2^{3/2}$	(0.)	0.029	0.091	0.082	0.084
Excess, κ_4 / κ_2^2	(-0.65)	9%	180%	26%	32%
Dust Term	(0.)	0.	(0.)	0.0015	0.0025

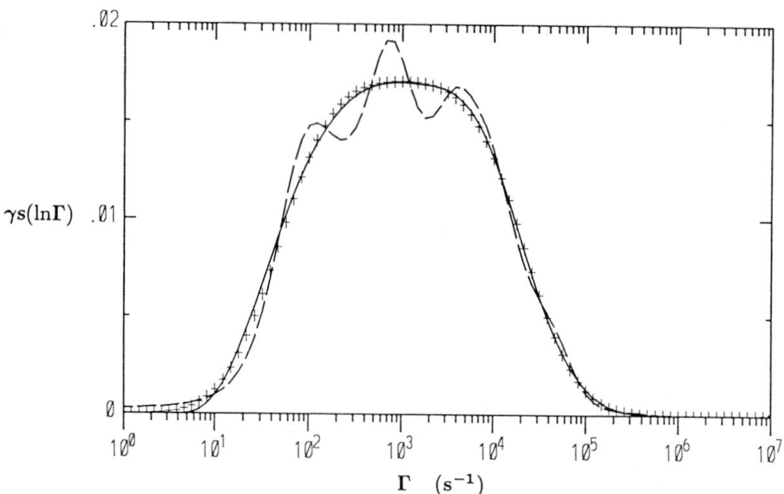

Figure 1. Failure of MaxEnt. In the analysis of a broad unimodal distribution (+++), MaxEnt (broken line) consistently produces artifact peaks and CONTIN (solid line) does not.

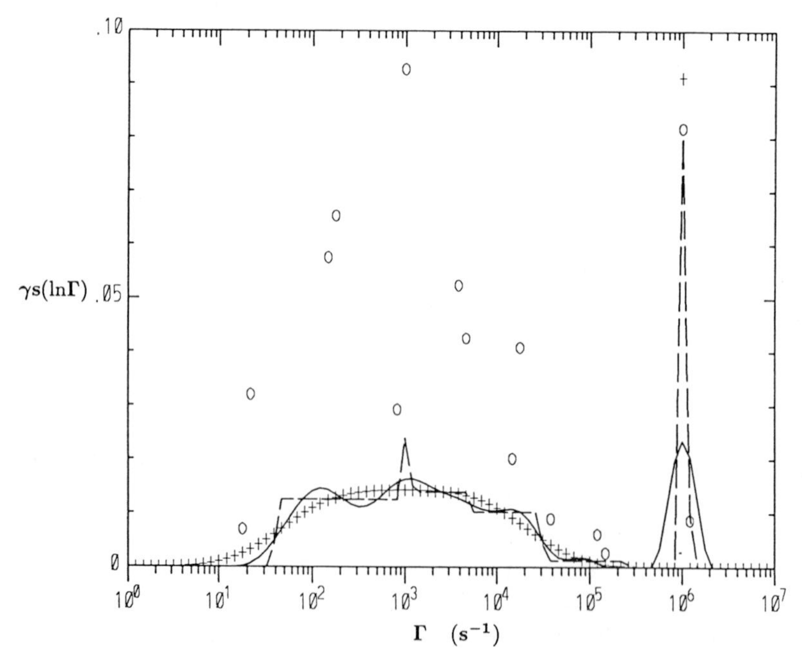

Figure 2. Failure of CONTIN. The simulated distribution (+++) is the same as in Fig. 1, except for an added isolated δ-function. Now CONTIN (solid line) consistently produces artifact peaks similar to MaxEnt (not shown). The two-peak MinPeak analysis (broken line) and the 15 non-zero points in the Reference Solution (circles) are also shown.

determined peak. MaxEnt has difficulty with the standard deviation, excess, and higher-degree statistics because its solution is sometimes biased upward toward m in eq. (8) at the ends of the z-grid, where there is less information in the data.

As predicted[1], MaxEnt is susceptible to artifact peaks, because its regularizor, eq. (8), takes no account of the relative positions of the grid points on the z-axis or of possible correlations between neighbours. It simply biases the solution toward m. This lack of extra bias toward correlation is sometimes considered an advantage, but the price paid is that it is insufficient to suppress artifact peaks, even with this high-quality data. CONTIN imposes a bias toward smoothness. In this simulation, s(z) happens to agree well with this bias, and CONTIN produces excellent results.

Failure of CONTIN

This smoothing bias can sometimes cause problems, however. To see this, an isolated δ-function at $\Gamma = 10^6\ s^{-1}$ was simply added to the s(z) of Fig. 1. In the renormalized s(z), the isolated δ had $^1/_6$ of the total amplitude. Fig. 2 shows a typical analysis of one of the ten simulated data sets. Although the extra δ is about a factor of 10 in Γ from the main peak, and therefore practically uncorrelated, it causes a drastic deterioration in the CONTIN solution. In order to satisfy eq. (7), CONTIN reduced the average α by a factor of 35. This is a compromise, in which the main peak is now undersmoothed but the added δ still oversmoothed. Similar effects have been observed with CONTIN with single asymmetric peaks with one very steep side and one broad side[13,7]. However, the simulated data had no added noise, which is necessary if the PROB1 criterion is to properly work. Similar problems occur in the related method of fitting functions with rapidly changing second derivatives with cubic splines[14]. However, this "δ-Effect" is very clear and occurs at attainable noise levels. It seems to have been first discovered and explained by Jakeš [15].

As expected[16], MaxEnt (not shown) yielded results similar to those in Fig. 1 for the main peak. They were slightly worse, probably because of the reduction in the long-lived components. The results of

CONTIN and MaxEnt are also similar, except that CONTIN consistently oversmoothes the added δ slightly more, with an average standard deviation ($\sqrt{\kappa_2}$) of 0.27, compared with 0.25 for MaxEnt. When there is no main peak, only a δ, the solutions (point-spread functions) with CONTIN and MaxEnt are remarkably similar, with standard deviations agreeing to within 1%.

MinPeak and MinClass resolve the δ much better, with an average $\sqrt{\kappa_2}$ of 0.092 and 0.075, respectively. For the main peak, MinClass actually had slightly smaller errors than those shown in Table 1, and MinPeak slightly larger, so that both methods had approximately the same accuracy. MinPeak strongly rejected all ten one-peak analyses with PROB1 exceeding 0.999 and accepted the two-peak analyses with an average PROB1 = 0.066. All ten two-class MinClass analyses correctly separated the two peaks, although the Reference Solutions, $s_0(z)$, themselves are uninterpretable (see Fig. 2). The staircase structure of the MinPeak solution in Fig. 2, with many binding inequality constraints, is typical and unaesthetic. However, they yield good low-bias peak statistics, unambiguous indications of which CONTIN and MaxEnt peaks are artifact, and a good stopping criterion for MinClass.

The CONTIN and MaxEnt analyses shown here are directly comparable because they have the same variance of fit to the data. The criterion in eq. (7) is relatively conservative, generally choosing larger values of α than other criteria. Smaller α values would have led to even stronger artifacts.

4. DISCUSSION

Regularization Methods

Regularization methods have been justified with Bayesian arguments[18,19,1,9] and in terms of the reduced variance of biased estimators[17,1,5] and their moments. Regardless of the justification, it is always necessary to specify in effect a prior probability distribution for the unknown s(z). When this prior is known, then a useful improved *a posteriori* estimate can be made using a rigorous formula (*Bayes' theorem*) from the rules of conditional probability. In practice, it is almost always necessary to guess at this prior (often as a uniform distribution), and the formula is then sometimes called *Bayes' postulate.* I would refer to the necessity of guessing at this prior as *Bayes' curse.* Not only is the guess difficult, but users tend to forget that it has been made and place far too much faith in the results.

With CONTIN, this guess is the form or order of the regularizor and with MaxEnt, the m_i in eq. (8). In addition the choice of α is critical. These bias the analysis; *i.e.* they introduce systematic errors into the solution in that the expectation value of the solution is no longer the true value. When the bias happens to agree with the truth (as with CONTIN in Table 1), then much greater accuracy can be achieved than with unbiased methods. If not, then serious artifacts can occur, as with MaxEnt and CONTIN in Fig. 2.

Many attempts have been made at *a posteriori* corrections to the regularizors. The δ-Effect can be reduced by suppressing the regularization in the region of the z-axis where a sharp peak is observed[15] or by subtracting the contribution of the sharp peak to the data and re-analyzing the data[20]. Similarly, one of the first applications of "Classic MaxEnt" produced poor results in a simple image restoration problem, because the uniform prior, $m_i = m$, was poor. "Classic MaxEnt" was then superseded by "New MaxEnt," in which the preceding solution is used to modify the prior for the next iteration, introducing a correlation-length parameter and a pre-model[21] to impose an extra bias toward spatial correlation, which is ordinarily prohibited by MaxEnt. I have multiplicatively modified the weighting of the regularizor for each half-peak (the segment between two successive extrema) until the penalty in the regularizor of each was equal, effectively charging each half-peak the same Bayesian price for admission to the solution. This worked well for a small set of tests, but the convergence and uniqueness properties are probably poor. In special cases, these methods may be useful when carefully applied, but they usually involve subjective decisions and added bias, and this can be dangerous in the analysis of experimental data.

The standard error estimates for the regularized solution also contain an unknown bias term[1] that is often ignored. In the limit of large α, these error estimates become meaningless and approach zero[1]. Thus, they cannot be used to unambiguously detect artifact peaks or to guide the above subjective modifications. Bayesian methods have also been used[18,21,8] to choose α, but, as always, the results depend on the choice of the prior probability distribution that must be specified for α.

At least as a stopgap measure, CONTIN can still be useful. In the vast majority of cases, it produces no artifact peaks and yields reasonably accurate moments. Peak-constrained solutions can be used if the question concerns one or two peaks. ALPST should then be used to gradually increase α from a very small value to a value where no diagnostics for numerical instability occur. The speed is adequate with one peak, but poor with two. Without peak-constrained solutions, bias can be reduced by choosing the smallest α that yields a solution with no more peaks than the solution with PROB1 nearest 0.5. The correct order of magnitude for error estimates of the moments can then usually be obtained from those for the solution with PROB1 nearest 0.5. Sometimes polymerization or aggregation mechanisms might even provide an excuse for considering a solution biased toward smoothness as a possibility.

Another *a posteriori* regularizor correction with CONTIN could be made using independent α values for each class determined by MinClass or MinPeak. The α's could then be independently increased until each class had only one peak. This would yield very impressive results for Fig. 2, but it is only a visual aid, showing what s(z) might look like if it were smooth.

MinPeak

Regularization methods generally make statements about one extreme member of Ω, *e.g.* the member with the maximum smoothness measure or entropy. MinPeak makes a statement about *all* members of Ω, *i.e.* "*all* members of Ω have at least n peaks." This is a very direct way to guard against artifact peaks.

In less ill-posed problems, where shoulders and higher-order critical points can be reliably determined, MinPeak could in principle be generalized to minimize the number of higher-order critical points. This would yield a smoother solution, but would increase the bias and the complexity of the algorithm. With QLS, this is probably seldom worthwhile. The main use of MinPeak is to provide a reliable estimate of the minimum number of peaks for MinClass, not a smooth microscopic solution.

The bias due to the peak constraints seems to be small, probably because of the poorly determined higher-order shape properties with QLS data. A few tests with one-peak analyses of two barely unresolvable δ-functions, resulted in only slightly worse peak statistics with MinPeak than with the nearly unbiased MinClass. Another advantage of MinPeak and MinClass is that they can be applied beyond the conservative stopping criterion of PROB1 = 0.5, if prior information dictates this. Thus, they both yielded good two-peak analyses of the two δ's, which were not resolved by CONTIN or MaxEnt using eq. (7). This could be useful, for example, for systems known to have two or three closely-spaced peaks.

When there are binding constraints, the PROB1 criterion contains added approximations[1]. This simply means that an n-peak solution with a PROB1 less than, say, 0.99 (and not strictly 0.5) should still be considered a possibility along with the (n+1)-peak solution. If one is forced to choose between the two, then the following (expensive) criterion could be used in such rare borderline cases to decide whether an n-peak solution, $\tilde{s}$, should be rejected. Let ρ_0 be the ratio of VAR for $\tilde{s}$ to VAR for the Reference Solution. Use $\tilde{s}$ to compute a set of noisy replicate data sets (as for MinClass). For each data set, ν, compute VAR for the n-peak solution and for the Reference Solution, and the ratio of the two, ρ_ν. A majority with $\rho_\nu < \rho_0$ indicates that constraining the original solution to n peaks caused a larger relative increase in VAR than for most of the true (replicate) n-peaked distributions, and that n should probably be increased. About ten replicates might be adequate. It is important to use variance ratios for these criteria, because of uncertainties in the absolute noise levels in practice.

The number of critical points, including extrema, is not necessarily invariant to changes of variable. For most reasonably smooth distributions and changes of variable, the number of extrema remain constant, but it is still important to define the problem so that the distribution of interest is obtained directly and not by a drastic change of variables afterwards. For example, one should never solve for the distribution in Γ and then convert it to s(z) by multiplying by Γ. For the grid in Fig. 1, this would magnify one end of the solution 10^7 times relative to the other end, and lead to false conclusions[15], possibly based on round-off error only.

MinClass

All the solutions in Fig. 2 (and infinitely more) are in Ω in that they fit the data to within experimental error. Therefore a single s(z) is insufficient to represent all members of Ω. MinClass makes no attempt at a meaningful microscopic solution, s(z), and concentrates on macroscopic statistics. It is attractive because it has no bias except that inherent in the non-negativity constraints. In addition, with the criterion used here, it is invariant to changes of variable. If G(Γ), the distribution in Γ, were used, eq. (9) would not change, since $G(\Gamma)\, d\Gamma = s(z)\, dz$.

For each possible partition, eq. (10) yields error estimates. In this sense, *every* partition is correct in that the estimated total density in each class and their correct error estimates are given. Other statistics and their estimated errors could also be computed. However, with a partition that splits a peak in two, the replicate solutions have larger fluctuations in density across the boundary, and the estimates have larger expected errors and stronger negative correlations. The estimates are still correct, but less informative. This is the basis of the criterion of minimizing the sum of the σ_{ik}^2.

This criterion has worked well in about 300 tests, but there is no reason to believe that it is optimal. This could even depend on the questions being posed in the experiment. There are numerous possible alternatives, *e.g.* involving correlations, covariances divided by the I_v, higher-degree moments, and Fisher information[22]. Since a new partition produces essentially new parameters, care is necessary in scaling and consistency. Some quantities, such as correlation coefficients, become data-independent, and therefore useless, when there are no non-negativity constraints.

One interesting peak-oriented possibility is to search all partitions for replicate solutions having alternating local maxima and minima in the mean value of $s_0(z)$ in each class. Since continuous functions, at least, always attain their mean value somewhere on the interval, this might be a useful way of searching for peaks. The partition(s) for which the maximum number of replicates satisfy the inequalities could be used for MinClass and for a starting point for MinPeak. This criterion has guaranteed convergence and is distribution independent, since it works only with frequencies. If necessary, something like the criterion of this paper might be used as a tie-breaker.

It is vital to explicitly impose non-negativity on the way to estimating the macroscopic statistics. Without this constraint, the accuracy and resolution drastically deteriorate[5]. It is remarkable that the Reference Solution, which is obtained by globally convergent and numerically stable linear algebra, is also a global solution to the difficult nonlinear problem of fitting sums of exponentials with positive amplitudes[23]. CONTIN, MinPeak, and MinClass can be directly applied to problems where s(z) can go negative, but there is the inevitable drastic loss in resolution.

MinClass attempts to partition a distribution into well-defined, possibly independent classes. In QLS and other areas, this has become the traditional strategy with CONTIN and its output of peak moments. However, in some areas, such as studies of compact structures involving less ill-posed problems, it might be necessary to modify MinClass to address more relevant questions, or to remain with biased methods. In trying to estimate fractional compositions from poor data[19], biased methods seem unavoidable.

Necessary Improvements

Experimental errors in B produce systematic errors in the y_j when eq. (3) is solved for the y_j. Ruf[24] proposed a useful and convenient linear correction for CONTIN. Jakeš[20] modified CONTIN to directly analyze the Y_j in eq. (3) using nonlinear optimization. This approach will be used with MinPeak and MinClass, because it avoids additional approximations for the w_j and bias from linearizing negative Y_j data points[15]. It also has the advantage that multiple data sets, *e.g.* from several scattering angles, can be analyzed together. Interchannel correlations can also be properly treated.

It is usually easy to count until B > 10^7, but great care is then necessary to keep systematic errors below this noise level. All four methods discussed here can rigorously include a nondiagonal covariance matrix for the Y_j in the analysis, and this will probably be necessary, since correlations between the Y_j can be significant[25,26].

Systematic errors are probably now the main limitation in QLS. When they are not eliminated by accounting for them in the model, they produce a data set corresponding to an incorrect distribution, s'(z). Any correct analysis method will yield incorrect information reflecting this s'(z) . Any method claiming to be insensitive to such systematic errors is so insensitive that it would not come close to s'(z) even if it *were* the correct solution. Only constraints based on prior knowledge can help. This problem becomes increasingly acute as the random noise level decreases.

ACKNOWLEDGEMENTS

I thank P. Štěpánek for calling my attention to the δ-Effect and J. Jakeš, Z. Kojro, H. Ruf, and K. Schätzel for making their manuscripts available to me prior to publication.

REFERENCES

1. S.W. Provencher, *Comput. Phys. Commun.*, 1982, *27*, 213.
2. S.W. Provencher, *Makromol. Chem.*, 1979, *180*, 201.
3. S.W. Provencher, J. Hendrix, L. De Maeyer, and N. Paulussen, *J. Chem. Phys.*, 1978, *69*, 4273.
4. S.W. Provencher, *Comput. Phys. Commun.*, 1982, *27*, 229.
5. S.W. Provencher, 'Photon Correlation Techniques in Fluid Mechanics,' E.O. Schulz-DuBois, ed., Springer, Berlin, 1983, p. 322.
6. A.K. Livesey, P. Licinio, and M. Delaye, *J. Chem. Phys.*, 1986, *84*, 5102.
7. S.-L. Nyeo and B. Chu, *Macromolecules*, 1989, *22*, 3998.
8. R.K. Bryan, *Eur. Biophys. J.*, 1990, *18*, 165.
9. J. Skilling, 'Maximum Entropy and Bayesian Methods,' J. Skilling, ed., Kluwer, Dordrecht, 1989, p. 45.
10. D.W. Marquardt, *SIAM J.*, 1963, *11*, 431.

11. M.G. Kendall, 'The Advanced Theory of Statistics,' Griffin, London, 1947, Vol. 1.
12. D.E. Koppel, *J. Chem. Phys.*, 1972, *57*, 4814.
13. R.S. Stock and W.H. Ray, *J. Polym. Sci. Polym. Phys. Ed.*, 1985, *23*, 1393.
14. C. de Boor, 'A Practical Guide to Splines,' Springer, New York, 1978, p. 303.
15. J. Jakeš, *Czech. J. Phys. B*, 1988, *38*, 1305.
16. J.E. Shore and R.W. Johnson, *IEEE Trans. Inform. Theory*, 1980, *IT-26*, 26.
17. A.E. Hoerl and R.W. Kennard, *Technometrics*, 1970, *12*, 55.
18. V.F. Turchin, V.P. Kozlov, and M.S. Malkevich, *Sov. Phys.-Uspekhi*, 1971, *13*, 681.
19. S.W. Provencher and J. Glöckner, *Biochemistry*, 1981, *20*, 33.
20. T. Nicolai, W. Brown, R.M. Johnsen, and P. Štěpánek, *Macromolecules*, 1990, *23*, 1165.
21. S.F. Gull, 'Maximum Entropy and Bayesian Methods,' J. Skilling, ed., Kluwer, Dordrecht, 1989, p. 53.
22. S.W. Provencher and R.H. Vogel, *Math. Biosci.*, 1980, *50*, 251.
23. A. Ruhe, *SIAM J. Sci. Stat. Comput*, 1980, *1*, 481.
24. H. Ruf, *Biophys. J.*, 1989, *56*, 67.
25. Z. Kojro, *J. Phys. A: Math. Gen.*, 1990, *23*, 1363.
26. K. Schätzel, *Quantum Optics*, 1990, *2*, in press.

8

Mesoparticle Diffusion in Biopolymer and Polymer Solutions

By George D.J. Phillies

DEPARTMENT OF PHYSICS, WORCESTER POLYTECHNIC INSTITUTE, WORCESTER, MASSACHUSETTS 01605, U.S.A.

1. INTRODUCTION

The motion of small particles through polymers is important to understanding a wide range of chemical and biochemical processes, ranging from the austerely scientific to the utterly applied, including protein transport within and between living cells, photographic film development, and the mixing of paints. An understanding of mesoparticle diffusion in matrices is also central to the rational design of large-scale facilities for the preparation, purification, and enzymatic post-isolation processing of biological extracts in biotechnology. The purpose of this paper is to discuss circumstances under which quasi-elastic light scattering spectroscopy can be used to measure particle diffusion in complex fluids.

The remainder of this paper is divided into four major sections:

First, a general treatment of light scattering spectra of multi-macrocomponent mixtures is presented. QLS spectra of ternary systems are generally quite complex. Under special conditions, though, spectra may simplify, greatly easing the difficulty of interpreting them. A particularly useful set of simplifying conditions applies to the diffusion of dilute, intensely-scattering probe particles through solutions of a perhaps-concentrated but weakly-scattering matrix.

Secondly, the general behaviour of the probe diffusion coefficient D_p for probes in macromolecule solutions is discussed. A simple scaling law that accurately describes most data is given. Temperature studies show that D_P obeys Walden's Rule, scaling in temperature inversely as the solvent viscosity. Solute binding by the probe particles potentially causes anomalies, exemplified here by bridging behaviour between probes in protein solutions, and by Langmuir adsorption-isotherm adsorption onto probes in solutions of some synthetic polymers.

Thirdly, critical experiments that reveal the dependence of the probe diffusion coefficient on the shapes of the probe and matrix polymers are examined. Results using spherical and random coil probes, diffusing either through globular or random-coil matrix polymers, are noted. These experiments serve to constrain the nature of acceptable models for polymer dynamics in concentrated solution.

Fourthly, we present a picture consistent with these experiments, namely the hydrodynamic scaling model. Experimental results compare favourably with the model's predictions. Comparison is made with other less successful models of polymer dynamics in solution.

2. QLS OF MULTICOMPONENT SOLUTIONS

Formal theoretical treatments of QLS spectra of concentrated and multicomponent systems have been made by many authors[1-5]. While there are still disagreements about the fine details on some issues, in broad outline this problem is now well-understood. Here I present major results from these calculations, leaving detailed derivations to the original papers.

In extremely dilute solution, the centre-of-mass diffusion of macromolecules is adequately described by a single *translational diffusion coefficient* D_T, whose value for spheres is given by the Stokes-Einstein equation

$$D_T = \frac{k_B T}{6 \pi \eta R} \tag{1}$$

Here k_B is Boltzmann's constant, T is the absolute temperature, and R and η are the sphere radius and the solvent or solution viscosity, respectively. If the concentration of the diffusing solute is sufficiently large that the solvent and solution viscosity differ significantly, the Stokes-Einstein equation is not applicable to the problem.

At elevated solute concentrations, a single number is inadequate to describe translational diffusion. Instead, two distinct diffusion coefficients are necessary. One of these, the self- or tracer-diffusion coefficient D_s, describes the motion of single macromolecules through a uniform solution. The other diffusion coefficient, the mutual or cooperative diffusion coefficient D_m, characterizes the relative motion of pairs of macromolecules; D_m determines how rapidly concentration gradients relax toward equilibrium.

D_s, which can be measured by pulsed-field-gradient NMR[6], fluorescence recovery after photobleaching[7], or fluorescence correlation spectroscopy with low labelling levels[2], is simply related to the single-particle mobility μ_{ii} or drag coefficient f_s, namely

$$D_s = \frac{k_B T}{f_s} \equiv \mu_{ii} \, k_B T \tag{2}$$

Quasi-elastic light scattering from a binary macromolecule:solvent mixture measures the mutual diffusion coefficient. D_m, which is determined by a balance of hydrodynamic and direct (*e.g.* electrostatic) forces, is described by the generalized Stokes-Einstein equation

$$D_m = \frac{\left[\frac{\partial \Pi}{\partial c}\right]_{P,T}}{f_m} (1 - \phi) \tag{3}$$

ϕ and Π being the solute volume fraction and osmotic pressure, respectively, while f_m is the drag coefficient for mutual diffusion.

In a dilute 2-macrocomponent solution, a QLS spectrum is typically composed of 2 exponentials, each exponential corresponding to the diffusion of one species of macromolecule. At elevated macromolecular concentrations, a gradient in the concentration of one solute species causes diffusive flows of all solutes; the motions of different macromolecular species are therefore coupled. The spectrum of a non-dilute 2-macrocomponent system is still composed of 2 exponentials, but each exponential refers to motions of both solutes. In particular, even if only one species scatters light, the spectrum generally still contains two exponentials[1].

However, *in the special case that the scattering species is dilute*, while the other solutes (which may be concentrated) do not scatter light significantly, the spectrum simplifies: one observes a single exponential, whose decay time is determined by the tracer diffusion coefficient of the dilute species through the solution. As first shown by this author[2] for the case of the fluorescence correlation method, fluctuation spectroscopies can measure self as well as mutual diffusion coefficients of macromolecular solutes. This result has been extensively developed. Especially noteworthy are theoretical papers by Jones[3] and Benmouna *et al.*[5], who treat quasi-elastic scattering by polymer mixtures.

The special case in which one dilute component dominates scattering by a mixture can often be attained experimentally. If one combines an isorefractive polymer:solvent pair with a compatible, dilute, intensely-scattering random-coil polymer, one has a polymer:polymer:solvent system in which the concentrated polymer is non-scattering, while the dilute polymer dominates the QLS spectrum. This approach has been extensively exploited by Lodge[8] and Nemoto[9], among others. As an alternative probe, mesoparticle species such as polystyrene latices typically scatter much more light than most biopolymer solutions, even when particle concentrations are very low. Studies of particle:macromolecule:solvent mixtures are discussed below.

3. PROBE DIFFUSION: PHENOMENOLOGY AND ARTIFACTS

Let us turn to the behaviour of probe particles in polymer/biopolymer solutions. This section will present representative data, and a simple equation that reduces the data to a few parameters. Most biochemists will recognize the importance of temperature in biological functions; the temperature dependence of D_p - a quantity relevant to the rate at which diffusion-limited reactions proceed - is discussed. Finally, the adsorption of matrix polymers by probe particles, an important potential artifact, is treated.

Our laboratory has acquired extensive data on the motion of polystyrene latex spheres in a variety of polymer solutions, including those of polyethylene oxide, polyacrylic acid (neutralized and non-neutralized), dextran, and bovine serum albumin. Polystyrene spheres in water have very large scattering cross-sections, so light scattering from these spheres easily dominates scattering from any of the matrix polymers that we use. Representative results are shown in Fig. 1, which plots D_p of 208 Å radius polystyrene spheres in 60% neutralized poly-acrylic acid of nominal molecular weight 450 kDa[10] D_p falls substantially with increasing polymer concentration c (in all these measurements, the probes were highly dilute, $\leq 10^{-3}$ by volume.) D_p also depends strongly on the ionic strength I (from added NaCl) in the solution, D_p increasing substantially at large I.

The lines in Figure 1 are *stretched exponentials*, functions of the form

$$D_p = D_0 \exp(-\alpha c^\nu) \tag{4}$$

Here D_0 is D_p at low concentration c, while α is a scaling prefactor. Stretched exponentials differ from regular exponentials in the presence of the scaling exponent ν, which raises the independent variable c to a (usually non-integer) power. α and ν both change systematically with increasing I.

Results similar to Fig. 1 describe most of our data on probe diffusion in polymer solutions. It is found that D_p generally has a stretched exponential[10,11,12] dependence on its independent parameters

$$D_p = D_0 \exp(-a\, c^\nu M^\gamma R^\delta I^\beta A^\varpi) \tag{5}$$

Here c and M are the matrix concentration and molecular weight, R is the probe radius, I is the solution ionic strength, and A is the degree of polymer neutralization. ν, γ, δ, β, and ϖ are scaling exponents. The term $I^\beta A^\varpi$ is only present if the matrix polymer is a polyelectrolyte. ν and β depend on M; β also depends on c[18,24]. Phenomenologically, ν is 0.5 -1.0, $\gamma = 0.8 \pm 0.1$, $\delta = 0 \pm 0.2$[10,12], β is in the range -0.1 to -1, and $\varpi \approx 0.25$ [12]. Note that D_p depends strongly on matrix M, even when the matrix polymer is in the semidilute ("entangled") concentration regime.

The dependence of D_p on temperature and viscosity is important in some biological and biotechnological problems. Also, the reproducibility of D_p after an extended temperature cycle - the return of D_p to its initial value on returning the system to its initial temperature - is a useful test of the probe's stability. Fig. 2 gives D_p

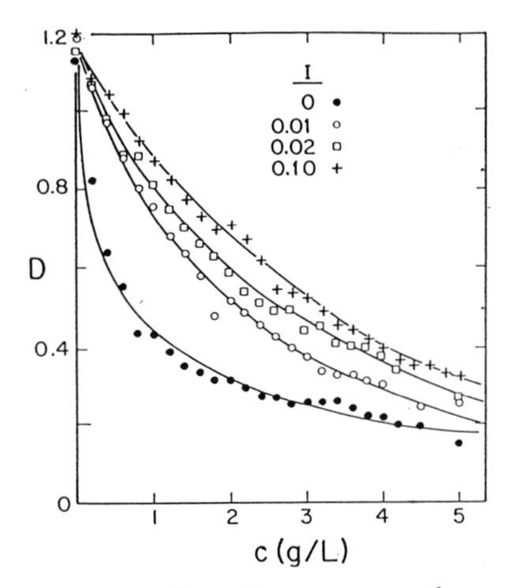

Figure 1. D_p (units 10^{-7} cm^2/s) of 208 Å polystyrene spheres in 450 kDa poly-acrylic acid, 60% neutralized, as a function of polymer c (g/L). Points correspond to different concentrations of added NaCl, namely 0 (filled circles), 0.01 M (open circles), 0.02 M (squares) and 0.10 M (crosses). After ref. 10.

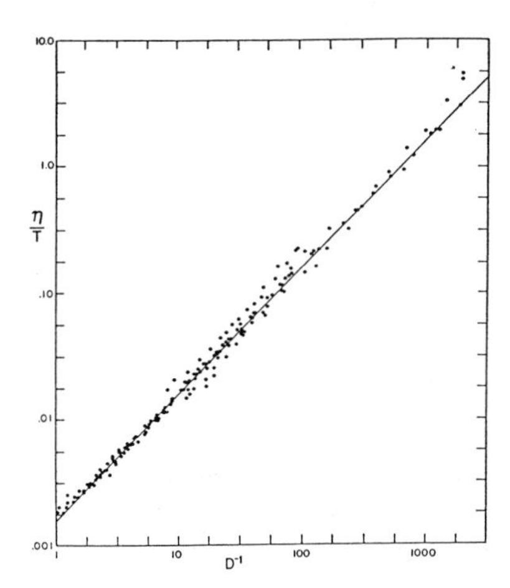

Figure 2. D_p^{-1} (units 10^{-7} cm^2/s) against η/T (units cP/K) for 0.091 μm polystyrene spheres in water:glycerol at different temperatures and glycerol concentrations. After ref. 13.

as a function of T/η for polystyrene spheres in the small-molecule solvent water:glycerol:sodium lauryl sulphate[13], η being the solution viscosity. η may be extended over three orders of magnitude by varying temperature ($0°C \leq T \leq 50°C$) and glycerol concentration (from 0 wt% to 95 wt%). In small-molecule solvents, D_p follows Walden's rule

$$\frac{D_{p1}}{D_{p2}} = \frac{T_1\eta_2}{T_2\eta_1} \tag{6}$$

the subscripts 1 and 2 referring to systems of different composition or temperature.

Fig. 3 represents measurements on carboxylate-modified polystyrene spheres in 2/3 neutralized 450 kDa poly-acrylic acid. For a system of fixed composition, the temperature dependence of D_p follows Walden's rule, eq. 6. D_p is linear in T/η, η being the *solvent* viscosity. However, if η is increased by increasing c of the polymer at fixed T, Walden's rule need not apply[14]. Large polymers are more effective at increasing the solution viscosity than at retarding probe diffusion. Equivalently, in concentrated solutions of high-molecular-weight polymers the Stokes-Einstein equation usually fails, $D_p\eta$ increasing with increasing c. With very-high-molecular weight ($M \sim 1 \times 10^6$ Da) polyacrylic acid, $D_p\eta$ can increase by four orders of magnitude over its value for probes in pure water[14].

It is well known that polystyrene latices are often effective at adsorbing polymers and biopolymers onto their surfaces. Such adsorption would increase the hydrodynamic radius of the latex (or perhaps cause it to aggregate!) and retard its diffusion. Surface adsorption leads to apparent failures of the Stokes-Einstein equation in which D_p falls excessively when c and η are increased, the product $D_p\eta$ therefore decreasing with increasing c. Surface adsorption gives changes of $D_p\eta$ in the wrong direction to explain the non-Stokes-Einstein diffusion effects that we have observed in many polymer solutions.

Surface adsorption is, however, an important possible artifact, potentially obscuring the physical interpretation of D_p. As an example of adsorption effects, Figure 4[15] presents D_p of 517 Å radius polystyrene spheres in pH 7.0 bovine serum albumin (BSA):0.15 M NaCl. The filled circles correspond to samples made by adding spheres to protein solutions of the indicated

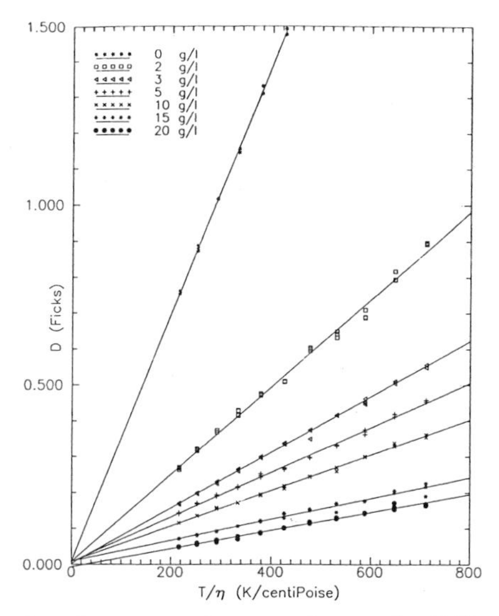

Figure 3. D_p of polystyrene latex spheres in 2/3 neutralized poly-acrylic acid (M = 450 kDa, I = 0) as a function of T/η. Each set of points and matching linear-least-squares fit line represents a different polymer concentration, η being the (temperature-dependent) solvent viscosity.

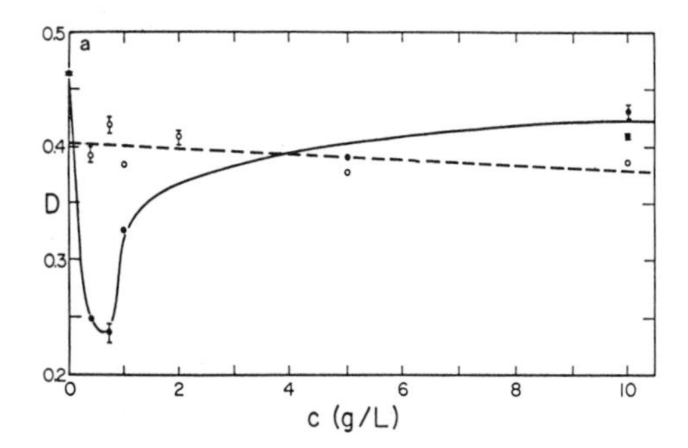

Figure 4. D_p (units 10^{-7} cm^2/s) of 517 Å polystyrene spheres against concentration c of bovine serum albumin. Solid points: samples prepared by mixing at the indicated c. Open points: samples prepared by mixing spheres into concentrated BSA and then diluting to the indicated c. After ref. 15.

concentrations; there is a peculiar dip and recovery in D_p between 0 and 4 g/L. The origin of the dip is revealed by the open circles, which correspond to samples made by adding spheres to concentrated BSA solutions and then diluting to the indicated concentrations. With the diluted samples, the dip is absent, but D_p extrapolated to $c \to 0$ is significantly smaller for the diluted samples than D_p for spheres in 0.15 M NaCl: no BSA.

All the odd features of Fig. 4 can be explained as surface adsorption phenomena. As is well known, polystyrene latices bind many proteins irreversibly. Spheres added to dilute protein solutions acquire a partial protein coat. Before the coat completes itself, spheres encounter other spheres, leading to *bridging*, in which one protein molecule binds to two spheres to form aggregates. The protein coats eventually fill in, so that no further bridging occurs; D_p is then time-independent. Contrarily, if spheres are added to a more concentrated protein solution, protein adsorption is faster, so spheres can acquire a complete protein coat before encountering other spheres. Bridging does not occur. Since adsorption is irreversible, dilution does not affect the coat. Spheres in diluted protein solutions remain coated and do not aggregate, so D_p of diluted spheres does not show a dip. Because they are protein-coated, in the limit $c_{BSA} \to 0$ diluted spheres are bigger than uncoated spheres. From the difference between the $c = 0$ intercepts of the solid and dashed lines, the surface layer of adsorbed serum albumin is roughly 100 Å thick, a value compatible with known dimensions of serum albumin and a plausibly-expected unimolecular layer of adsorbed protein[15].

Figs. 5 - 7 show a second example of polymer adsorption, this time by polyethylene oxide onto polystyrene latices[16]. Data interpretation is here complicated by the failure of the Stokes-Einstein equation in polymer solutions, so that D_p, η_M (the viscosity obtained macroscopically - in this case, with capillary viscometers), and eq. 1 do not give the true radii of the probes. Fig. 5 gives apparent sphere radii r for spheres in polyethylene oxide:water, calculated from eq. 1 by using D_p and η_M. Sometimes r increases with increasing polymer c, consistent with adsorption; in other cases r falls.

The behaviour of r is clarified by observing spheres in water:polyethylene oxide solutions that contain trace amounts (0.1 wt%) of the anionic detergent Triton X-100 (Sigma). Triton X-100

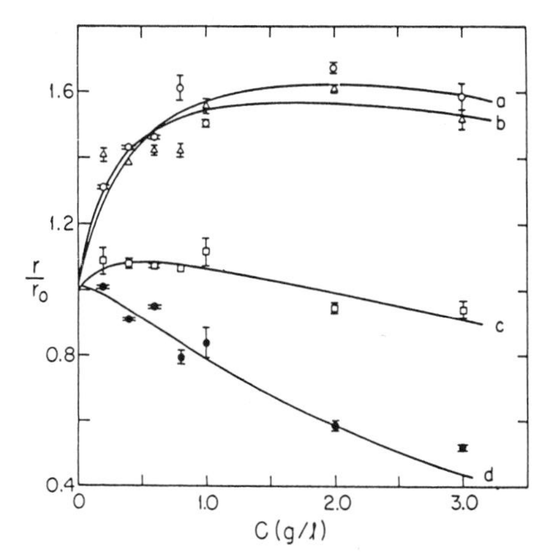

Figure 5. Apparent hydrodynamic radii of (a) 208 Å, (b) 517 Å, (c)0.322 μm and (d) 0.655 μm polystyrene spheres in water:polyethylene oxide (M = 300 kDa), as calculated from D_p and the macroscopic solution viscosities η_M. r_0 is r for a sphere in pure water. Solid lines are Langmuir adsorption isotherms, corrected for non-Stokes-Einsteinian diffusion effects. After ref. 16.

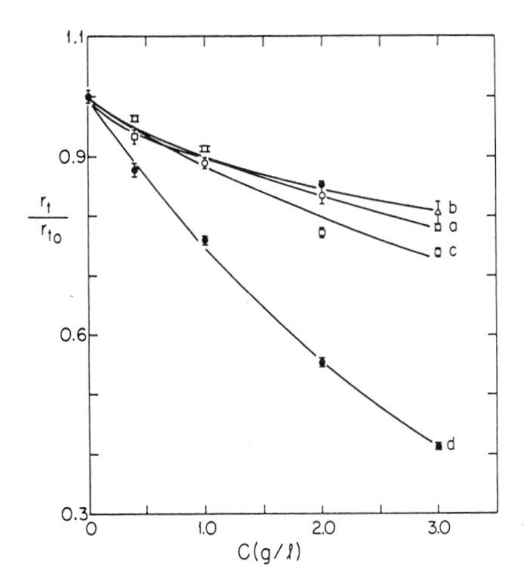

Figure 6. Normalized apparent hydrodynamic radii r_t of (equivalently, normalized microviscosity η_μ encountered by) (a) 208 Å, (b) 517 Å, (c)0.322 μm and (d) 0.655 μm polystyrene spheres in water:polyethylene oxide (M = 300 kDa): 0.1 wt% Triton X-100, as calculated from D_p and the macroscopic solution viscosities η_M. r_{to} is r_t for a sphere in water:0.1 wt% Triton X-100. Solid lines are stretched exponentials. After ref. 16.

is strongly bound by spheres; a competitive binding study indicates it is not displaced by polyethylene oxide. Fig. 6 shows r_t, the apparent radius of spheres in detergent-containing solutions, against polymer concentration c. These spheres have physically constant radii; the fall in r_t relative to its c = 0 value r_{t0} is a hydrodynamic effect. Since r_{t0} is known, Fig. 6 effectively supplies calibrating curves, allowing calculation of true sphere radii from D_p at any c by dividing out the non-Stokes-Einstein behaviour. Fig. 7 shows true sphere radii, for spheres in systems that contain no detergent, as computed from their D_p and r_t/r_{t0}. Figs. 5 and 7 refer to the same physical samples. Solid curves of Figs. 7 and 5, the latter with non-Stokes-Einstein effects folded in, are the expected Langmuir adsorption isotherms. Polymer adsorption first increases with increasing c and then saturates.

This section has treated the phenomenology of probe diffusion in protein and random-coil polymer solutions. Typical data on $D_p(c)$, and a scaling law that is usually effective at describing how D_p depends on major independent variables, were presented. D_p changes with temperature inversely as the solvent viscosity. A significant effect, potentially dependent both on temperature and on matrix concentration, is matrix adsorption. Figs. 4-7 and the discussion show how adsorption phenomena may be identified and interpreted.

4. EFFECTS OF PROBE AND MATRIX ARCHITECTURE

In a living system, the total macromolecular concentration is high, but most species are individually dilute. The diffusion of a molecule through complex inter- and intra-cellular environments is therefore well-described by the molecule's probe diffusion coefficient. Biopolymer solutions typically contain macromolecules having a variety of physical architectures, ranging from the tightly folded forms of globular proteins to the near-random-coil structures of some linear polysaccharides. This section treats the effect of probe and background architecture on D_p. We ask, for example, whether a solid particle should be expected to diffuse more or less rapidly than a random coil that has the same radius of gyration.

The significance of diffusant architecture will be demonstrated by contrasting D_p from three key sets of experiments. The experiments show what happens if we change the probe architecture, if we change the matrix architecture, or if we replace a polymer

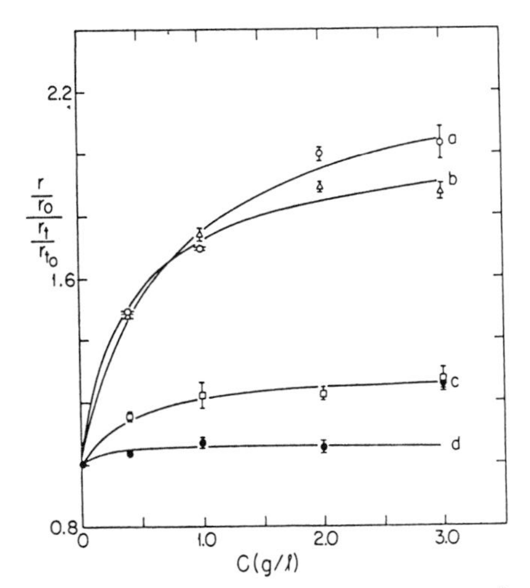

Figure 7. Hydrodynamic radii of (a) 208 Å, (b) 517 Å, (c) 0.322 μm and (d) 0.655 μm polystyrene spheres in water:polyethylene oxide (M = 300 kDa) solutions containing no Triton X-100, as calculated from D_P and the solution microviscosity η_μ. Solid lines are Langmuir adsorption isotherms. After ref. 16.

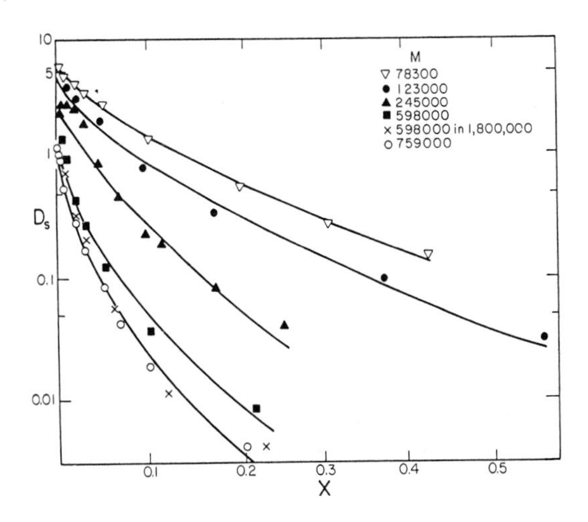

Figure 8. D_S of polystyrenes of various M (and D_S of 598 kDa polystyrene in an 1800 kDa matrix) in tetrahydrofuran, taken from Leger *et al.*[17]. D_S in 10^{-7} cm^2/s, x in g/g. Solid lines are stretched exponentials. After ref. 27.

solution with a covalently cross-linked gel.

Effect of Probe Architecture

Fig. 1 gave typical data on D_P of particles through polymer coils. There is an extensive literature on the diffusion of polymer coils through other coils, representative sections of which appear in Figs. 8 and 9. Fig. 8[12] shows self-diffusion coefficients D_S measured by Leger *et al.*[17] using forced Rayleigh scattering on polystyrene:benzene, while Fig. 9[12] shows pulsed-field-gradient NMR data of Callaghan and Pinder[6] on polystyrene:CCl_4. Solid lines are stretched exponentials (eq. 4), α and ν being obtained from non-linear-least-squares fits. As apparent from the figures, D_S for each system follows closely a stretched exponential in polymer concentration. A systematic re-analysis[18] of the modern literature confirms the generality of this result; eq. 4 is appropriately described as a universal scaling equation for D_S.

Leger *et al.*'s results[17] are sometimes interpreted as confirming the reptation-scaling picture, by showing that D_S follows power laws $D_S \sim c^x M^{-2}$. Fig. 8 matches D_S with stretched-exponential behaviour. Comparison of Fig. 8 with Leger *et al.*'s [17] original figure reveals that measurements which fit power laws over a limited concentration range can also be fit by stretched exponentials over wider concentration ranges.

Callaghan and Pinder's work[6] is unusual among earlier studies of polymer:polymer:solvent systems in that it included small polymers (2000 Da) as well as large ones (350 kDa), and in that it incorporated extensive work on both dilute and concentrated polymer solutions. Fig. 9 shows that stretched exponentials describe findings on low-molecular-weight "non-entangling" polymers and on high-molecular-weight "entangled" polymers. In Fig. 9, vertical arrows denote the overlap concentration c^*. There is no sign of a break in the curves at or near c^*. A single function with constant parameters is adequate to describe D_S at all concentrations from the most dilute to the highly concentrated (here 300 g/L). The factors that control diffusion in concentrated solutions are apparently already active at $c \leq c^*$.

A few experiments of Brown and Zhou[19] cover random-coil and spherical probes of the same size diffusing through the same

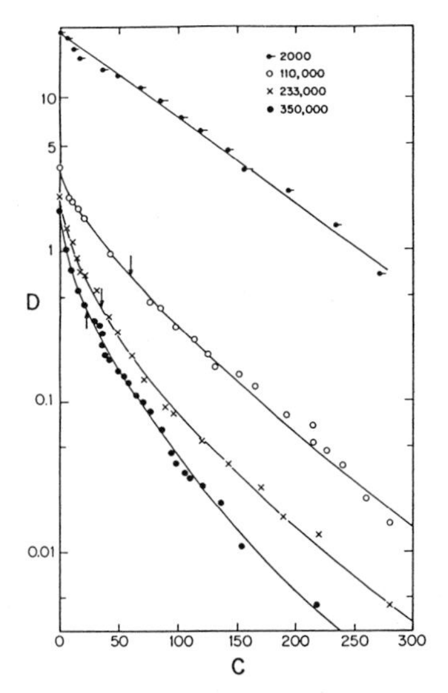

Figure 9. D_s of polystyrene in CCl_4, from ref. 6. Arrows marked estimates of c^*; solid lines are stretched exponentials. 2000 Da polystyrene has $M < M_e$ and hence no nominal c^*. After ref. 27.

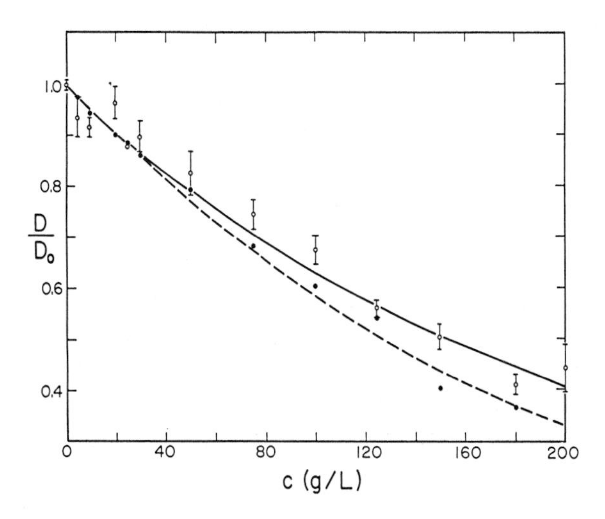

Figure 10. D_p/D_{p0} (open points) and fluidity η_0/η of 0.322 μm carboxylate-modified polystyrene spheres in bovine serum albumin:0.15 M NaCl. Solid and dashed lines are stretched exponentials. After ref. 16.

matrix polymer. In published work, Brown and Zhou compared D_p of 1600 Å silica spheres and 4.9 x 10^6 Da polyisobutylene ($R_g \approx$ 1400 Å) in 6.1 x 10^5 polyisobutylene:$CHCl_3$, matrix concentrations ranging from dilute to highly concentrated ($c \gg c^*$). The two probe species had the same D_p at all c, to within experimental error. More recent unpublished work[20] on silica spheres having the same radius as the matrix polymer show that the probe D_p and matrix D_s are approximately equal at all matrix concentrations.

How is polymer self-diffusion correlated with solution viscosity? Numasawa *et al.*[21] measured D_s and η for polystyrene in a polymethylmethacrylate:benzene matrix, varying c of the matrix and M of both matrix and probe polymers. Numasawa *et al.* found that $D_s\eta$ can increase dramatically (up to one-hundred-fold relative to its value in pure solvent) at large c and M. The $D_s\eta$ phenomenology seen by Numasawa *et al.*[21] for polystyrene in polymethylmethacrylate:benzene is qualitatively similar to the $D_p\eta$ phenomenology seen by Lin and Phillies[14] for polystyrene spheres in polyacrylic acid:water.

From the above, globular and random-coil probe species share a common stretched-exponential dependence on c, and a qualitatively similar failure of Walden's rule (with c as the controlled variable) at large matrix M and c. Limited measurements find that polymer matrices retard probes of the same size to the same extent, regardless of the probe architecture. D_p is thus indifferent to differences between coil and sphere probes. It is extremely difficult to account for this behaviour on the basis of models that ascribe modes of motion (such as snake-like slitherings) to polymer chains, when those modes are not available to spherical particles.

Effect of Matrix Architecture

Results of Ullmann *et al.*[16] on probe diffusion through bovine serum albumin (BSA) solutions are described by Fig. 10, which also presents values for the fluidity η^{-1} of these solutions. The curves are stretched exponentials; probe spheres in BSA solutions and in random-coil polymer solutions have concentration dependences of this same form. Furthermore, as c increases D_p/D_{p0} rises above the normalized fluidity η_0/η; probes in concentrated BSA solution diffuse faster than expected from the solution viscosity. This non-Stokes-Einstein behaviour, which occurs even though the probes are much larger (0.3 μm radius) than the serum albumin molecules ($R \leq 50$ Å), is qualitatively highly similar to the results shown in Fig. 6. Probes diffuse through a background composed of globular molecules much as they would diffuse though a background containing long polymer chains. Probe diffusion is not substantially affected by matrix architecture.

Effect of Matrix Cross-Linking

From eq.5, D_p is correlated with the probe radius R as

$$D_p = D_0 \exp(-b R^{\delta})$$

D_0 depends on R, but δ is nearly zero for probes in solution. Non-dilute polymer solutions are sometimes described as containing *pseudogel lattices* or *transient gel networks.* It is interesting to compare probe diffusion in solutions that might contain transient gels with probe diffusion in true, covalently cross-linked gels. Hou *et al.*[22] observed ficolls in F-actin gels by means of fluorescence recovery, finding $\delta \approx 0.7$. Similar results are reported by Bansil *et al.*[23].

Based on Hou *et al.*[22] and Bansil *et al.*'s[23] work, on time scales to which probe diffusion is sensitive true gels are size filters, retarding the motion of large particles while letting small particles pass. If a polymer solution contains a transient lattice, the lattice would trap particles with radii exceeding the lattice spacing. It can be shown[24] that probe diffusion involves distance and time scales appropriate to detect the pseudogel lattice, if it existed. However, macromolecule solutions do not behave like gels; they are not size filters. Polymer solutions retard the diffusion of large and small mesoscopic particles equally. (Molecules so small that they are comparable in size with solvent molecules behave differently.) Probe diffusion is substantially perturbed by the insertion of matrix cross-links. Polymer solutions therefore do not contain transient gel lattices.

Conclusions from Critical Experiments

Measurements on probes in macromolecule solutions find D_p depends on the probe and matrix sizes, but is largely indifferent to probe and matrix architecture. The diffusion of polymer chains through other chains appears to resemble closely the diffusion of colloidal spheres through other spheres. Results of Hou *et al.*[22] and Bansil[23] show that probe diffusion can sense polymer crosslinks, if they are physically present. Diffusion through gels is qualitatively different from diffusion through solutions. The hypothesized "pseudogel lattice" - a network of transient chain entanglements in solution - must therefore not be fundamental to polymer dynamics, even on short time scales.

This final conclusion must be viewed as a favourable development. The results of many workers - for example, Pusey and Tough[28] at the Royal Signals and Radar Establishment - have converged to describe how spherical colloids diffuse at intermediate and large concentrations. By applying the principles already tested for colloid diffusion to the polymer diffusion problem, an approximate description of polymer dynamics should be obtained. The description is provided by the hydrodynamic scaling model, which incorporates hydrodynamic and excluded volume forces in a systematic way.

5. HYDRODYNAMIC SCALING MODEL

Let us consider a model for polymer diffusion. The theoretical underpinnings of the model have appeared elsewhere[25,26,24,27]. Here I sketch the primary assumptions of the model, its major predictions, and compare with experiment and with competing models, notably reptation-scaling.

Model Assumptions

The hydrodynamic scaling model invokes three major assumptions, namely (i) chain contraction with increasing concentration, (ii) self-similarity, and (iii) identification of the dominant forces in chain motion. These assumptions suffice to determine the concentration and molecular weight dependences of D_p, numerical values for the principal scaling parameters, and correlations with other transport coefficients.

By *chain contraction*, I refer to the demonstration of Daoud *et al.*[29] that the radius of gyration of a polymer chain in solution scales as

$$R_g^2 \sim M c^{-x} \qquad (8)$$

For short chains, and for large chains at very low concentration, x ≈

0. For large chains, $x \approx {}^1/_4$. Experimentally[29], chain contraction begins at quite small c, is perhaps one third complete at c^*, and continues all the way out to the melt.

The purpose of the *self-similarity* approximation is to bootstrap the calculation from the low-concentration linear regime up to higher concentrations. A plausibility argument for the approximation has been advanced[25], but is omitted here. The mathematical effect of the approximation is to rewrite a linear concentration dependence as

$$D_0 (1 - k_D\, c) \rightarrow D_0 \exp(-k_D\, c) \tag{9}$$

k_D being the (weakly c-dependent) linearized concentration dependence of D_p. Eq. 9, read right to left, would be a Taylor series expansion.

The heart of the theory lies in identifying the forces that control k_D. The fundamental assumption of the model is that *polymer dynamics at elevated concentrations are dominated by hydrodynamic interactions. Topological constraints - chain crossing restrictions - provide at most secondary corrections.* By hydrodynamic interactions, I refer to the bead-bead interactions treated in the Kirkwood-Risemann picture of polymer chains[30]. It should be recognized that Kirkwood's original work used the Oseen tensor approximation to the hydrodynamic interaction tensor $\mathbf{T}_{ij}$. More recent work on colloid dynamics[31,32] has emphasized the significance of short-range (r^{-3}, r^{-7}) hydrodynamic forces. Short range forces cause the motion of nearby Brownian objects to become correlated, so that they diffuse parallel to each other. Objects suffering parallel displacements do not collide; short-range hydrodynamic interactions therefore suppress dynamic consequences of chain entanglements, perhaps explaining why chain entwinements are not important to polymer diffusion.

Model Predictions

Major predictions of the hydrodynamic scaling model are[24]:

(i) A polymer's self-diffusion coefficient should depend on its concentration *via* a stretched exponential

$$D_s = D_0 \exp(-\,\alpha\, c^{\nu}) \tag{10}$$

D_0 is the diffusion coefficient of a hypothetical chain that has, in pure solvent, the hydrodynamic radius that the real chain has in solution at concentration c. Chain radii depend on polymer concentration *via* eq. 8 D_0 is therefore weakly sensitive to c.

(ii) α and ν are determined by polymer molecular weight. For random-coil polymers

$$\alpha \sim M^1 \tag{11}$$

Treating bead-bead interactions at the Kirkwood-Risemann level allows a numerical estimate of α; for 1×10^6 Da polystyrene, $\alpha \approx 2$.[26] The exponent ν is determined by chain contraction, namely $\nu = 1 - 2x$. For large polymer chains, $\nu \approx {}^1/_2$; for small chains (and large chains at small c), $\nu \approx 1$.

(iii) The translational diffusion of linear and star polymers through a linear polymeric matrix employ the same physical mechanisms, so α and ν for linear and star polymers of the same molecular weight should be similar. A dependence of α and ν on f is expected because f-armed stars have a smaller radius than do linear chains (hence, smaller α for stars), and because the compressibilities (eq. 8) of many-armed stars and linear chains may differ[27].

(iv) Spheres and random coils that have the same size should diffuse through polymer solution at roughly the same rate.

(v) Treatments of different transport coefficients should use the same mathematical apparatus, so a range of transport parameters (for example, zero-shear viscosity η_0) should use the same functional form, namely $\exp(-\alpha c^\nu)$, for concentration scaling.

(vi) Since ν is determined by static (equilibrium) properties of polymer chains, ν determined from different transport coefficients should show its transition from small- to large-M behaviour over the same range of M.

Comparison of Model with Experiment

The hydrodynamic scaling model successfully predicts a wide range of experimental behaviour. Treating the above predictions *seriatim*:

(i) Self- and tracer-diffusion coefficients in random coil solutions uniformly follow eq. 10.

(ii) Figure 10 shows α values taken[24] from a substantial part of the polymer literature. Over three orders of magnitude in both variables, α and M are correctly correlated by eq. 11. In this figure, the dashed line refers to a numerical prediction[26] of α, no free parameters being employed. The prediction, from the lead term of a series expansion, is in reasonable agreement with somewhat scattered data.

Fig. 11 shows (filled circles) ν for polymer self-diffusion. ν is close to $^1/_2$ for large polymers, and roughly 1 for polymers with $M < 10^4$, as predicted.

(iii) Experiments[8] on linear and star polystyrenes diffusing through polyvinylmethylether:toluene find that D_s of the polystyrenes systematically follows eq. 10, including the expected weak c-dependence of D_0. For f-armed stars of fixed molecular weight, α (while scattered) tends to fall with increasing f, as predicted by the model.

(iv) Measurements of Brown and Zhou[19,20] confirm that spheres and random coils having equal R_g have the same D, as predicted.

(v) Phillies and Peczak[33] analysed published data on η_0, sedimentation coefficient s, and rotational diffusion coefficient D_r, finding for a large number of systems that the transport coefficients all follow stretched exponentials in c, as predicted. At very large c and M (corresponding to $\eta \geq 10^4$ Poise) a transition to melt-like behaviour, seen as a crossover to a power-law dependence of η on c, is apparent in some work.

(vi) Fig. 12 gives ν from η and D_p as well as ν from D_s. The shaded area in the Figure refers to ν as predicted by the Martin equation, an early (pre-1950) correlation of η and c. While one could wish for data covering a wider range of molecular weights, ν measurements from D_s, D_p, and η agree with each other (about as well as ν measurements on one

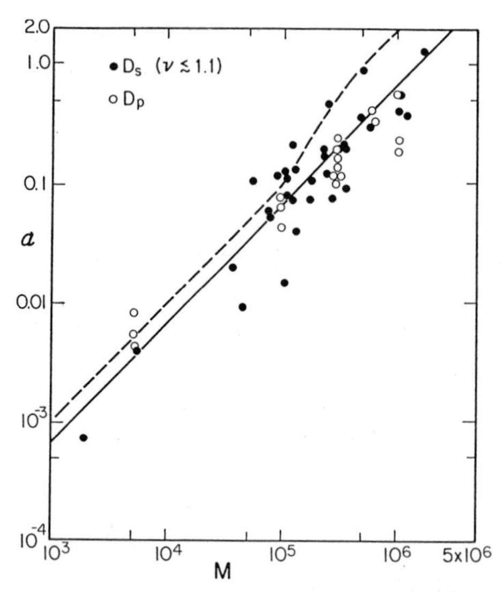

Figure 11. Scaling prefactor α from probe and self-diffusion coefficients D_p and D_s, against matrix M, for systems with $\nu \leq 1.1$. Solid line is the scaling prediction $\alpha \sim M^1$. Dashed line (*no free parameters!*) is predicted by the hydrodynamic scaling model[26]. After ref. 27.

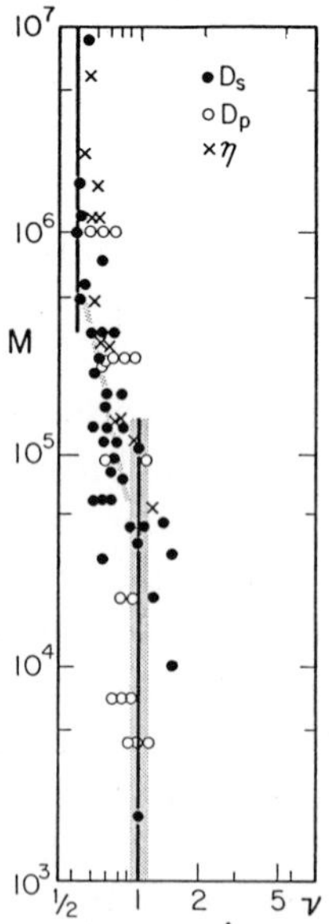

Figure 12. Scaling exponent ν against molecular weight M of the matrix polymer, including values taken from fitting stretched exponentials to D_s, D_p and η. Shaded region is the Martin equation prediction $\nu = 1$. Solid lines are the small- and large-M predictions $\nu = 1$ and $\nu = 0.5$. Transition region (narrow shaded line) is $\nu \sim M^{-1/4}$. After ref. 27.

transport coefficient from different laboratories agree with each other), in accord with the prediction that the transition in ν is the same for D_S, D_P, and η.

Alternative Treatments of Polymer Dynamics

A variety of models of polymer dynamics presently have at least some defenders. A widely used picture is the reptation-scaling model of de Gennes[34]; the entanglement model of Doi and Edwards[35] is a close relative. For polymer melts, Hess[36] used Mori theory to advance a non-scaling reptation-based picture; Schweizer[37] used mode coupling and a generalized Langevin equation to treat polymer fluid dynamics. Computer modelling led Skolnick and collaborators[38] to a dynamic entanglement model. Also noteworthy are the Ngai-Rendell coupling scheme[40] and Oono's renormalization group treatment[39] of mutual diffusion.

Many of the above are primarily intended as models of polymer melts, not solutions. The Ngai-Rendell scheme is more nearly a series of sum rules than a physical model, though Ngai *et al.*[41] have recently shown that it correctly correlates D_S of linear and star polymers in a linear chain matrix. Oono's model[39], while detailed, refers to a single physical parameter.

Of the above theories, the reptation-scaling picture is unique in the range and detail of predictions it makes about polymer dynamics in solution. As important as the range of predictions made by the model is the correctness of the predictions. A comparison[24] of the reptation-scaling model (RSM) with the phenomenology sketched above is highly unfavourable to this model. For example:

(i) The RSM divides solutions into entangled ($c > c^*$) and non-entangled ($c < c^*$) regimes, with a crossover regime separating them. Different nodes of motion are predicted to be important for self-diffusion in different concentration regimes. Experimentally, a single function with constant parameters predicts D_S, there being no sign of a crossover from dilute to semidilute behaviour in the experimental data.

(ii) The RSM predicts scaling (power-law) dependences of variables on each other. On a log-log plot, a power law may always be said to provide a straight-line tangent to a curve;

however, on a log-log plot stretched exponentials are always curves, not straight lines. While the stretched exponential form has some flexibility, a stretched exponential can never fit well to data that is correctly described by a power law, and *vice versa*. Extensive analysis of the literature[33,24] shows that most real data has stretched-exponential concentration dependence, and hence does not follows power (scaling) laws.

(iii) Sphere and chain dynamics: The RSM predicts for semidilute solutions that large spheres diffuse nearly in accord with the Stokes-Einstein equation, while random-coil chains with $M > M_e$ reptate, therefore diffusing faster than spheres of the same size. Contrary to this prediction, spheres and probe chains of equal size diffuse at nearly the same speed. Furthermore, contrary to the RSM, in semidilute solutions of large chains the Stokes-Einstein equation fails, η being decoupled from D_p.

(iv) The reptation model is based on an assumed transient gel (pseudolattice) structure in solutions. The pseudolattice is predicted[42] to act as a filter, selectively obstructing the diffusion of large probe particles. Experimentally, solutions are equally effective at retarding diffusion by large and small particles.

(v) The RSM predicts for the zero-shear viscosity $\eta \sim M^3$; the HSM implies $\eta \sim \exp(aM^\gamma)$. Solutions follow the HSM prediction[27], even when consideration is focused on experiments[43] that are identified elsewhere[35] as having confirmed reptation.

At a wide range of points, experiment thus rejects the reptation-scaling model of polymer solutions (melts are not being considered here), but supports the hydrodynamic scaling model.

6. SUMMARY

The feasibility of light scattering experiments on ternary mixtures was discussed, emphasis being placed on systems in which a single dilute species dominates the spectrum. The diffusion of spherical probes through a polymer solution adheres to Walden's rule, eq. 6. In appropriate cases, the consequences of matrix adsorption by the probe can be seen. When adsorption and bridging effects were suppressed, D_p followed a universal scaling equation in matrix concentration. D_p is substantially independent of probe and matrix architectures, but probe diffusion through solutions differs markedly from probe diffusion through true gels. A picture of polymer dynamics that is consistent with these findings, namely the hydrodynamic scaling model, was presented and argued to be superior to major competing models.

ACKNOWLEDGEMENTS

The partial support of this work by the National Science Foundation under Grant DMR89-43855 is gratefully acknowledged.

REFERENCES

1. G.D.J. Phillies, *J. Chem. Phys.*, 1974, *60*, 976; 1974, *60*, 983.
2. G.D.J. Phillies, *Biopolymers*, 1975, *14*, 499.
3. R.B. Jones, *Physica*, 1979, *97A*, 113.
4. T.P. Lodge, *Macromolecules*, 1983, *16*, 1393.
5. M. Benmouna, H. Benoit, R. Borsali and M. Duval, *Macromolecules*, 1987, *20*, 2620.
6. P.T. Callaghan and D.N. Pinder, *Macromolecules*, 1984, *17*, 431.
7. J.A. Wesson, I. Noh, T. Kitano and H. Yu, *Macromolecules*, 1984, *17*, 782.
8. T.P. Lodge, P. Markland and L.M. Wheeler, *Macromolecules*, 1989, *22*, 3409.
9. N. Nemoto, S. Okada, T. Inoue and M. Kurata, *Macromolecules*, 1988, *21*, 1509.
10. G.D.J. Phillies, C. Malone, K. Ullmann, G.S. Ullmann, J. Rollings and L.-P. Yu, *Macromolecules*, 1987, *20*, 2280.
11. G.D.J. Phillies, G.S. Ullmann, K. Ullmann and T.-H. Lin, *J. Chem. Phys.*, 1985, *82*, 5242.
12. G.D.J. Phillies, H. Inglefield, M. Kiss, L. Li, D. Maclung, M. Maguire, M. Malone, T. Pirnat, A. Rau, N. Teasdale, P. Chinigo, J. Rollings and L.-P. Yu, *Macromolecules*, 1989, *22*, 4068.
13. G.D.J. Phillies, *J. Phys. Chem.*, 1981, *85*, 2838.
14. T.-H. Lin and G.D.J. Phillies, *J. Colloid Interface Sci.*, 1984, *100*, 82.
15. K. Ullmann, G.S. Ullmann and G.D.J. Phillies, *J. Colloid Interface Sci.*, 1985, *105*, 315.
16. G. Ullmann and G.D.J. Phillies, *Macromolecules*, 1983, *16*, 1947.

17. L. Leger, H. Hervet and F. Rondelez, *Macromolecules*, 1981, *14*, 1732.
18. G.D.J. Phillies, *Macromolecules*, 1986, *19*, 2367.
19. W. Brown and P. Zhou, *Macromolecules*, 1989, *22*, 4031.
20. W. Brown, private communication.
21. N. Numasawa, K. Kuwamoto and T. Nose, *Macromolecules*, 1986, *19*, 2593.
22. L. Hou, F. Lanni and K. Luby-Phelps, *Biophys. J.*, in press.
23. I. Nishio, J.C. Reina and R. Bansil, *Phys. Rev. Lett.*, 1987, *59*, 684.
24. G.D.J. Phillies, *J. Phys. Chem.*, 1989, *93*, 5023.
25. G.D.J. Phillies, *Macromolecules*, 1987, *20*, 558.
26. G.D.J. Phillies, *Macromolecules*, 1988, *21*, 3101.
27. G.D.J. Phillies, *Macromolecules*, 1990, *23*, 2742.
28. P.N. Pusey and R.J.A. Tough, in 'Dynamic Light Scattering' edited by R. Pecora, Plenum, New York, 1985, Chapter 4.
29. M. Daoud, J.P. Cotton, B. Farnoux, G. Jannink, G. Sarma, H. Benoit, R. Duplessix, C. Picot and P.-G. de Gennes, *Macromolecules*, 1975, *8*, 804.
30. J.G. Kirkwood and J. Risemann, *J. Chem. Phys.*, 1948, *16*, 515.
31. C.W.J. Beenakker and P. Mazur, *Physica*, 1983, *120A*, 388; 1984, *126A*, 349.
32. J.M. Carter and G.D.J. Phillies, *J. Chem. Phys.*, 1985, *89*, 5118.
33. G.D.J. Phillies and P. Peczak, *Macromolecules*, 1988, *21*, 214.
34. P.-G. de Gennes, *Scaling Concepts in Polymer Physics*, Cornell University Press, Ithaca, NY, 1979.
35. M. Doi and S. F. Edwards, *The Theory of Polymer Dynamics*, Clarendon Press, Oxford 1986.
36. W. Hess, *Macromolecules*, 1986, *19*, 1395; 1986, *20*, 2587.
37. K.S. Schweizer, 1990 International Discussion Meeting on Relaxations in Complex Systems, Heraklion, Crete, June 1990.
38. J. Skolnick and R. Yaris, *J. Chem. Phys.*, 1987, *88*, 1407; *88*, 1418.
39. Y. Oono and P.R. Baldwin, *Phys. Rev. A*, 1986, *33*, 3391.
40. K.L. Ngai, A.K. Rendell and S.T. Rajagopal, *Annals N.Y. Acad. Sci.*, 1984, *484*, 150.
41. K.L. Ngai, A.K. Rajagopal and T.P. Lodge, *J. Polymer Sci.: Polymer Physics Ed.*, 1990, *28*, 1367.
42. Ref. 34, p. 80, cited there as H. Benoit, private communication (1975).
43. S. Onogi, S. Kimura, T. Kato, T. Masuda and N. Miyanaga, *J. Polym. Sci.* 1986, *C*, 381.

9

The Diffusion of BSA Molecules in Poly(*N*-Isopropyl Acrylamide) Gels: A Forced Rayleigh Scattering Study

By William G. Griffin and Mary C.A. Griffin

AFRC INSTITUTE OF FOOD RESEARCH, READING LABORATORY, SHINFIELD, READING. RG2 9AT U.K.

1. INTRODUCTION

The restricted diffusion of colloidal particles in chemically cross-linked gels is of widespread technological importance. For example, poly(acrylamide) gels are commonly used as bioseparation media in gel permeation chromatography and electrophoresis. The physical properties of many cross-linked gels often show strong dependence on temperature/solvent quality, ionic strength and polymer concentration[1]. It is possible to produce linear, water-soluble, polymers which exhibit a critical temperature *above* which the polymer precipitates from solution. This is called the lower critical solution temperature (LCST) and gels derived from linear polymers exhibiting LCST behaviour can also undergo such phase transitions. Typically, such behaviour in a gel is characterized by the development of opacity and shrinkage of the gel above the LCST. Aqueous poly(acrylamide) solutions do not appear to exhibit an LCST. Although the aqueous gels of poly(acrylamide) have been shown to undergo phase transitions and the phase diagram (composition-temperature) for such gels is rather complex[1,2,3] the cross-linked aqueous gels of poly(acrylamide) also do not show LCST phenomena. Some polymers of N-substituted acrylamides, however, do exhibit an LCST. Linear poly(*N*-isopropyl acrylamide) (poly(NIPAAM)) is an example of such a polymer[4]. The LCST property of poly(NIPAAM) is reflected in the behaviour of poly(NIPAAM) gels, prepared by polymerizing NIPAAM in the presence of the cross-linker, *N,N'*- methylene bis-acrylamide. These gels have been found to shrink reversibly on heating above the LCST[5]. Poly(NIPAAM) gels can absorb small molecules and macromolecules, showing a selectivity for molecular size that depends on poly(NIPAAM) and cross-linker concentration[6,7]. A further level of control of gel properties arises from manipulation of the LCST by inclusion of other acrylamide derivatives as co-monomers[8,9]. The change in gel pore size which occurs when the gel crosses the LCST (at approximately 32°C for poly(NIPAAM)) can also be used to modulate the activity of enzymes trapped in the gel[9,10]. The diffusion of particles in gels and solutions of polymers is also of importance from a more fundamental viewpoint. The reduction in probe diffusion rates caused by an immobile matrix or an entanglement network is thought to be explained in part by hydrodynamic screening effects (ref. 11 and references therein) but other physical mechanisms such as entrapment and adsorption may also be important. It is therefore of considerable interest to obtain accurate measurements of self-diffusion coefficients (D_S) for probe molecules of known shape and size in gels where the gel pore size can be controlled *via* polymer concentration and temperature. In the present study the forced Rayleigh scattering (FRS) technique has been used to measure D_S for dye-labelled bovine serum albumin (BSA) molecules in poly (NIPAAM) gels.

2. MATERIALS AND METHODS

All chemicals used were BDH Analar grade, except where otherwise stated. Water was purified using a MilliQ system (Millipore).

Labelling of BSA with 4-dimethylaminophenylazophenyl-4'-maleimide (DABMI)

BSA (Miles Ltd., Slough, Pentex ® monomer grade) was treated with DABMI, followed by a work-up procedure including dialysis against activated charcoal. The method has been described in detail elsewhere[12].

Preparation of poly(NIPAAM) gels

(a) Samples prepared for photography A 20% w/w aqueous solution of *N*-isopropyl acrylamide (as supplied by Kodak) was mixed with *N,N'*-methylene bis-acrylamide (bis) as cross-linker. The ratios of cross-linker to NIPAAM are given below. The NIPAAM/bis solution was diluted to the desired concentration with water before adding 2 μl/ml *N,N,N',N'*-tetramethyl ethylene diamine (TEMED) (Aldrich) as catalyst and 1 mg/ml ammonium persulphate (Sigma) as initiator. The pre-gelling solutions were mixed by repeated inversion of the containers. To minimize large scale inhomogeneities which might arise by convection effects the solutions were kept as nearly as possible at constant temperature (either 4°C or 35°C) throughout the gel preparation.

(b) Samples prepared for FRS or fluorescence All samples were prepared in a phosphate buffer, pH 7.0, at approximately 75 mM. The polymerization was carried out at 4°C as described in (a) but in fused silica cuvettes (Hellma) (5 mm and 10 mm path length for FRS and fluorescence measurements respectively). The ratio of cross-linker to NIPAAM was 0.02 w/w. BSA or DABMI-labelled BSA (DABMI-BSA) was incorporated into the gels by inclusion in the polymerization mixture. This was done by making up the NIPAAM solution in buffer containing BSA or DABMI-BSA. FRS measurements were made with DABMI-BSA present in the gels at a concentration of approximately 7 $mg.ml^{-1}$.

3. FORCED RAYLEIGH SCATTERING

Forced Rayleigh scattering is a technique for measuring self-diffusion coefficents of tracer particles over a wide range of values (*ca.* 10^{-6} cm^2 s^{-1} - 10^{-14} cm^2 s^{-1}). The tracer particles are labelled with a photochromic dye whose optical properties are changed by exposure to blue (488 nm) light. The sample is illuminated with a diffraction pattern created at 488 nm. A transient grating is thus formed in the sample and the depth of modulation decreases as the dye-labelled molecules diffuse. The relaxation time of the transient grating is measured by following the intensity of a red (reading) laser beam scattered by the transient grating at the Bragg angle. The three-dimensional nature of the transient grating is depicted in Fig. 1. The FRS apparatus used has been described elsewhere[12]. A schematic diagram of the apparatus is given in Fig. 2. Data analysis was carried out as follows. For a monodisperse system, such as DABMI-BSA in buffer the decays are of the form:-

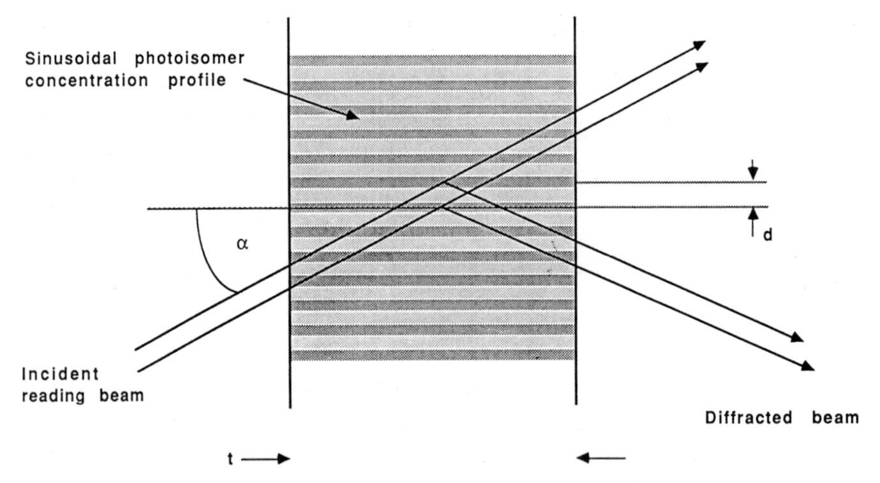

Figure 1. Schematic diagram of FRS interference fringes. t is the grating thickness, d is the grating spacing and α is the Bragg angle.

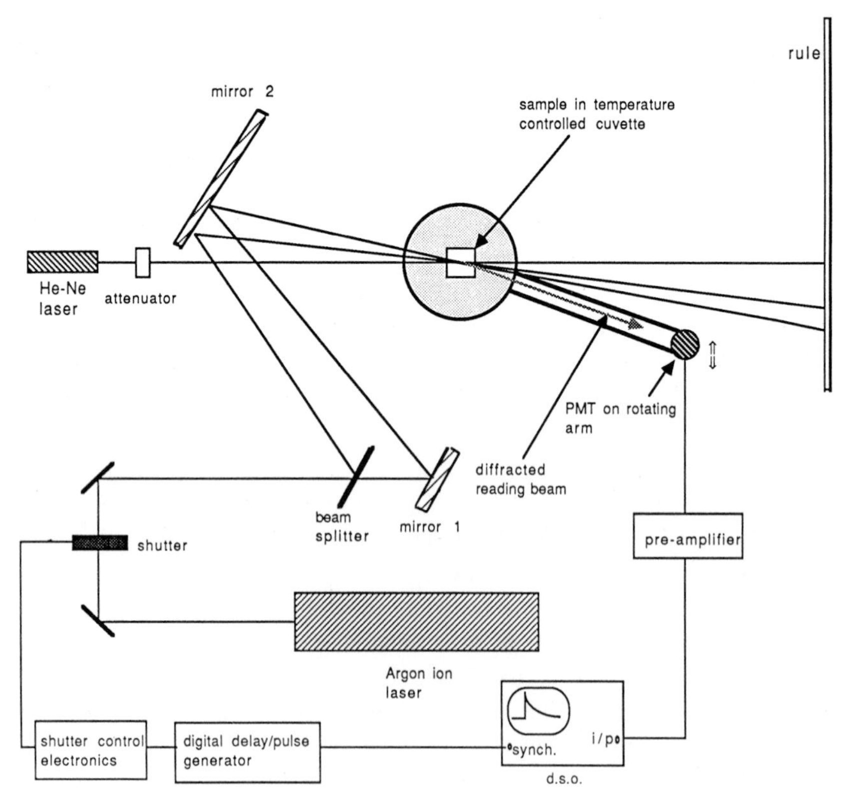

Figure 2. Schematic diagram of FRS apparatus.

$$V(t) = (A \exp(-t/\tau) + B)^2 + C^2 \quad (1)$$

where V is the FRS signal amplitude, τ is the signal decay time, B^2 and C^2 are proportional to the coherent and incoherent background levels respectively. $(B^2 + C^2)$ is equal to the total baseline, measured as the output voltage before the writing beam excitation. The data sets were fitted to eq. (1) using a non-linear fitting routine. The three parameters A, B and τ, were varied such that a minimum was obtained in the sum of squares of the residuals weighted by $[V(t)]^{-1/2}$. This corresponds to Poisson statistics for the noise[12].

Eq. (1) gave excellent fits to the measured FRS decays from DABMI-BSA in buffer[12]. By contrast, the FRS decays from DABMI-BSA incorporated into poly(NIPAAM) gels did not give good fits to eq. (1), indicating that it did not provide an adequate description of the decays. For analysing data from photon correlation spectroscopy (PCS), Chu[13] suggested obtaining a mean decay time by fitting the data to the sum of two exponentials, if the polydispersity is greater than 0.1. In a similar way, the FRS decays from DABMI-BSA in poly(NIPAAM) gels were fitted to the following equation:

$$V(t) = \{A_1 \exp(-t/\tau_1) + A_2 \exp(-t/\tau_2) + B\}^2 + C^2 \quad (2)$$

which describes the FRS decay for two diffusing species, and may also represent the FRS decay from a polydisperse system. B was found to be close to zero, and was therefore set to zero, to reduce the number of variable parameters, and simplify the optimization. Excellent fits to eq. (2) were obtained for the FRS decays of DABMI-BSA in poly(NIPAAM) gels of concentration up to 9.9% w/v.

From these fits we obtain a mean decay time, τ_{av}, (an average determined by the number of dye groups attached to each diffusing species) from:

$$\frac{1}{\tau_{av}} = \frac{A_1/\tau_1 + A_2/\tau_2}{A_1 + A_2} \quad (3)$$

where τ_1 and τ_2 are the decay constants varied in the fitting procedure and A_1 and A_2 are fitting coefficients. Measurements of the FRS decay were made for different grating spacings and $1/\tau$ or $1/\tau_{av}$

was plotted against q^2, defined by $q=4\pi\sin(\theta)/\lambda_{write}$, where θ is the writing beam crossing angle, and λ_{write} is the wavelength of the writing beam. A straight line plot of this kind was taken to indicate that a diffusive process was being studied. D_s, the self-diffusion coefficient, is obtained as the slope. The plot of $1/\tau_{av}$ against q^2 for BSA in 9.9% poly(NIPAAM) is shown in Fig. 3.

4. RESULTS AND DISCUSSION

The physical properties of poly(NIPAAM) gels depend on the temperature at which the polymerization is carried out. To illustrate this, Fig. 4 shows photographs of poly(NIPAAM) gels and sols polymerized at (A) 4°C and (B) 35°C. The concentration of NIPAAM and the ratio of cross-linker to NIPAAM in each sample is indicated by the side of the grid in Fig. 4. At 4°C all the mixtures formed clear gels. At 35°C the 2 %w/v reaction mixtures formed cloudy sols; cloudy gels were formed from the higher NIPAAM concentrations. After the polymerization was complete the gels were placed in an incubator at 37°C to equilibrate for 5 hours. All the samples became turbid at this temperature. They were then brought out of the incubator and allowed to cool to room temperature (26°C). Photographs (A1) and (B1) were taken shortly after removal of the gels from the incubator. The other photographs were taken subsequently. The gels prepared at 4°C clarified within 6 minutes, the more highly cross-linked gels clearing more rapidly. The gels prepared at 35°C became less turbid much more gradually. After holding these gels at room temperature for approximately 2 hours, they were kept in a 4°C room overnight. The following morning the gels were allowed to warm to room temperature and a final photograph (B3) was taken. The photographs show that the polymers are held in a network structure characteristic of their temperature of polymerization. The gels formed at 35°C showed a tendency to retain the phase separated structure characteristic of poly(NIPAAM) above the LCST. The more highly cross-linked samples and those at higher NIPAAM concentration were slower to clarify. The 10% gels did not clarify over a period of weeks.

The fluorescence emission spectra were recorded at 25°C of poly(NIPAAM) gels, both with and without incorporated BSA. The spectrum of BSA in poly(NIPAAM) was then corrected for the fluorescence shown by impurities in the poly(NIPAAM). Fig. 5

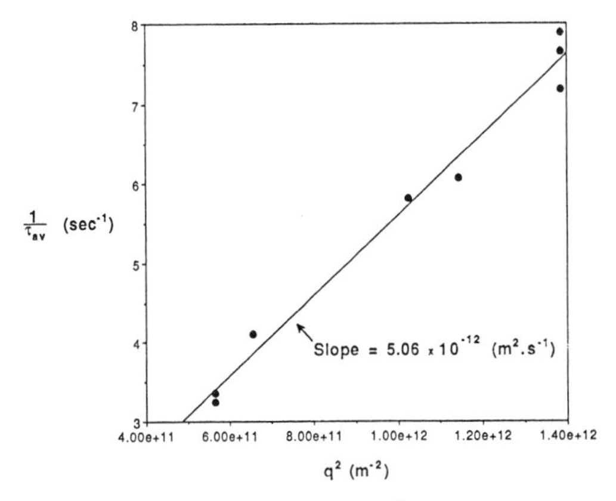

Figure 3. Plot of $1/\tau_{av}$ *versus* q^2 for DABMI-BSA in a 9.9% poly(NIPAAM) gel.

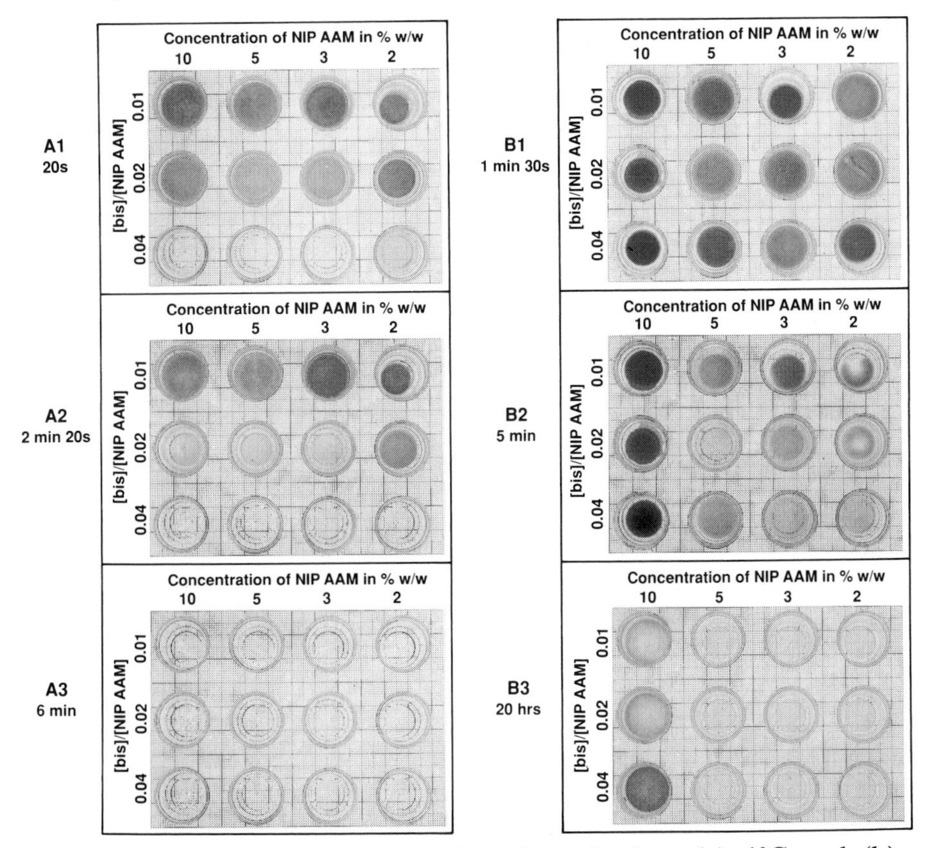

Figure 4. Poly(NIPAAM) gels polymerized at (a) 4°C and (b) 35°C. Details of the experiment are given in the text. Each photograph is annotated with the corresponding time after removal of the gel from the incubator.

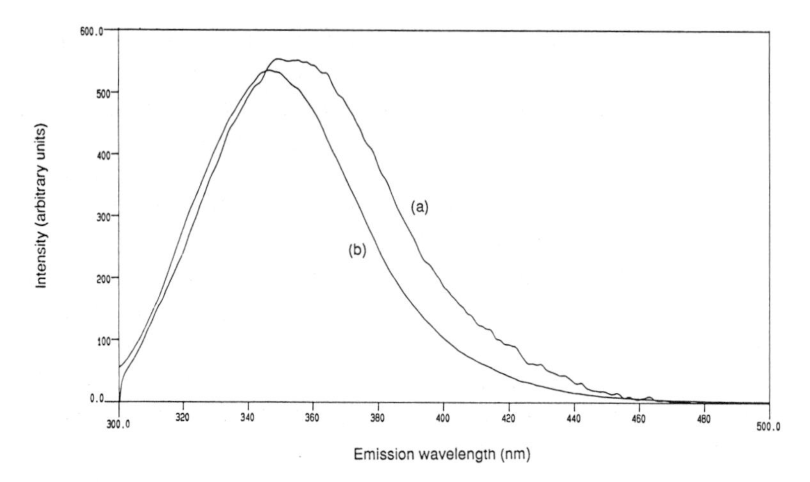

Figure 5. Fluorescence spectra of BSA in (a) 3 % poly(NIPAAM) gel and (b) pH 7.0 buffer. The excitation wavelength was 290 nm.

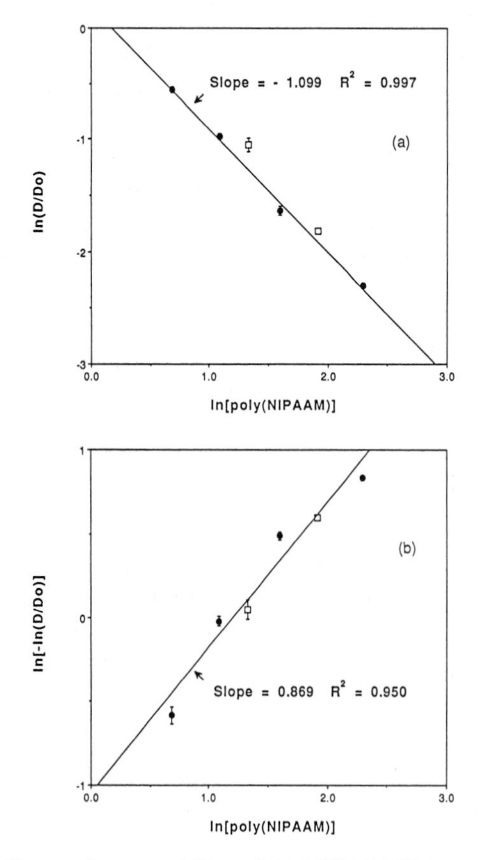

Figure 6. Dependence of D_S of DABMI-BSA on poly(NIPAAM) concentration.

curve (a) shows the fluorescence emission spectrum, obtained from excitation at 290 nm, of BSA (0.075 mg/ml) in a 3% w/v poly(NIPAAM) gel after appropriate correction. The emission spectrum of BSA (0.069 mg/ml) in buffer is also shown Fig. 5 curve (b), recorded for the same temperature and excitation wavelength. Spectrum (a) has been multiplied by 4.5 to bring it on to the same scale as spectrum (b). Poly(NIPAAM) gel causes a reduction in intensity of BSA fluorescence and a shift in the emission maximum of about 10 nm to longer wavelength. The reduction in intensity may be due to quenching by residual NIPAAM monomer. The red shift may be due to binding of residual monomer (NIPAAM), poly(NIPAAM) or impurities to a hydrophobic pocket on the BSA. It is planned to investigate the quenching and the spectral shift by making further measurements on gels made from more extensively purified NIPAAM.

The mean value of D_S of 5.06 x 10^{-11} m^2 s^{-1} for DABMI-BSA in buffer at 18°C is in good agreement with the values obtained by PCS (see reference 12). Fig. 6 shows the dependence of D_S for DABMI-BSA, on the concentration of poly(NIPAAM) at the same temperature. The graphs show data from two experiments. The filled circles represent data from samples which were not mixed after the onset of gelation. The open squares represent data from two samples which were mixed as gelation proceeded. The fitted straight lines were obtained by linear regression to the data represented by the filled circles. R is the regression coefficient. It has been proposed by various authors (*e.g.* refs. 11,14,15) that the effect of a polymer network on the diffusion of hard spheres is described by the stretched exponential equation:

$$\frac{D}{D_0} = \exp(-\alpha\, C^{\delta}) \qquad (4)$$

where D_0 is the diffusion coefficient of the probe in the absence of the network and C is the polymer concentration. The parameters α and δ can be determined by fitting curves of $\ln(-\ln(D/D_0))$ *versus* ln C. The value of α will, in general, depend on the size of the probe molecule.

We have also considered the possibility that the variation of D/D_0 with C could be described by a power law:

$$\frac{D}{D_0} = A\,C^{\beta} \qquad c > c^* \qquad (5)$$

where c^* represents the NIPAAM concentration below which the gel will not form. The plot of $\ln(D/D_0)$ *versus* ln(C) (Fig. 6(a)) gave a better fit to our data than the plot of $\ln(-\ln(D/D_0)$ *versus* ln(C) (Fig. 6(b)). The difference in fit is not great but there seems no reason to prefer the stretched exponential form on the basis of the available evidence. The straight line fit of $\ln(D/D_0)$ is based on samples with four different gel concentrations, prepared under identical conditions. The diffusion coefficients for two samples prepared under slightly different conditions, do not fall on the same straight line. It seems very important to control the thermal history and mixing conditions for the gels in order to achieve reproducibility in probe diffusion coefficients.

5. CONCLUSIONS

Our results suggest that the variation of $(D/D_0)_{BSA}$ with poly(NIPAAM) gel concentration over the range 2% - 10% w/v can be described by a power law. The fit to the stretched exponential formula is not quite so good. Other workers studying self-diffusion of probe particles in polymer networks have reported good fits to the stretched exponential form. A review of such data for a variety of macromolecules diffusing in macromolecular solutions has been presented by Phillies[16]. He concludes that the stretched exponential formula gives a better fit than power laws to the experimental data for spherical probes and flexible polymers diffusing in entanglement networks. The electrophoretic mobility of probe molecules in gel matrices has also been shown to vary with gel concentration according to the stretched exponential form[17]. However, the presence of an electric field may affect the gel matrix properties and may also cause flow of solvent so that comparisons of friction coefficients derived from electrophoretic mobility measurements with those from simple diffusion measurements must be made with caution. Leloup *et al.*[18,19] have described experiments in which the diffusion of BSA in amylose and amylopectin gels was measured by a classical diffusion cell technique. Their plot of $\ln(-\ln(D/D_0))$ against ln(c) for BSA in amylose gels is linear (R=0.98) and the slope, δ, given in the later paper[19], has a value of 0.73. Literature values of D/D_0 of ABITC-BSA (azobenzene-*p*-isothiocyanate labelled BSA) in a poly(acrylamide) gel[20] fit a stretched exponential (Fig. 7(b)) somewhat better than a power law (Fig. 7(a)), though the difference is not marked. The value of δ is given as 0.92, significantly different from the value reported by Leloup *et al.*[19] It should be noted that the data from reference 20 include two points at concentrations above 7% w/v, which do not follow the stretched exponential equation. Bansil and Gupta[3] showed a phase diagram for the co-polymerization of acrylamide and bis at 25°C. For the ratio of bis to acrylamide used by Park *et al.*[20] in their polymerization mixtures (5% w/w), Bansil and Gupta indicate that above a total monomer concentration of approximately 5% w/v opaque gels are formed. At concentrations between 2% and 5% the mixtures form clear gels. The gels described by Park *et al.* were polymerized at a slightly different temperature (22°C). It seems possible that their observation of different behaviour in gels of concentration above 7% reflects a phase transition. It is also possible that the simple power law we observe is due to changes in the structure of the poly(NIPAAM) gel which may occur with changing gel concentration. It should be noted that there were no indications of changing solvent content in the gels over the period of measurement. Alternatively, the power law may reflect some kind of pre-transitional behaviour in the gel. In this context it should be noted that we observed a dramatic decrease in the FRS signal amplitude at a temperature (*ca.* 22°C for the 10% gel) well below the LCST for the gel at a given concentration. This diminution in signal amplitude coincided with the development of increased scattering from the gel suggesting that some increase of heterogeneity was occurring. It may be that the appearance of such inhomogeneities in the gel reflects the formation of hydrophobic regions which could cause some segregation of the probe molecules. In this case the azo dye groups attached to the protein molecules may find themselves in an environment unfavourable to the *trans-cis* isomerization which is required for the FRS grating to form. This point is currently under investigation. Another possible explanation for deviations from the stretched exponential form would be the reversible binding of the probe molecules to the gel matrix. Our fluorescence results do indicate that there is an interaction between constituents of the gel and the probe but we are not yet able to decide whether this would be sufficient to account for our results. Other FRS studies[21,22] have shown that reversible binding of small molecules to a gel matrix can markedly affect the measured diffusion coefficients.

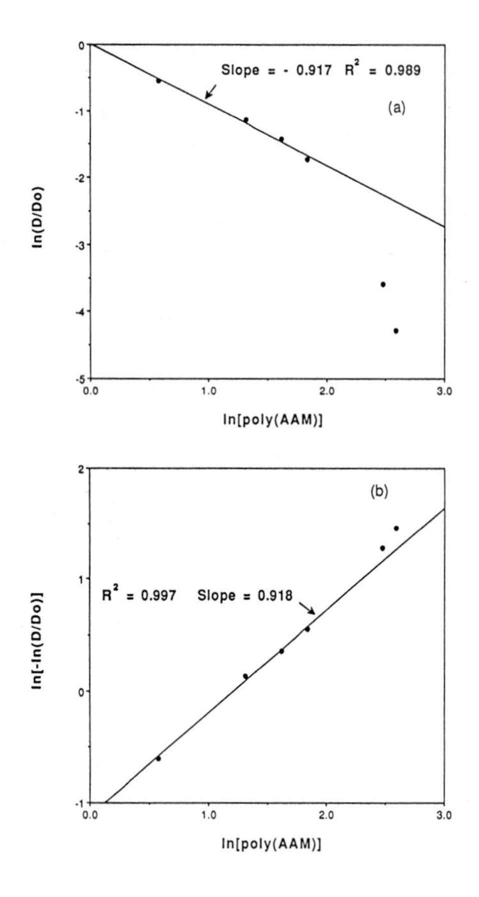

Figure 7. Dependence of D_S of ABITC-BSA on polyacrylamide (poly(AAM)) concentration (data taken from ref. 20).

Further experiments are needed to test the power law description of probe diffusion data in poly(NIPAAM) gels and establish whether, for example, binding effects or dependence of gel structure on poly(NIPAAM) concentration are significant. However, it is worth remarking that the theoretical validity of eq. (4), as derived by Cukier[11], rests on the approximations inherent in the effective medium description of diffusion of a probe particle in a Brinkman[23] fluid. If the structure (*e.g.* the statistical distribution of gel matrix density) and/or dynamics (*e.g.* the gel matrix flexibility) of the gel are such that an effective medium approximation is not valid then different approximate solutions of the Brinkman equation will need to be examined. The theoretical explanation of the retardation of protein molecules in poly(*N*-isopropyl acrylamide) gels, and possibly in other gels formed from acrylamide and its *N*-alkyl derivatives, is still open to further development.

6. SUMMARY

Measurements have been reported of the self-diffusion coefficients of bovine serum albumin (BSA) molecules in chemically cross-linked gels of poly(*N*-isopropyl acrylamide). The gels exhibited a lower critical solution temperature above which they became opaque and shrank. It was found that the self-diffusion coefficient of the BSA varied with polymer concentration approximately according to a power law. This behaviour was contrasted with results for probe diffusion in other gel systems. Some indications of gel-probe interactions have been discussed and possible explanations for the power law behaviour were examined.

ACKNOWLEDGEMENTS

We thank Mr John Price for assistance in preparing the DABMI-labelled BSA. This work was supported by the Agriculture and Food Research Council.

REFERENCES

1. *e.g.* T. Tanaka, *Scientific American*, 1981, *244*, 110, and references therein.
2. J. Baselga, I. Hernandez-Fuentes, R.M. Masegosa, and M.A. Llorente, *Polymer Journal*, 1989, *21*, 467
3. R. Bansil and M.K. Gupta, *Ferroelectrics*, 1980, *30*, 63.

4. M. Heskins and J.E. Guillet, *J. Macromol. Sci. Chem.*, 1968, *A2*, 1441.
5. Y. Hirokawa and T. Tanaka, *J. Chem. Phys.*, 1984, *81*, 6379.
6. R.F.S. Freitas and E.L. Cussler, *Sep. Sci. Technol.*, 1987, *22*, 911.
7. A.S. Hoffman, A. Afrassiabi, and L.C. Dong, *J. Contr. Rel.*, 1986, *4*, 213.
8. L.D. Taylor and L.D. Cerankowski, *J. Polymer Sci., Polym. Chem. Ed.*, 1975, *13*, 2551.
9. L.C. Dong and A.S. Hoffman, *J. Contr.Rel.*, 1986, *4*, 223.
10. T.G. Park and A.S. Hoffman, *Appl. Biochem. Biotechn.*, 1988, *19*, 1.
11. R.I. Cukier, *Macromolecules*, 1986, *17*, 252.
12. W.G. Griffin and M.C.A. Griffin, in 'Food Polymers, Gels and Colloids', Royal Society of Chemistry Special Publication *82*, ed. E. Dickinson, 1991, p. 47.
13. B. Chu in 'The application of laser light scattering to the study of biological motion', ed.J.C. Earnshaw and M.W. Steer, NATO A.S.I., 1982, *A59*, 53.
14. A.G. Ogston, B.N. Preston, and J.D. Wells, *Proc. Roy. Soc.*, 1973, *A333*, 297.
15. D. Langevin and F. Rondelez, *Polymer*, 1978, *19*, 875.
16. G.D.J. Phillies, *Macromolecules*, 1986, *19*, 2367.
17. D. Rodbard and A. Chrambach, *Proc. Nat. Acad. Sci.*, 1970, *65*, 970.
18. V.M. Leloup, P. Colonna and S.G. Ring, *Food Hydrocolloids*, 1987, *1*, 465.
19. V.M. Leloup, P. Colonna and S.G. Ring, *Macromolecules*, 1990, *23*, 862.
20. I.H. Park, C.S. Johnson and D. Gabriel, *Macromolecules*, 1989, *23*, 1548.
21. W.G. Griffin and M.C.A. Griffin, *Makromol. Chem., Macromol. Symp.*, 1991, *45*, 271.
22. J.A. Lee and T.P. Lodge, *J.Phys.Chem.*, 1987, *91*, 5546.
23. H.C. Brinkman, *Appl.Sci.Res.*, 1947, *A1*, 27.

10

Developments in Electrophoretic Laser Light Scattering and some Biochemical Applications

By Kenneth H. Langley

DEPARTMENT OF PHYSICS & ASTRONOMY, UNIVERSITY OF MASSACHUSETTS, AMHERST, MASSACHUSETTS 01003, U.S.A.

1. INTRODUCTION

Electrophoretic light scattering (ELS) provides a direct measurement of the mobility of charged particles moving in solution under the influence of an applied electric field. The mobility, defined as the ratio of the particle velocity V to the electric field strength E,

$$U = V/E \tag{1}$$

is a useful quantity as it can, under suitable conditions, be related to the density of surface charge on the particle or to the zeta potential. The zeta potential has become widely used to characterize the stability and transport properties of colloidal systems and is therefore of obvious interest in many commercial applications such as waste water treatment, processing and paper industries, and the suspension and deposition of paints, inks, and other coatings. While not nearly so numerous, biological and biochemical applications of electrophoretic mobility[1] have included the characterization and modification of various particles found in the blood and the study of particles for which the attachment of smaller entities for purposes of drug delivery or analysis might be important. The concentration of ions at the particle-solution interface may also play a significant role in the surface interaction of proteins, viruses and enzymes. Several instruments have recently become available which routinely obtain the particle velocity needed for mobility measurements by utilizing the Doppler frequency shift in laser light scattered from the particles undergoing electrophoresis. I shall describe one such instrument and illustrate its capability by presenting a few measurements that are relevant to biochemistry and biology.

2. FUNDAMENTALS OF ELECTROPHORETIC LIGHT SCATTERING

Electric Charge and Zeta Potential

The separation of charge and the existence of the electrical double layer at the interface between the surface of a particle and the surrounding medium are phenomena which have been studied for many years. The only aspects of the situation which will concern us here are that the particle itself, together with nearby ions which move with the particle, form a system with some net effective surface charge density which moves through the solution under the influence of an electric field, while effectively separated from the surrounding solution by an imaginary shear plane. The electrical potential at this shear boundary relative to the potential far from the particle (assuming the double layer thickness is small compared to the particle size) is known as the zeta potential and may be related to the mobility defined in equation 1:

$$U = \varepsilon\zeta/\eta \tag{2}$$

where ε is the permittivity of free space ε_0 times the dielectric constant of the solvent, ζ is the zeta potential and η is the viscosity of the solvent. Thus, in many situations of experimental interest, U and ζ differ only by a multiplicative constant. In general, the larger the zeta potential, the more stable the colloid.

Light Scattering Detection of Electrophoretic Velocity

Laser Doppler detection of particle velocity, the distinguishing feature of ELS, is illustrated in Fig. 1. Light from the main laser beam scattered by particles through some angle θ falls on a photodetector simultaneously with unscattered light from a reference beam derived from the same laser source. The scattered light is shifted in frequency relative to the reference beam because of the Doppler effect by an amount given by

$$\Delta\nu = \frac{\mathbf{K.V}}{2\pi} = \frac{2nV}{\lambda_0}\sin\left\{\frac{\theta}{2}\right\} \tag{3}$$

where $K = (4\pi n/\lambda_0)\sin(\theta/2)$ is the magnitude of the scattering wavevector $\mathbf{K}$, n is the index of refraction of the solution, λ_0 is the vacuum wavelength of the light, and θ is the scattering angle. The range of frequency differences typically encountered in ELS is only *ca.* 1 kHz, but can easily be measured because the detector is sensitive to the frequency difference between the scattered light and the reference beam (heterodyne detection). Measurements performed at a series of angles will yield a plot of frequency *vs.* angle which is very nearly linear for angles up to 30° - 40°.

The mobility may be expressed from equations 1 and 3 as

$$U = \frac{\Delta\nu\ \lambda_o}{[2nE\ \sin(\theta/2)]} \tag{4}$$

A frequency spectrum of peaks corresponding to species of particles of different mobilities may thus be converted to a spectrum of mobilities or of zeta potentials by appropriate scaling, but we must be careful to remember that the amplitude of a peak represents only the intensity of scattered light at that value of the abscissa, not the number or weight fraction. Fig. 2 illustrates these points: measurements at 4 different scattering angles of a polystyrene latex are plotted as a function of frequency in Fig. 2a, and replotted as a function of measured mobility in Fig. 2b. The four peaks in Fig. 2a are simply four representations of the single mobility peak shown in Fig. 2b.

Diffusion Broadening

The spectral width of the peaks shown above can arise from two distinct mechanisms (aside from instrumental resolution which is always present, and which may, in fact, be the dominant effect). In addition to the effect of possible dispersion of the mobility within the population of particles sampled, there is always present a finite peak width due to the distribution of velocities of random Brownian motion. The contribution of Brownian diffusion to the peak width $\delta\nu$ is given by

$$\delta\nu = \frac{D\ K^2}{\pi} \tag{5}$$

here D is the diffusion coefficient which for spherical particles of radius R is given by

$$D = \frac{kT}{6\pi\eta R} \tag{6}$$

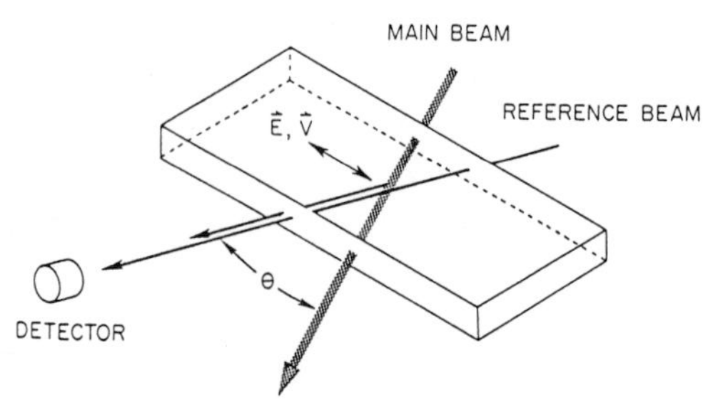

Figure 1. Scattering geometry in the ELS instrument. E and V a the electric field and particle velocity, respectively. K is the scattering wavevector.

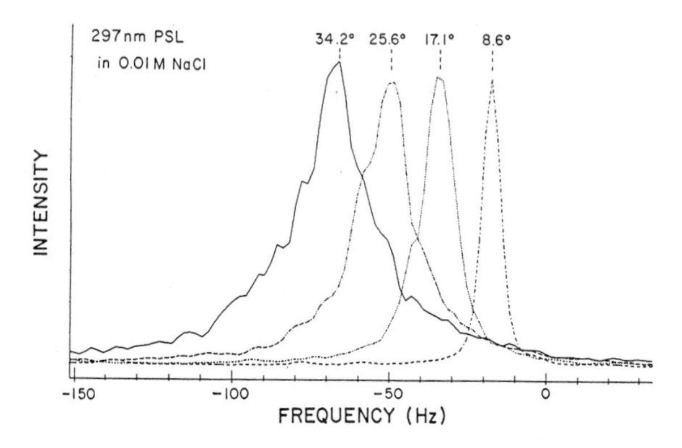

Figure 2a. Frequency spectrum of a polystyrene latex at four scattering angles. The frequency shift and linewidth depend on scattering angle.

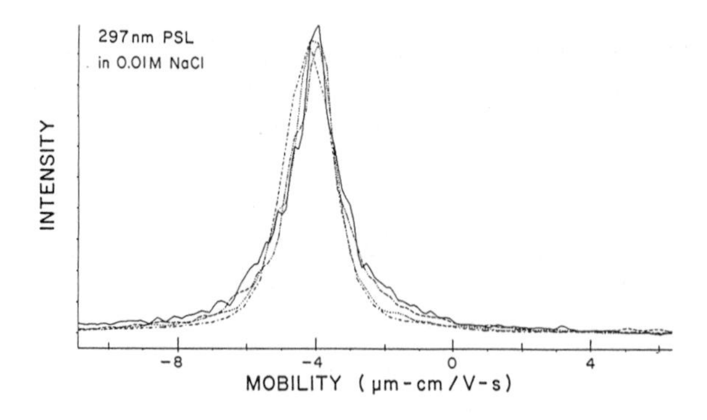

Figure 2b. The same PSL data as in Fig. 2a, rescaled and plotted on the same mobility axis. Measurements at all four angles now coincide as expected.

Eqs. 5 and 6 are well known as the basis of quasielastic or dynamic light scattering determination of particle diffusion coefficient or size. The dramatic effect of diffusion is shown in Fig. 3. At the same angle, the spectral diffusion broadening of 21 nm diameter PSL is enormous while 1000 nm PSL produces a narrow sharp spike. The diffusion limited resolution of a spectral peak *i.e.* the ratio of the frequency shift to the peak width is thus

$$r = \frac{KV}{DK^2} = \frac{V}{DK} \tag{7}$$

showing that for small particles it is desirable to perform measurements at small angles, although this will result in small frequency shifts which are inherently more difficult to measure because of the instrumental limits of frequency resolution and the presence of low frequency noise or vibration. Equation 7 confirms that higher electric field strength (and therefore velocity) also leads to increased resolution, although one must be cautious about excessive Joule heating of the sample during the course of the experiment.

3. AN ELECTROPHORETIC LIGHT SCATTERING INSTRUMENT

The above principles have found expression in a new generation of instruments designed to perform ELS measurements quickly and accurately. I would like to review some details of one such instrument[2], the Coulter® DELSA 440, and in the next section display some results obtained with this instrument in order to illustrate the current level of capability.

The heart of the instrument, the scattering chamber, is shown in Fig. 1. The liquid sample is contained within a rectangular cross section channel through a glass block. Most of the laser power is concentrated in the main beam which intersects the much weaker reference beam at angle θ. The detector signal (centred at frequency Δν given by equation 3) is processed by a 256 channel autocorrelator and Fourier transformed to obtain the frequency spectrum. In fact, there are four reference beams at nominal scattering angles of 7.5°, 15°, 22.5° and 30°, and a detector and correlator corresponding to each angle. Thus, frequency spectra are obtained at four angles simultaneously, providing redundancy which is useful for the elimination of artifacts and essential for extracting qualitative information about particle size. The main beam is continuously advanced or retarded in phase in order to determine the sign of the frequency shift and hence the mobility. The electric field, transverse to the laser beams, as shown, is applied through electrodes which form spherical caps to the liquid spaces at either end of the rectangular channel. Large surface area of the electrodes avoids high current densities and the attendant bubbling problems. The entire glass scattering cell and attached electrode assembly may be removed from the instrument for easy filling, and can be raised or lowered with respect to the laser beams and detectors in order to place the scattering volume within the stationary layer. The electro-osmotic effect initiates a flow countered by a gravitationally induced flow such that the net solution velocity is zero at only two heights within the rectangular channel. The velocity detected is therefore the true electrophoretic velocity only if the cell height is adjusted such that the scattering volume is placed within one of the two stationary layers.

4. ELS MEASUREMENTS ON VARIOUS SYSTEMS

Multimers of Bovine Serum Albumin

The electrophoretic mobility spectrum of a preparation of BSA is shown in Fig. 4. Three frequency spectra were obtained simultaneously at three different scattering angles in one experimental run, and the results scaled and plotted on the same mobility axis. At least five species of different mobility are resolved and the correspondence in peak positions for the three scattering angles shows that these peaks are not just noise, but do represent species of different mobility. Independent work revealed that none of the individual mobility populations represents monomeric BSA.

Stability of a Vesicle Preparation

While developing a series of liposome based drug delivery systems, a small commercial research facility discovered that some formulations aggregated with time. It was thought that perhaps a surface charge density change accompanied this aggregation, yielding a species of different mobility than the initial 68 nm particles. A sample which had been originally analyzed as monodisperse particles of 68 nm diameter was analyzed two weeks later in several ways on the ELS instrument. First, frequency spectra were obtained at four scattering angles in zero applied electric field as shown in Fig. 5a. One would expect to see diffusion broadened peaks centred at zero frequency. The presence of a sharp "spike" at the centre of the diffusion broadened peak indicated the existence of some slowly diffusing larger species. Analysis of the entire peak yielded an average size of approximately 400 nm. Analysis of only the central portion of the peak indicated the presence of a 620 nm diameter species; exclusion of the central spike from the analysis yielded about 88 nm. Mobility analysis with a 0.1 mA current passing through the scattering cell produced the mobility spectra shown in Fig. 5b. Indeed there are two species, one of -4.2 and one of -2.6 µm cm/(V s) mobility. Further, we can see that the right hand peak of -2.6 µm cm/(V s) mobility is the species of larger size. We know this from the fact that large particles scatter light much more strongly in the forward direction than at higher angles, while small particles scatter nearly isotropically. At scattering angle 8 degrees, the left peak is nearly invisible; as the scattering angle increases, the left peak becomes stronger and stronger relative to the right. An independent photon correlation analysis confirmed the bimodality of the sample, indicating a weight distribution of 93% 67 nm particles and 7% 623 nm particles.

Lysosyme

The ability to analyze particles as small as 6 nm is shown in Fig. 6, a mobility spectrum of free lysosyme. The need to overcome diffusion broadening in small particles by performing the ELS measurement at small scattering angle (see equation 7) is vividly demonstrated: at 34° or 25°, the mobility peak is practically nonexistent. At 17° and 8.6° the peak is quite apparent although the effect of limited resolution is rather noticeable, particularly at 8.6°.

Bacterial Cells With Differing Amounts of Surface Antigen

ELS is a sensitive method of detecting the surface charge condition of membrane bounded biological particles. Fig. 7 depicts the changes in mobility associated with different degrees of expression of a specific antigen on the surface of a bacterial cell. Four different isolates of the same strain of bacteria in which the amount of surface antigen is genetically determined were analyzed by ELS. The group of six curves making up the rightmost peak is derived from two of the isolates (three scattering angles each) which lack the antigen

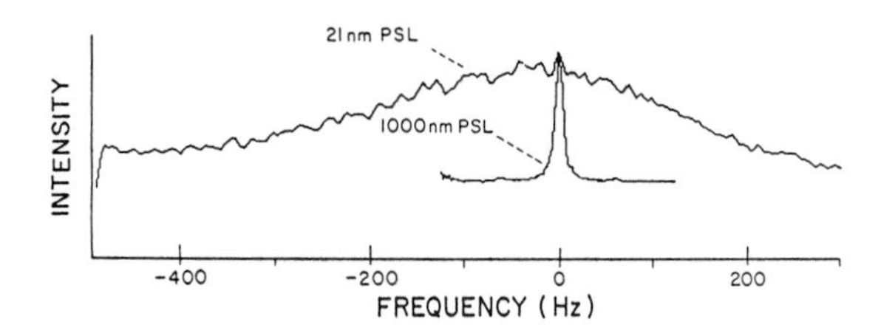

Figure 3. Diffusion broadening of light scattered at 34° from 21 nm and 1000 nm PSL. There is no applied electric field.

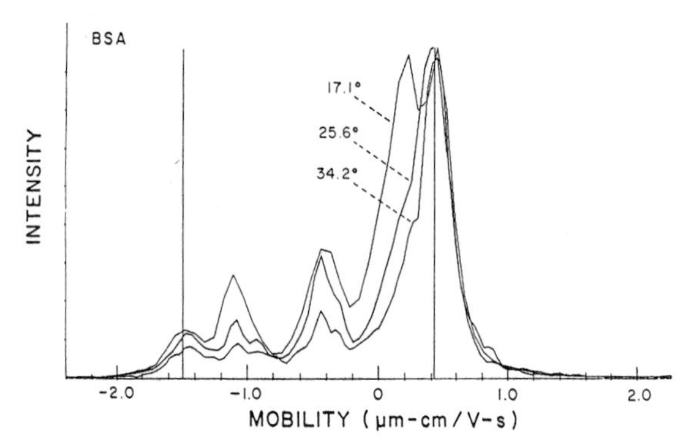

Figure 4. Resolution of multimers of bovine serum albumin by ELS. Sample was prepared by dissolving 0.32 g of BSA in 20 mL of filtered (0.22 μm pore) 20 mM KCl, pH 7.0.

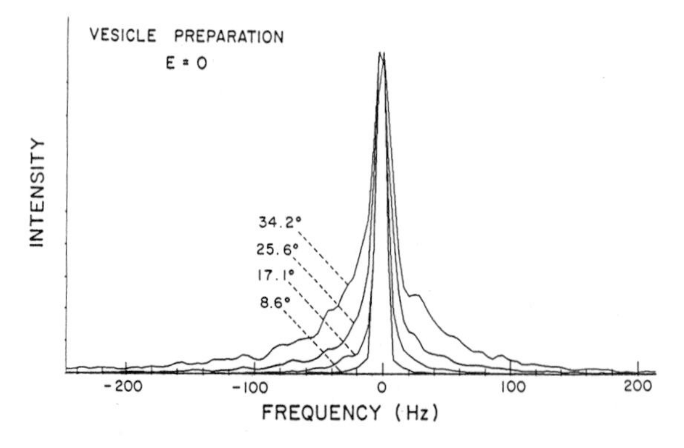

Figure 5a. Zero field diffusion analysis of vesicle preparation after two weeks aging. The sharp central spike indicates that a larger more slowly diffusing species has formed in the original sample of monodisperse 68 nm particles. Size analysis of the wings and central parts of the spectrum yielded particles sizes of 88 and 618 nm, respectively in good agreement with dynamic light scattering analysis.

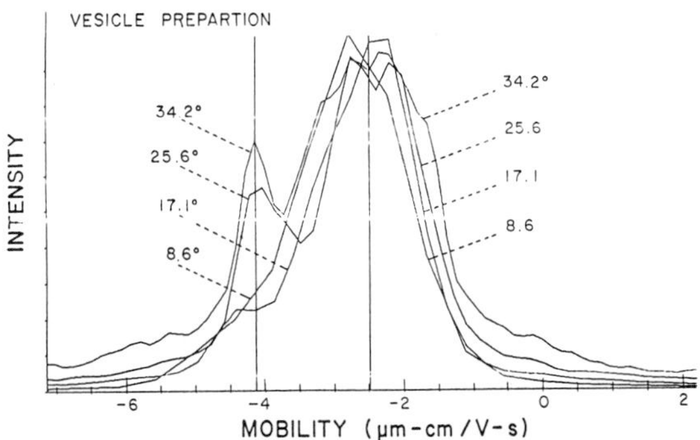

Figure 5b. Mobility analysis of the same vesicle sample shown in Fig. 5a. A bimodal distribution is clearly evident: the species responsible for the left peak is smaller in size as is indicated by the greater intensity relative to the right peak at 25° and 34° scattering angles.

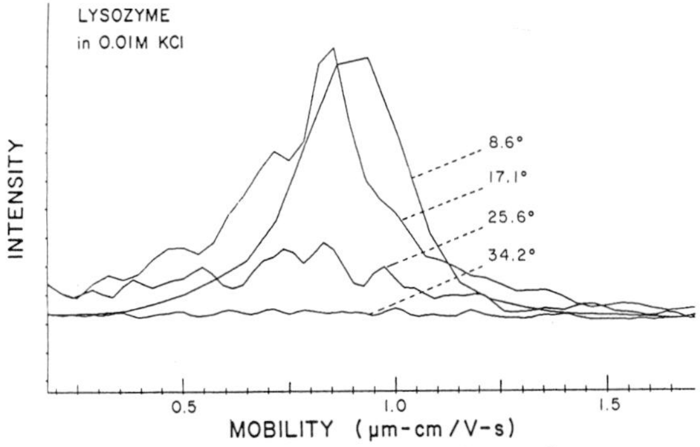

Figure 6. Mobility spectrum of free lysosyme. Diffusion broadening smears out the peak except at the two smaller scattering angles.

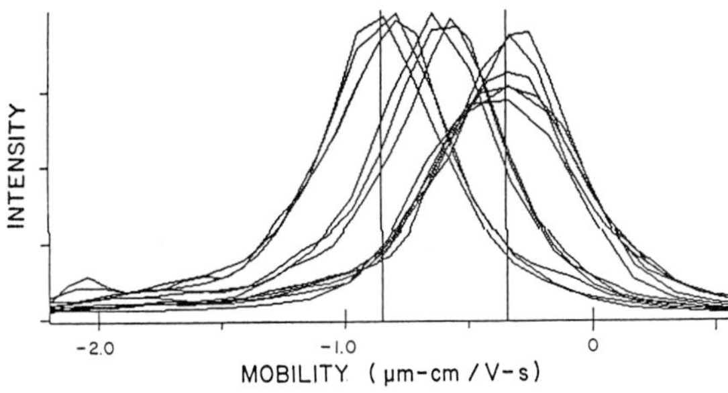

Figure 7. Bacterial cells with differing amounts of specific surface antigen display different mobility spectra. The cluster of curves forming the right peak are obtained from two isolates (three angles each) of bacteria in which the antigen is absent. The left and centre clusters (three angles each) are from bacteria genetically altered to express two different amounts of the antigen.

entirely. The leftmost peak (three scattering angles) and centre peak (three scattering angles) are from two isolates which express different amounts of the surface antigen.

5. CONCLUSIONS

As far as the biologically oriented user of ELS is concerned, there are two significant advances in recent instrumentation. First, by virtue of low angle measurement, one can overcome the drastic diffusion broadening effects that previously prevented users of commercial instruments from analyzing small proteins. Secondly, high resolution enables one to separate electrophoretic subspecies effectively and to quantitate the low mobilities that obtain in physiological salt solutions.

ACKNOWLEDGEMENTS

My thanks to Stan Sugrue of Coulter Electronics for his collaboration in selecting material for this paper.

REFERENCES

1 B. R. Ware and D. D. Haas in *Fast Methods in Physical Biochemistry and Cell Biology*, R. I. Sha'afi and S. M. Fernandez, Eds., Elsevier, 1983.
2 S. Sugrue, T. Oja, and S. Bott, *American Laboratory*, Feb 1989.

11

An Investigation of Rigid Rod-like Particles in Dilute Solution

By P. Johnson and W. Brown[1]

CAVENDISH LABORATORY, MADINGLEY ROAD, CAMBRIDGE. CB3 0HE U.K.

[1]INSTITUTE OF PHYSICAL CHEMISTRY, UNIVERSITY OF UPPSALA, BOX 532, S-751 21, UPPSALA, SWEDEN.

1. INTRODUCTION

In previous work[1] turnip yellow mosaic virus particles (TYMV) were chosen as suitable monodisperse and largely non-interacting spheres for detailed study by a combination of physical methods. Thus using sedimentation (both velocity and equilibrium) methods together with dynamic light scattering, it was possible not only to confirm the molecular weight, and to measure the second virial coefficient but from a detailed study of the effects of concentration on sedimentation and diffusion coefficients, a value for the effective specific volume of the particles was derived. With the partial specific volume this made possible the calculation of the hydration of the particles, and their volume fraction in solution at a given weight concentration. This allowed an experimental check of Batchelor's coefficient[2] for the concentration-dependent term in both sedimentation velocity and diffusion. Although the experimental values were of the correct order, they were usually some 15% - 20% smaller than calculated, an effect due probably to residual electrostatic interactions.

A natural extension of the above work would be to rigid monodisperse rod-like particles, and tobacco mosaic virus (TMV) would seem to be a suitable model substance. With its long RNA spiral upon which some 2130 interacting corpuscular protein subunits are attached, the parent particle appears as rigid rods of length approximately 3000 Å and diameter 150 Å - 180 Å, though, as discussed below, aggregation of the basic particle can and does occur. Clearly the asymmetry of the rods introduces new features as compared with spherical particles. Thus rotation of rods is meaningful and measurable, and orientation will affect any property involving flow. Coupling of rotational and translational motion has also to be borne in mind. The large effective volume of a long rod will also produce considerable excluded volume effects and therefore modified virial coefficients. Several groups of workers have tackled these problems, notably Fujime[3], Schaefer, Benedek *et al.*[4], Newman and Swinney[5] and more recently Wilcoxon and Schurr[6]. Though substantial progress has been made, discrepancies remain. In particular, measured translational diffusion and sedimentation coefficients are not compatible with the most accurate molecular weight determinations, and attempts to measure the anisotropy of the translational diffusion coefficient (*i.e.* $D_{\parallel} - D_{\perp}$) have led to improbable results.

Aggregation has been mentioned as a complicating factor and it seems likely that this may have led to low values of the translational diffusion coefficient and erroneous rotational diffusion coefficients in past work. Sedimentation coefficients would not be so affected since aggregated material sediments ahead of the normal component and the presence of significant aggregation is readily discernible from the schlieren diagram. Accordingly, in the present work,the sedimentation velocity method has been used not only to determine sedimentation coefficients over a range of TMV concentrations and ionic strengths but also as a control. Thus solutions for dynamic and integrated light scattering were examined only if no significant aggregation 'peak' was observable (*i.e.* <5% of total).

It seems clear that, biologically, the basic infective particle is 3000 Å long and that shorter particles, derived by damaging the basic particle (*e.g.* at pH >11), are non-infective[7]. However certain solution conditions lead to aggregation, particularly end-on, and increased sedimentation velocity. Schachman[8] using viscosity and electron microscope observations attempted to define desirable solution conditions for no aggregation. He concluded that chloride ions promote aggregation but that higher ionic strengths of other ions (*e.g.* 0.1M phosphate, pH 7) cause reversal of aggregation.

More recently Steere[7] considering previous work as well as his own findings mentioned three conditions for the avoidance of end-to-end aggregation: (1), pH > 7.2, (2) Ionic strength as low as possible, (3) Absence of certain salts, particularly phosphates. The latter condition clearly conflicts with Schachman's findings. Steere also used 0.5 M EDTA in extracting and purifying the virus, whilst other workers (Witz, unpublished) recommend absence of EDTA. Accordingly, in the present work, a compromise solution has been used, where chloride and EDTA have been avoided, and a pH of 7.5 has been achieved by the use of phosphate at ionic strengths of 0.05, 0.0125 and 0.00625. As shown below, under these conditions, no significant aggregation of a monodisperse material occurred during storage for up to two weeks. Accepting the monodisperse state of the TMV particles, which are visualized as rotating continuously in a moving sphere of radius 1500 Å, the concentration, C^*, at which such spheres account for the whole volume is calculated to be approximately 0.5g/100ml. In the work described, concentrations were mainly between 0.01 and 0.25g/100ml.

Sedimentation analysis was crucial in confirming the absence of aggregated material. With this knowledge, a detailed study of both sedimentation and diffusion coefficients (translational and rotational) against concentration was performed. By combination of the appropriate translational regression coefficients, the value of the second virial coefficient was obtained and this magnitude was confirmed by integrated light scattering measurements.

2. MATERIALS AND METHODS

Phosphate buffer salts were of A.R. grade and solutions were made up in water which had passed through an ion-exchange column prior to distillation from a pyrex glass still. Ionic strengths of 0.05, 0.0125 and 0.00625, all at pH 7.5, were achieved by the relevant proportions of Na_2HPO_4 and KH_2PO_4 according to Green[9]: minor pH adjustment being performed by the addition of dilute NaOH or HCl.

The purified TMV samples used (Common Vulgari or UI Strain), gifts from Dr. M. Wilson of the John Innes Institute, Norwich, were obtained as concentrated solutions in phosphate buffers and were stored at 4°C. Small quantities were removed by micro-pipette, diluted appropriately and dialyzed at 4°C, with several changes, against large volumes (x 500) of the same buffer. In most cases (always for dynamic light scattering), filtration of 1 ml quantities by syringe through small Millipore type HA, 0.45 μm or 0.22 μm pore size was performed and if necessary repeated. Similar filters of polypropylene did not allow passage of TMV particles. Concentrations were obtained by spectrophotometry using 1 mm-pathlength cells at 260 nm and an absorption coefficient of 3.0 litre g^{-1} cm^{-1}.

Sedimentation

A Beckman model E analytical ultracentrifuge, equipped with an accurately calibrated (± 0.02°C) RTIC temperature-measurement system and accurately focused schlieren optics was used. The speed control mechanism was accurately adjusted so that the coarser control rarely operated - thus speed variation was of the order of 2 or 3 parts per 1000, and actual speeds uncertain to a similar extent. To eliminate possible temperature gradients across the cell, the rotor temperature was not controlled but at the speeds most frequently employed (15,220 r.p.m.) the temperature rise was approximately 0.2°C/hr. Mean temperatures around ambient, accurate to ±0.05°C, were thus obtained and used in correcting measured sedimentation coefficients s_c to water at 20°C, s^o_{20}. A partial specific volume, $\bar{v}$, of 0.73 was assumed[10]. In correcting s_c to s^o_{20} values, the properties (viscosity and density) of the buffer solvent (rather than the TMV solution) were utilized as recommended by Harding and Johnson[1].

In keeping with a pronounced s upon c dependence, the sedimenting boundaries were very sharp and, using 8-12 frames measured on a micro-comparator, well-defined linear plots of ln r *versus*. time (t) were obtained, r being the radial distance of the maximum refractive index gradient.

An attempt was also made to examine TMV solutions by sedimentation equilibrium in a Beckman ultracentrifuge equipped with interference optics. However to produce a measurable concentration gradient over a 3 mm column required rotational speeds lower than 2.000 r.p.m. and it was not found possible to achieve vibration-free rotation at such speeds.

Dynamic Light Scattering

Most of the dynamic light scattering measurements were made on a laboratory-made instrument which was based upon the classical light scattering instrument of Johnson and McKenzie[11] in which extreme precautions had been taken in the scattering housing to eliminate stray light. By introducing suitable blackened baffles, the path of the incident light through the thermostatting water was shielded from the detector (EMI 9863) and after passage through the 1 cm square cell, the light was effectively removed from the system by a simplified Tyndall cone (for details see Godfrey *et al.*[12]). It was thus possible to use an angular range of 5° - 140°. The water surrounding the cell was regularly filtered and, though stationary, was thermostatted at 25.0°C by annular tanks placed above and below and fed by water pumped through long flexible tubes from a thermostat accurate to 0.01°C. Temperature constancy at the cell was better that ±0.02°C.

The above housing was placed centrally on a rotary milling table, itself fixed at one end of a stout 2 metre rectangular cross section optical bench. A 15 mW He-Ne laser (Spectra Physics - model 124A) giving light at 632.8 nm was placed with suitable adjustments at the other end of the bench. The milling table carried a stout 27" length dural arm upon which the detector and its pinhole system could be rotated about the central axis of the milling table. The whole optical bench plus its attachments was placed upon a vibration-proof table supported by anti-vibration mounts (supplied by Anti-Vibration Methods Ltd., Warminster, Wilts.) driven by low pressure compressed air. Thus the most sensitive parts of the apparatus were unaffected by external disturbance.

Output from the detector was passed to a 128-channel Malvern K7025 Multibit correlator *via* a Malvern Amplifier-Discriminator (RR63). Output from the correlator was monitored continuously on an oscilloscope and directed *via* a BBC microcomputer to floppy discs, capable of storing 30 runs. Batches of runs were then transferred to the main university IBM 3084Q computer for various forms of processing described below. Such batches contained runs at 5 or 6 different angles between 15° and 130°, with sample times at each angle covering a 4-5 fold range, being chosen to give approximately the same percentage decay of the correlation function.

Satisfactory operation depended upon the careful alignment of the laser with respect to the scattering housing and the detecting system which was described by Godfrey *et al.*[12] However electronic 'noise' proved to be a further considerable problem. Thus, without precautions, spurious counts at the correlator appeared when control equipment (switches, thermostats *etc.*) in the neighbourhood were operated and generally if electric mains surges occurred. Mains filters (*e.g.* Radiospares 238-671) were used on all electric supplies and the amplifier-discriminator, which was particularly sensitive, was supplied by a large constant voltage (magnetic saturation) transformer. It was essential also to check that the laser with its power supply was working within specification. A suitable test was the operation of the whole apparatus with the detector completely screened when the correlator should show dark current count levels only. Occasionally, however, the monitoring oscilloscope showed obvious irregularity in which case the data was rejected. A further possibility was to adjust the correlator to near overflow. If spurious counts and overflow counts *then* occurred, rejection of data was required.

Data Processing

Several stages of data processing were employed. The first essentially treated the decay of the correlation function in terms of a single exponential and the program employed produced an accurate plot of ln $[g^2(t) - 1]$ *versus* time, t, where $g^2(t)$ is the normalized intensity correlation function. The best least-squares fit to this plot was obtained, treating it in turn by linear, quadratic and cubic procedures and deriving the Z-average translational diffusion coefficient (D_T) from the limiting slope. The 'polydispersity factor' or Z-averaged normalized variance of the distribution of diffusion coefficients was also obtained as shown by Pusey[13]. This form of processing is strictly applicable for rods of length comparable with wavelength only at low scattering angles ($\theta < 30°$) where only translational diffusion contributes. For rigid rod-like particles, rotation is also involved in the decay of the correlation function at higher θ values. Thus if coupling between translational and rotational motions can be ignored, then the following expression applies[14]:

$$(g^2(t) - 1)^{1/2} = g^1(t) = e^{-D_T K^2 t} (A_0 + A_1 e^{-6D_R t} + ...) \quad (1)$$

where $g^1(t)$ is the field correlation function, $K = (4\pi n/\lambda) \sin \theta/2$ is the Bragg vector with n the refractive index of the solution, λ the wavelength *in vacuo*, and D_R is the rotational diffusion coefficient. A_0, A_1 and higher coefficients are functions of K which can be calculated for a particular model structure. For 3000 Å rods, higher coefficients may be ignored and for $\theta < 30°$, A_1/A_0 is negligibly small. Thus for TMV in dilute solution with $\theta < 30°$, one exponential term only ($\exp(-D_T K^2 t)$) is involved and this allows the determination of D_T by the procedure already described. It should be mentioned, however, that since dust and foreign material show up markedly under such conditions, their thorough removal is imperative (see below). A_1/A_0 increases with scattering angle and for $\theta \geq 90°$, the second term in eq. (1) begins to contribute significantly. Before dealing with the computation of D_R from eq. (1) it was desirable to check its general correctness experimentally. Eq. (1) may be written

$$(g^1(t)\, e^{D_T K^2 t} - A_0) = A_1 e^{-6D_R t} \tag{2}$$

and in logarithmic form

$$\ln (g^1(t)\, e^{D_T K^2 t} - A_0) = \ln A_1 - 6D_R t \tag{3}$$

A plot of the L.H.S. against t essentially describes the rotational contribution to the decay of the correlation function (Fig. 1) and its linearity demonstrates the correctness of eq. (1).

In using eq. (1) for the determination of D_R, it should be realized that over the useful range of θ values, $6D_R$ is of similar magnitude to DK^2 (at $\theta = 106°$, $6D_R = D_T K^2$). Accordingly attempts merely to separate the correlation decay by computer into two exponential terms with their coefficients were not profitable. However in addition to D_T values, obtained at low angles, A_1/A_0 could be calculated for rigid rods. Accordingly the calculation (using a Harwell routine) was incorporated into the programming along with various NAG curve-fitting procedures. Further, by use of $(g^1)^2$ as a weighting factor, experimental points involving fewer counts were given less weight and occasionally the last few noisy channels were rejected. Fig. 2 demonstrates the quality of fit obtained. A subroutine of the program applied eq. (3) producing plots like Fig. (1) and approximate values of D_R. However the most accurate values of D_R were obtained from the parameters of the best fit to the curve of $g^1(t)$ against channel number.

The maximum entropy method[15] was also considered as a

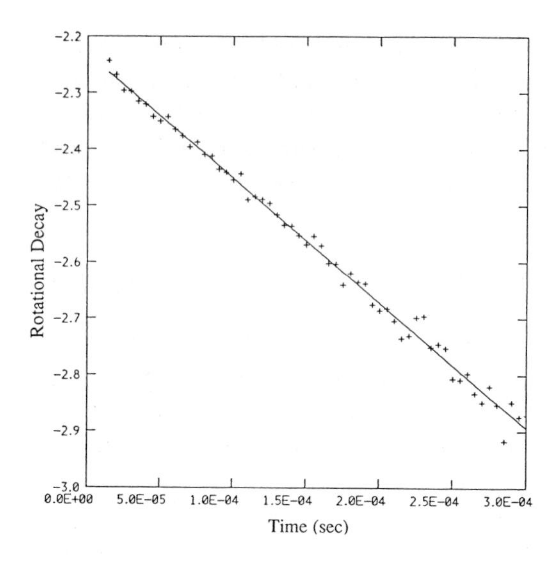

Figure 1 A plot of L.H.S. of eq. (3) against time for TMV at 0.067g/100ml and I = 0.0125, demonstrating the exponential nature of the rotational decay. Sample Time = 5 μs. Scattering angle = 90°.

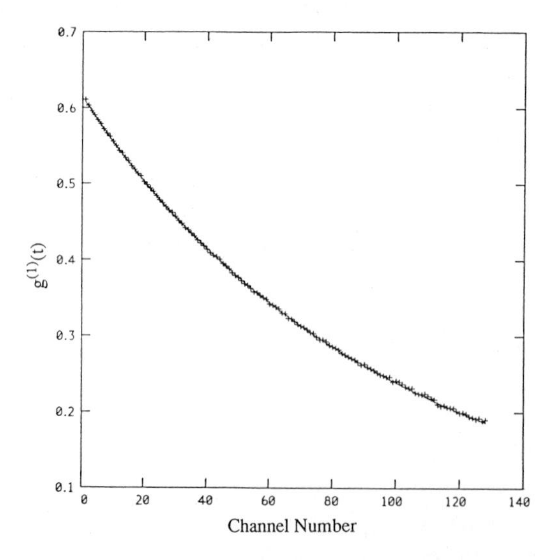

Figure 2 Experimental (+) and calculated (full line) plot of $g^1(t)$ against channel number for 0.0673g/100ml solution of TMV in phosphate buffer (I = 0.0125) at pH 7.5 and 25.0°C. Sample Time = 5 μs. Scattering angle = 90°. For this fit D_T = 4.70 cm^2 sec^{-1} and D_R = 324 sec^{-1}.

possible treatment of the experimental data. However this is a very general approach which ignores the almost monodisperse character of the TMV preparation. Even for $\theta = 15°$, very poor precision in the determination of D_T was observed and for $\theta > 30°$, the exponents of the exponential terms were too close to one another for separation.

Integrated Light Scattering

The Malvern System 4700 with 40 mW He-Ne Siemens laser was used to perform integrated light scattering measurements designed to measure the second virial coefficient. An angular range of 15° - 140° was available and 1 cm square fluorimeter-type cells were employed. The instrument was programmed to take readings at 10° angular intervals, care being taken to avoid viewing the cell corners. At each angle the time of observation (usually 20 secs) was divided up into short intervals (usually 1 sec) and count rates very different from average were automatically discarded. Standard deviations in the count rates were also evaluated and used in the weighting procedure. Calibration of the instrument was performed by the use of dilute polystyrene latex suspensions (*e.g.* ICC spheres, diameter = 38 nm, Portland, Oregon, U.S.A.) whose optical density, measured in an accurate spectrophotometer, was used to calculate Rayleigh's Ratio for the suspension. The cell was surrounded by filtered water thermostatted to 25.0 ± 0.05°C. Background and solvent scatter were always measured and appropriately subtracted from solution scatter. TMV solutions were handled as for dynamic light scattering.

3. RESULTS AND DISCUSSION

The State of Aggregation of TMV

When diluted stock solution was examined in the ultracentrifuge at a concentration of approximately 0.2g/100ml without filtration, the TMV sedimented mainly as a hyper-sharp peak with a sedimentation coefficient of s^0_{20} of 180S - 184S but with a significant smaller faster peak with s^0_{20} of *ca.* 210S (*e.g.* see Fig. 3) The latter peak corresponds closely with an end-to-end dimer. Filtration through a 0.45 μm filter gave a solution of unchanged concentration but possessing only the single 180S peak. This experiment was repeated at I = 0.05 and I = 0.0125 with similar results. It must be concluded that hydrostatic stretching in or near the membrane is responsible for converting the dimer to monomer. It follows also that the forces causing aggregation are relatively weak and probably not covalencies. No significant aggregation occurred after storage at 4°C for up to 2 weeks. This allowed a range of concentrations to be examined without further filtration but in light scattering work, filtration was routinely performed at each dilution.

Sedimentation of TMV

In view of the asymmetry of TMV particles, the possibility of orientational complications existed. The Peclet Number, Pe (*e.g.* Kay and Nedderman[16]), which is concerned with such effects may be written for the ultracentrifuge cell as

$$Pe = \frac{\text{Sedimentation Flux}}{\text{Diffusion Flux}} \approx \frac{c \cdot \frac{dr}{dt}}{D_T \cdot \frac{c}{w}} \approx \frac{w \cdot \frac{dr}{dt}}{D_T} \tag{4}$$

where r is radial distance of the sedimenting boundary at which a concentration change c occurs, w is the radial width of the boundary and t is time. Pe can be evaluated semi-quantitatively as a mean value for a particular experiment. Accordingly a series of runs were performed at a constant concentration of TMV (0.156g/100ml) and *ca.* 25°C in which the rotational speed varied from 8210 r.p.m. to 24,630 r.p.m. Under these conditions the boundary remained sharp and w = 0.1 cm was assumed with $D_T = 4.2 \times 10^{-8}\ cm^2\ sec^{-1}$ (see later). Table 1 contains results obtained. It is clear that no dependence of sedimentation coefficient on speed is observed whilst the Peclet No., which is always » 1, varies over a ten-fold range. The absolute values of Pe are not significant but provide a demonstration of the condition during TMV sedimentation of the usual requirement in applying the sedimentation velocity method, that the rate of sedimentation shall be much greater than that of diffusion. It should be mentioned however that since no velocity gradients occur in the cell (apart from very thin layers at the walls), orientation of the sedimenting rods is not involved.

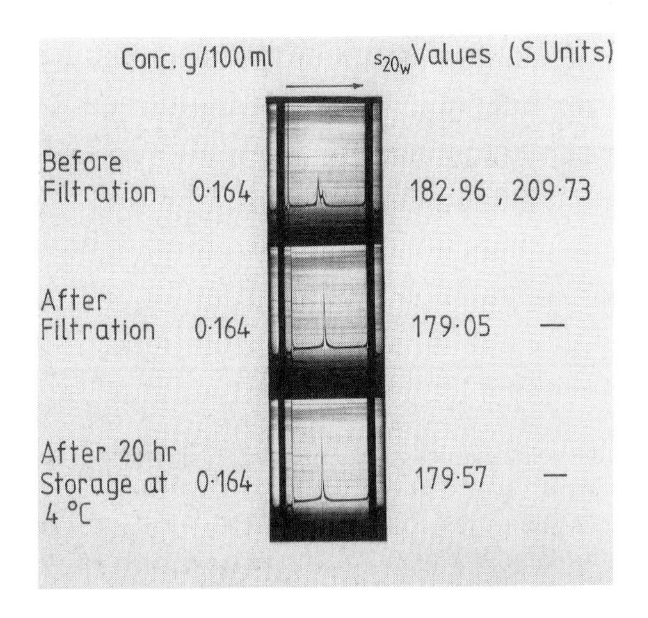

Figure 3 Sedimentation patterns for TMV at a concentration of 0.164g/100ml in phosphate buffer (I = 0.00625, pH 7.5) before filtration (see text), after filtration, and after further storage.

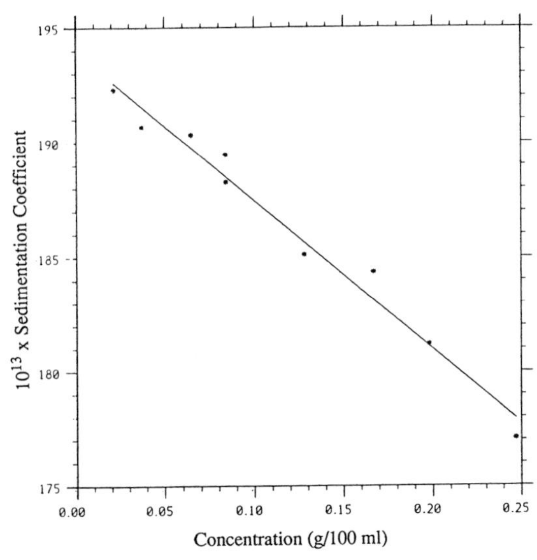

Figure 4 Plot of reduced sedimentation coefficient, s^0_{20}, against TMV concentration in phosphate buffer at I = 0.0125 and pH 7.5.

Table 1. Peclet number for sedimentation of TMV at 25°C. TMV concentration = 0.156 g/100ml; I = 0.00625; pH = 7.5.

Rotational Speed (r.p.m.)	Peclet Number	s^0_{20} (S)
8210	20.5	(175.9)
8660	20.5	177.9
12590	457	176.6
17250	900	178.0
24630	1915	177.9

Sedimentation velocity measurements were made at the three ionic strengths over the TMV concentration range 0.01-0.25 g/100ml. Fig. 4 shows results for I = 0.0125 but for I = 0.05 and I = 0.00625 also similar plots were obtained. In each case, no deviation from straight line behaviour was observed so that an equation of the following form is valid:

$$s_c = s\,(1 - k_s\,c) \tag{5}$$

where all of the s_c values refer to water 20°C (*i.e.* s^0_{20}) and $s \equiv (s^0_{20})_{c=0}$. Table 2 summarises the data for the three ionic strengths. The differences between the extrapolated sedimentation coefficients are thought to be real for the total error in the measured values is thought to be less than ±1S. On the other hand, the k_s values are subject to greater error and the differences might not be real. If charge effects are involved in the k_s values, then a larger k_s would be expected at lower ionic strengths. It should be mentioned that occasionally a sample of TMV with its dilutions would fit a line lower and almost parallel with those described by the parameters of Table 2. Apart from the possibility of bacterial contamination we were not able to explain this. Breakdown by the filtration procedures was not thought likely since it clearly did not always occur and repeated filtration did not increase the effect.

Table 2. Sedimentation velocity parameters for TMV. Solvent - phosphate buffer pH 7.5.

Ionic Strength	0.00625	0.0125	0.05
$(s^0_{20})_{c=0} \pm 1$ (S)	197.2	194.0	196.2
$k_s \pm 2$ (c in g/ml)	31.6	33.4	28.6

The Perrin equation may be modified for long ellipsoids of revolution into the form

$$s = \frac{2}{9}\left\{\frac{1 - \bar{v}\rho}{\bar{v}\,\eta}\right\} a^2 \ln\left(\frac{2b}{a}\right) \tag{6}$$

where b and a are the semi-major and semi-minor axes with η and ρ the viscosity and density of the solvent. Taking b/a to be 3000/180 = 16.7 and s = 195.8S (mean value), a is calculated to be 165 Å, a very reasonable value having regard to the assumption that TMV is an ellipsoid. If, however, the equation of Tirado and Garcia de la Torre[17] for capped cylinders is applied, then a b/a value of 13.6 is obtained.

Dynamic Light Scattering

The dynamic light scattering measurements were performed chiefly at I = 0.0125 and TMV concentrations from 0.02 to 0.12 g/100ml. In the first place, the best fit to a single exponential term was obtained over the whole angular range, yielding an apparent diffusion coefficient, D_{App} (from the exponent - $D_{App}K^2t$). Table 3 summarizes the derived values. At each concentration and each angle, three or four sample times were selected to cover 90% - 95% of the total decay and the mean value, derived usually by treating $\ln(g^2(t) - 1)^{1/2}$ as a quadratic, is quoted. Treatment as a linear or cubic equation gave less consistent results. At each concentration D_{App} rises over the whole angular range from the low angle value with no sign of reaching constancy, behaviour similar to that reported by Wilcoxon and Schurr[6]. This is undoubtedly due to the increasing importance of the second exponential term (in eq. (1)) involving D_R at the higher angles.

Table 3. Apparent diffusion coefficients at 25°C. TMV(W): I = 0.0125; pH = 7.5; phosphate buffer.

	TMV Conc. (g/100ml)	10^7 x D_{APP} - Quadratic Fit							D_R (sec^{-1})
		15°	25°	75°	90°	110°	120°	130°	
C	0.119	0.467	0.482	0.560	0.581	-	0.593	0.598	350
D	0.0917	0.472	0.495	0.559	0.581	0.584	-	0.606	342
E	0.0673	(0.455)	0.484	0.553	0.581	0.585	-	0.593	312
F	0.0553	0.474	(0.457)	0.544	0.565	0.589		0.606	341
G	0.0413	0.481	(0.509)	0.568	0.583	(0.605)		0.608	342
H	0.040	-	0.492	0.569	0.582	0.606		0.613	338
I	0.0317	0.482	0.510	0.568	0.584	0.611		0.605	360
J	0.023	0.484	0.522	0.560	0.586	0.597		0.612	331
								Mean	339.5 ± 13

Brackets (...) indicate an inaccurate value.

At 15°, there is no contribution from D_R and thus a plot of $\ln(g^2(t) - 1)$ against channel number is accurately linear. Fig. 5 contains such a plot from which D_T (= D_{App}) values accurate to better than 1% were regularly obtained. Fig. 6 contains such D_T values plotted against TMV concentration. The equation for the best least square line is

$$D_{T,25} = 0.487\ (\pm 0.002) \times 10^{-7}\ (1 - 0.353\ (\pm 0.045)\ c) \qquad (7)$$

where 0.487×10^{-7} is the limiting value of D_T at c = 0, c being in g/100ml. Correcting to water at 20°C, $D^0_{20} = 4.27 \times 10^{-8}$. The variation of D_T with concentration is usually written

$$D_T = D\ (1 + k_D\ c') \qquad (8)$$

where c' has units of g/ml. Comparing with eq. (7), k_D becomes -35.3 (ml/g).

Combining $[D^0_{20}]_{c=0} = 4.27 \times 10^{-8}$ with $[s^0_{20}]_{c=0} = 194.0$ S and $\bar{v} = 0.73$, the molecular weight $M = 40.8 \times 10^6$, somewhat larger than Caspar's chemical estimate[18] of 39.4×10^6 but nearer the value 41.6×10^6 obtained in a special form of sedimentation equilibrium by Weber *et al.*[19] If $M = 40.8 \times 10^6$ is combined with $[s^0_{20}]_{c=0} = 194.0$ S we obtain a frictional ratio $f/f_0 = 2.2$. Interpreting as an unsolvated prolate ellipsoid of revolution this gives an axial ratio, b/a, of 22 which is higher than expected. But accepting hydration of the usual order (≈30%), the b/a value is reduced to 19.5. A further test of D_T is available from the equation for rigid rods given by Doi and Edwards[20]

$$D_T = \frac{\ln(b/a)\ kT}{6\pi\eta b} \qquad (9)$$

where the rod has length 2b and diameter 2a. With b/a = 20 and 2b = 3000 Å, $D_T = 0.488 \times 10^{-7}\ cm^2\ sec^{-1}$ at 25° in almost exact agreement with experiment. For b/a = 16.7, $D_T = 0.457 \times 10^{-7}\ cm^2\ sec^{-1}$.

Knowing D_T at each concentration, it was now possible by computing methods already outlined to calculate D_R. Table 4 illustrates the use of the method over a range of angles. Clearly A_1/A_0 is too small at 75° and 90° to allow any accuracy in the calculated D_R, but at the higher angles, a more definite value is obtained. Such D_R values in I = 0.00125 buffer at 25° obtained at the different TMV concentrations are given in the last column of Table

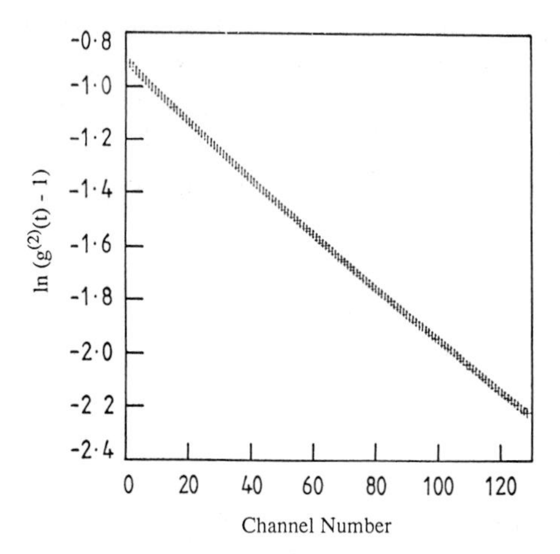

Figure 5 Typical plot of ln $(g^2(t) - 1)$ against channel number at 25°. Sample Time = 100 μs. TMV concentration = 0.119g/100ml in phosphate buffer at I = 0.0125, pH 7.5. Angle is 15°.

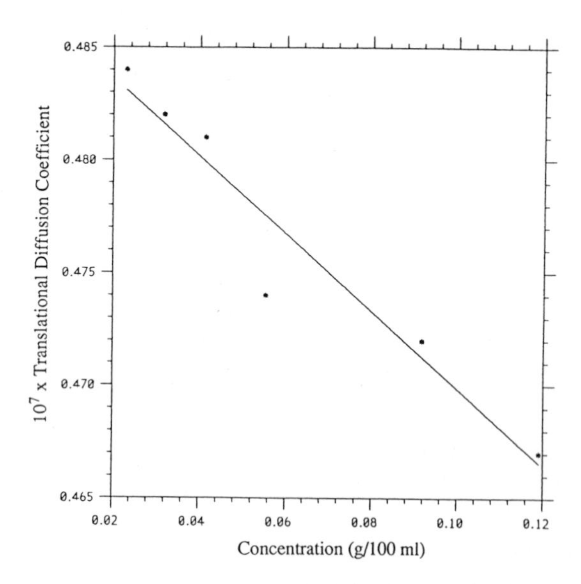

Figure 6 Plot of translational diffusion coefficient, D_T, in $cm^2 sec^{-1}$ against TMV concentration (g/100ml) in phosphate buffer at I = 0.0125, pH 7.5 and 25°.

3. The uncertainty in these values is such that no concentration dependence can be deduced but it cannot be excluded. Correcting to 25°C in water requires only a viscosity factor of 1.0045 (*i.e.* 341.0 sec^{-1}). Newman and Swinney[5] using electric birefringence studies obtained a value 318 sec^{-1} at 20°C which corresponds to 363 sec^{-1} at 25°C.

Table 4. Computer calculation of D_R. pH = 7.5; I = 0.0125; Phosphate buffer; temperature= 298.1 K; [TMV] = 0.092 g/100ml; $D_T = 4.75 \times 10^{-8}$ cm^2 sec^{-1}

Angle, θ	A_1/A_0	D_R (sec^{-1})	
75°	0.096	621	A_1/A_0 too small
		628	
90°	0.182	464	
		452	
		425	
110°	0.340	336	
		320	
		322	
130°	0.521	330	
		343	
		328	
	Mean value of D_R	329.8 ± 8	

Broersma[21,22] devised an equation for rods which may be written

$$D^R = \frac{3kT}{\pi\eta\,(2b)^3}\,\{\ln(2b/a) - \gamma\} \tag{10}$$

where γ varies weakly with b/a. Subsequently Newman, Swinney and Day[23] report conversations with Broersma in which the following more accurate form of γ is given:

$$\gamma = 1.45 - 7.5\,\{[\ln(2b/a)]^{-1} - 0.27\}^2 \tag{11}$$

Taking 2b = 3000 Å, we obtain b/a = 17.0, corresponding with a value for 2a of 176 Å close to the upper limit suggested by X-ray work (*e.g.* Stubbs, Warren, and Holmes[24]). In a more recent treatment, Tirado and Garcia de la Torre[17] give a modified expression for the end effect which results in a b/a value of 14.7, slightly lower than anticipated.

Concentration-Dependent Terms in Sedimentation and Diffusion

It was shown[1] that the coefficients of the concentration-dependent terms in sedimentation and diffusion (as defined in equations (5) and (8) are related by

$$k_D = 2BM - k_s \qquad (12)$$

where B is the second virial coefficient and M is the molecular weight of the macromolecule. Introducing $k_D = -35.3$ and $k_s = 33.4$ (from Table 2) and using $M = 40.8 \times 10^6$, we find $B = -0.233 \times 10^{-7}$ ml.mol.g^{-2}. It should be noted that 2BM is the small difference between two much larger quantities and is subject to large error (1.9 ± 2). Thus the possibility that B is a small positive quantity cannot be excluded (however, see later). For an asymmetric molecule like TMV a very significant contribution from excluded volume, B_{Ev}, would be expected. Tanford gives for rods[25]

$$B_{Ev} = \frac{bv}{aM} \qquad (13)$$

where v is the specific volume of the macromolecule. v differs from $\bar{v}$ in that it includes any hydrating or included solvent. Accepting the value v = 1.8 of Harding and Johnson[1] for the spherical turnip yellow mosaic virus, B_{Ev} is calculated to be $+0.88 \times 10^{-6}$ which is not only much larger than that obtained experimentally above but is also of opposite sign. To achieve the overall negative experimental B value, requires a negative contribution larger than B_{Ev}, probably of electrostatic origin. Thus Doty and Edsall[26] reported experiments in which mixtures of serum albumin and γ-globulin, of opposite electrical charge, gave negative B values which decreased in magnitude with increase in ionic strength. The pure components alone gave positive B values. Thus the charged state of the TMV particles requires consideration.

Integrated Light Scattering

An attempt to confirm the negative value of the second virial coefficient was made by the classical light scattering method. Fig. 7 contains a 'Zimm' plot derived from this data and Table 5 summarises the rather approximate B values obtained. However all were negative and numerically greater than estimated from eq. (12).

Table 5. Second virial coefficient (B) for TMV at pH 7.5.

Ionic Strength	10^5 x B (ml.mol.g^{-2})
0.00625	-0.11 (± 0.05)
0.0125	-0.6 (± 0.2)
0.05	-0.12 (± 0.05)

Neglect of Coupling

Separation into translational and rotational terms, as in eq. (1), involves the assumption of no coupling between the two motions. However since diffusion cofficients parallel ($D_{\parallel}$) and perpendicular ($D_{\perp}$) to the axis of the molecule must differ, this cannot be completely true. Schaeffer *et al.*[4] in a theoretical and experimental study of the spectrum of the light scattered from dilute TMV solutions were able to determine D_T as 0.39 x 10^{-7} cm^2 sec^{-1} at 25°, a value considerably lower than those reported in Table 3, and conceivably indicative of aggregation. They also derived the relation

$$-6.65\ (D_{\parallel} - D_{\perp}) + 0.84\ D_R L^2 = (3.10 \pm 0.30) \times 10^{-7} \qquad (14)$$

where L is the length of the rods. Unless $D_R \geq 415$ sec^{-1}, then ($D_{\parallel} - D_{\perp}$) is calculated to be negative which seems unlikely. The mean value for D_R from Table 3, like most published estimates, is also much smaller than 415 sec^{-1}. Wilcoxon and Schurr[6] developed a theoretical treatment for thin rods which linked the variation in D_{App} with changing KL values:

$$D_{App}\ (K) = D_T + 2\Delta\ (^1/_3 - F(KL)) + 2L^2 D_R\, G(KL) \qquad (15)$$

where $\Delta = (D_{\parallel} - D_{\perp})$ and F(KL) and G(KL) are tabulated functions.

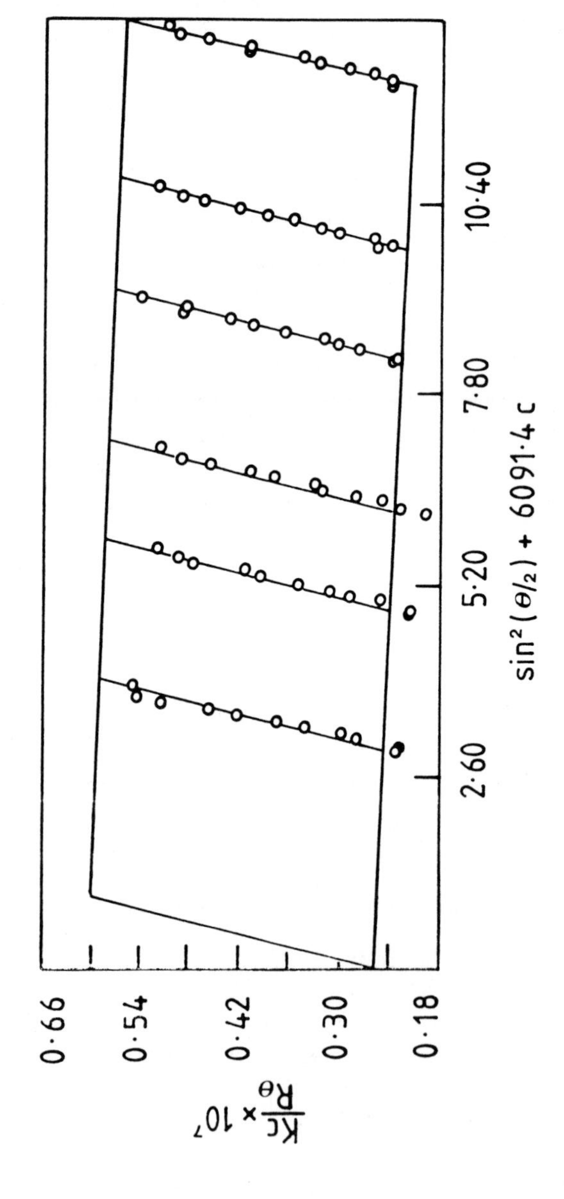

Figure 7 Biaxial 'Zimm' Plot of Kc/R_θ against $\sin^2(\theta/2) + k_{ARB}\,c$ for TMV at I = 0.00625 and pH 7.5. K is an optical constant, R_θ is the Rayleigh Ratio at angle θ and k_{ARB} is an arbitrary constant.

Their D_T value at 20°C is in reasonable agreement with that determined here but their Δ value of 1.8×10^{-8} $cm^2\,sec^{-1}$ is considerably larger than estimated in this work from the variation in D_{App} (*viz* 0.25 (±0.2) x 10^{-8} $cm^2\,sec^{-1}$). The latter value is considerably smaller than is calculated for the theoretical equations of Broersma and Tirado *et al.* but close to that reported by Schaeffer *et al.*

4. SUMMARY

TMV, with fundamental rod-like particles 3000 Å long by 150 Å - 180 Å, has been chosen as a model for the study of the behaviour of monodisperse long rigid rods in dilute solution. However, after storage, purified TMV contains aggregated particles, but ultrafiltration was found to convert them into normal TMV. Using the sedimentation method as a monitoring device, conditions for the maintenance of the monodisperse condition, required in physical studies were sought. The concentration c^*, at which the circumscribing spheres of the TMV particles account for the whole solution volume was calculated to be 0.5 g/100ml: sedimentation and light scattering studies were performed over a range of concentrations up to $c^*/2$.

Sedimentation coefficients, which were found independent of speed of rotation, were measured at three ionic strengths - a linear decrease with increasing TMV concentration was observed in each case. Dynamic light scattering at low angles gave translational diffusion coefficients which also decreased with increasing TMV concentration. Combining with sedimentation data, a molecular weight of 40.8×10^6 was obtained and a frictional ratio indicating an axial ratio of 19.5. $[D_T]_{c=0}$ was also in agreement with the value calculated for a rod of axial ratio 20. Knowing D_T values, D_R values accurate to ± 5% were now derived by curve-fitting procedures, but no concentration dependence was observed. The mean D_R value was in conformity with the equation of Broersma, indicating an axial ratio of 17 whilst the more recent equation of Tirado and Garcia de la Torre suggested a lower value. Combination of the regression coefficients in sedimentation and diffusion gave an estimate of the second virial coefficient B of -0.233×10^{-7} ml mol g^{-2}. The expected contribution to B from excluded volume effects (B_{Ev}) was positive and much larger ($B_{Ev} = +0.88 \times 10^{-6}$) and it would seem likely that considerable electrostatic effects are involved in the reversal of sign. The negative experimental B value was confirmed by integrated light scattering. Attempts to estimate the anisotropy in the translational diffusion gave values much smaller than expected theoretically.

ACKNOWLEDGEMENTS

We gratefully acknowledge the help of Miss Fiona Miller and Dr. John Horton with computing problems and the gift of TMV from Dr. Michael Wilson of the John Innes Institute, Norwich. We are indebted also to Neville Buttress for his meticulous ultracentrifuge and general assistance. We owe much to Professor Sir Sam Edwards, F.R.S. for the privilege of working in his laboratory.

REFERENCES

1. S.E. Harding and P. Johnson, *Biochem. J.*,1985, *231*, 543, 549.
2. G.K. Batchelor, *J. Fluid Mech.*, 1972, *52*, 245.
3. S. Fujime, *J. Phys. Soc. Jpn.*, 1971, *29*, 416.
4. D.W. Schaeffer, G.B. Benedek, P. Schofield and E. Bradford, *J. Chem. Phys.*, 1971, *55*, 3884.
5. J. Newman and H.L. Swinney, *Biopolymers*, 1976, *15*, 301.
6. J. Wilcoxon and J.M. Schurr, *Biopolymers*, 1983, *22*, 849.
7. R.L. Steere, *Science*, 1963, *140*, 1089.
8. H.K. Schachman, *J. Am. Chem. Soc.*, 1947, *69*, 1841.
9. A.A. Green, *J. Am. Chem. Soc.*, 1933, *55*, 2331.
10. M. Lauffer, *J. Am. Chem. Soc.*, 1944, *66*, 1188.
11. P. Johnson and G.H. McKenzie, *Proc. Roy. Soc. B*, 1977, *199*, 263.
12. R.E. Godfrey, P. Johnson and C.J. Stanley, (1982), in 'Biomedical Applications of Laser Light Scattering', Ed. Sattelle, Lee, and Ware, Elsevier BioMedical, p. 373.
13. P.N. Pusey, (1974) in 'Photon Correlation and Light Beating Spectroscopy' (Cummins, H.Z. and Pike, E.R., eds.), p. 387, Plenum Press, New York.
14. R. Pecora, *J. Chem. Phys.*, 1968, *48*, 4126.
15. A.K. Livesey, P. Licinio and M. Delaye, *J. Chem. Phys.*, 1986, *84*, 5102.
16. J.M. Kay and R.M. Nedderman, (1979) 'An Introduction to Fluid Mechanics and Heat Transfer', 3rd Edn., Cambridge, p. 210.

17. M.M. Tirado and J. Garcia de le Torre, *J. Chem. Phys.*, 1979, *71*, 2581; *J. Chem. Phys.*, 1980, *73*, 1986.
18. Caspar, D.L.D., in 'Advances in Protein Chemistry', 1963, *18*, 147, Academic Press.
19. F.N. Weber, R.M. Elton, H.G. Kim, R.D. Rose, R.L. Steere and D.W. Kupke, *Science*, 1963, *140*, 1090.
20. M. Doi and S.F. Edwards, 'The Theory of Polymer Dynamics', 1986, Oxford, p. 300.
21. S. Broersma, *J. Chem. Phys.*, 1960, *32*, 1626.
22. S. Broersma, *J. Chem. Phys.*, 1960, *32*, 1632.
23. J. Newman, H.L. Swinney and L.A. Day, *J. Mol. Biol.*, 1977, *116*, 593.
24. G. Stubbs, S. Warren and K. Holmes, *Nature*, 1977, *267*, 216.
25. C. Tanford, 'Physical Chemistry of Macromolecules', Wiley, 1961, p. 196.
26. P. Doty and J.T. Edsall, (1951) in 'Advances in Protein Chemistry', Vol.VI, Academic Press, p. 71.

12

Static and Dynamic Light Scattering by Semidilute Rodlike Polyelectrolytes

By Theo Odijk

DEPARTMENT OF POLYMER TECHNOLOGY, FACULTY OF CHEMICAL ENGINEERING, AND MATERIALS SCIENCE, DELFT UNIVERSITY OF TECHNOLOGY, P.O. BOX 5045, 2600 GA DELFT, THE NETHERLANDS.

1. INTRODUCTION

Even after many years of considerable effort there is still no unequivocal consensus concerning the structure of intrinsically flexible polyelectrolytes without salt. This is partly due to the fact that we are uncertain even about the precise structure of uncharged polymer solutions. So at this juncture it seems wise to focus on the structure factor of polyions that are rigid or slightly semiflexible rods. Here I shall review the current static and dynamic theory of isotropic, semidilute suspensions of uncharged rods and then discuss the effect of charge under excess salt conditions. Thus the theory applies solely to a narrow though important class of biopolymers.

I first discuss the screened electrostatic potential between two rodlike polyions which is quite well documented. Despite the twisting force between the two rods it is possible to view the rods as having an effective diameter when discussing the statistical thermodynamics of the solution. An important issue that I try to resolve here is whether the effective diameter remains a useful concept in describing light scattering, both static and dynamic.

2. INTERACTION BETWEEN TWO CHARGED RODS

We model the polyion backbones as cylinders of length L, diameter

D and charge spacing ν electrons per unit length. The electrostatic potential in the excess 1-1 electrolyte of concentration n_s is modelled within the Poisson-Boltzmann approximation. There are two important electrostatic scales: the Debye-Hückel screening length κ^{-1} and the Bjerrum length Q where $\kappa^2 = 8\pi Q n_s$, $Q = q^2/\varepsilon k_B T$, q is the elementary charge, ε the electric permittivity of the solvent, k_B is Boltzmann's constant and T the solution temperature. Consider two polyions skewed at an angle γ; the shortest distance between the centrelines is x. To a good approximation the electrostatic interaction can be written as

$$\frac{W_{el}}{k_B T} = \frac{A e^{-\kappa x}}{\sin \gamma} \qquad (1)$$

when x<L. (See ref. 1 for historical the background to eq.(1) and for an extensive discussion of the dimensionless quantity A in terms of the polyion parameters.) Eq.(1) exhibits not only a plausible screening term but also an inverse sin γ dependence which originates from the electrostatic torque between the rods: they want to twist away from the parallel configuration in order to minimize the repulsive force.

We want to address the effect of twist on the macroscopic properties of the polyelectrolyte solution. The relevant parameter turns out to be $A' \equiv Ae^{-\kappa D}$ instead of A. Few biopolymers are very weakly charged so that A'>2 under most practical conditions. Sometimes A' may be smaller which is interesting from a physical point of view (see Nyrkova and Khokhlov[2]) but this case will not be discussed here.

3. SECOND VIRIAL COEFFICIENT

Under excess salt conditions it is possible to express a thermodynamic quantity in terms of a virial expansion in the polyion number density c. For instance, the second virial coefficient at fixed angle γ is given by[3,4].

$$B_2(\sin\gamma) = L^2 \sin\gamma \int_0^\infty \{1 - \exp(-w_{tot}/k_B T)\}\, dx$$

$$\sim L^2 \sin\gamma\, \{D + \kappa^{-1}(\ln A' + C_E - \ln \sin\gamma)\} \qquad (2)$$

where $C_E = 0.5722...$ is Euler's constant, w_{tot} is the sum of the hard core and electrostatic potentials and the second line is valid when $A' > 2$ (Note that the factor 2 in eq.(46) of ref. 1 and in eq.(VI.2) of ref. 4 is erroneous; however the other equations of refs. 1 and 4 remain unchanged). The twisting effect gives rise to the $-\ln \sin \gamma$ term which becomes significant when small angles are predominantly sampled (as in the liquid crystal state[1]).

In the isotropic state the orientational average of eq.(2) yields:

the isotropic excluded volume

$$B_2 \equiv \langle B_2(\sin \gamma) \rangle = (\pi/4)\, L^2 D_{eff} \tag{3}$$

and the effective diameter

$$D_{eff} \equiv D + \kappa^{-1} (\ln A' + C_E + \ln 2 - {}^1\!/_2) \tag{4}$$

In ref. 5 an experimental test of eq.(3) is given for DNA solutions. Now in the second virial approximation, coefficients higher than the second are neglected which is allowed when the effective volume fraction is much smaller than unity ($(\pi/4)\, L\, D_{eff}^2\, c \ll 1$). For example, the osmotic pressure is then

$$\pi_p = k_B T\, c\, (1 + B_2 c) \tag{5}$$

It is important to note that in an experiment B_2c can be of the order unity while at the same time higher order terms may be deleted: the polyions must be sufficiently slender *i.e.* $L \gg D_{eff}$ so that the second virial term becomes appreciable even at low volume fractions. We focus on the semidilute region $B_2c = O(1)$ in this review. The thermodynamics depends on D_{eff} (or B_2) which collects together all other minor length scales. It is important to know whether a similar simplification occurs for light scattering.

4. STATIC STRUCTURE FACTOR

In order to describe the structure factor S(k) one must come to terms with the correlation between orientational and spatial degrees of freedom. As far as I am aware the problem was first stated correctly for rods albeit in the second virial approximation by Doi, Shimada and Okano[6]. Van der Schoot and the author[9] derived a closed analytical expression for the structure factor from their integral equation. To obtain S(k) I shall follow a slightly different line of argument[7] than that of ref. 6.

The coordinates $\mathbf{q} \equiv (\mathbf{r}, \mathbf{u})$ of a rod are specified in a Cartesian coordinate system: $\mathbf{u}$ is the unit vector along the rod axis and $\mathbf{r}$ is the vector distance from the origin to the rod centre. We let a hypothetical external field $\phi(\mathbf{q})$ act on each polyion rod. The rod configurations $\mathbf{q}$ are now described statistically by a local rod density $f(\mathbf{q})$ which is normalized to the total number of macromolecules N

$$\iint d\mathbf{q}\, f = N \tag{6}$$

The total free energy is given by[4]

$$F_{tot} = \text{const} + k_B T \int d\mathbf{q}\, f \ln \frac{4\pi f}{c}$$
$$+ \tfrac{1}{2} k_B T \iint d\mathbf{q}\, d\mathbf{q}'\, f(\mathbf{q})\, f(\mathbf{q}') \left\{1 - \exp\left(\frac{-w_{tot}(\mathbf{q},\mathbf{q}')}{k_B T}\right)\right\}$$
$$+ \int d\mathbf{q}\, f(\mathbf{q})\, \phi(\mathbf{q}) \tag{7}$$

which is the sum of an unimportant constant, an orientational entropy, a second virial term bilinear in f and an energy arising from the external field. The interaction w_{tot} depends on the configurations $\mathbf{q}$ and $\mathbf{q}'$ of two polyions. Note that when the external field is switched off ($\phi=0$) $f=f_{eq}=c/4\pi$ and one regains the usual virial expansion[3,4].

The structure factor is obtained according to the following recipe:

(i) The real distribution f is the one that minimizes F_{tot}: $\delta F_{tot}/\delta f = 0$.

(ii) Our interest is in small fluctuations away from equilibrium. Hence we linearize the resulting nonlinear integral equation by setting $f = f_{eq} + \delta f$ where the small perturbation δf should be linear in the presumably small external field ϕ.

(iii) We remark that the system is translationally invariant *i.e.*

$$w_{tot}(\mathbf{q},\mathbf{q}') = w_{tot}(\mathbf{r}-\mathbf{r}', \mathbf{u}, \mathbf{u}')$$

It is expedient to spatially Fourier transform the linear integral equation. The Fourier transforms are denoted by $\delta f(\mathbf{k},\mathbf{u})$, $\phi(\mathbf{k},\mathrm{u})$ and $w(\mathbf{k},\mathbf{u},\mathbf{u}')$, the latter being a kernel related to the Mayer function.

(iv) In order to derive the structure factor S(k) we choose $\phi(\mathbf{k},\mathbf{u}) = -\varepsilon\, s_{\mathbf{k}}(\mathbf{u})$ where ε is small and the rod interference factor is given by the well-known form

$$s_{\mathbf{k}}(\mathbf{u}) \equiv \frac{\sin \mathbf{K}.\mathbf{u}}{\mathbf{K}.\mathbf{u}} \quad ; \qquad \mathbf{K} \equiv \frac{\mathbf{k}\, L}{2}$$

The vector $\mathbf{k}$ is now the actual scattering vector in our experiment. Our choice for ϕ simulates the density fluctuations monitored by light scattering.
The small fluctuation in the distribution is given by

$$\delta f(\mathbf{k},\mathbf{u}) + f_{eq} \int d\mathbf{u}'\, w(\mathbf{k},\mathbf{u},\mathbf{u}')\, \delta f(\mathbf{k},\mathbf{u}') = \varepsilon\, f_{eq}\, s_{\mathbf{k}}(\mathbf{u}) \qquad (8)$$

(v) Yvon's theorem connects the fluctuation $\delta c(\mathbf{k})$, the Fourier transform of the local segment density, to the external field *per segment* which is simply equal to $-\varepsilon$.

$$\delta c(\mathbf{k}) = \chi(\mathbf{k})\, \varepsilon = c\, S(\mathbf{k})\, \varepsilon \qquad (9)$$

The response function χ is proportional to the structure factor $S(\mathbf{k})$. In view of the simple relation between rod and segment densities

$$\delta c(\mathbf{k}) = \int d\mathbf{u}\, s_{\mathbf{k}}(\mathbf{u})\, \delta f(\mathbf{k},\mathbf{u}) \qquad (10)$$

we get the structure factor in terms of the unknown function g $\equiv \delta f/\varepsilon f_{eq}$

$$S(\mathbf{k}) = \langle s_{\mathbf{k}}(\mathbf{u})\, g(\mathbf{k},\mathbf{u}) \rangle \qquad (11)$$

$$g(\mathbf{k},\mathbf{u}) + c \langle w(\mathbf{k},\mathbf{u},\mathbf{u}')\, g(\mathbf{k},\mathbf{u}') \rangle' = s_{\mathbf{k}}(\mathbf{u}) \qquad (12)$$

The orientational average $\langle ... \rangle$ denotes $(4\pi)^{-1}\int d\mathbf{u}$ with a similar prescription for the primed variable.

(vi) In practice, the double layer surrounding a polyion will be thin enough for no interference effects to arise within a Debye length. In that case

$$w(\mathbf{k},\mathbf{u},\mathbf{u}') = 2\, B_2\, (\sin\gamma)\, s_{\mathbf{k}}(\mathbf{u})\, s_{\mathbf{k}}(\mathbf{u}') \quad (13)$$

with $\gamma = \gamma(\mathbf{u},\mathbf{u}')$. This expression was also proposed by Maeda[8].

(vii) Even though the kernel (eq. (13)) is not definite the linear integral equation (12) can be addressed with the help of a variational principle[9] leading to

$$S(\mathbf{k}) = \frac{\langle s_{\mathbf{k}}^2(\mathbf{u})\rangle^2}{\langle s_{\mathbf{k}}^2(\mathbf{u})\rangle + 2c \langle\langle s_{\mathbf{k}}^2(\mathbf{u})\, B_2\, (\sin\gamma)\, s_{\mathbf{k}}^2(\mathbf{u})\rangle\rangle} \quad (14)$$

which can be simplified to[9]

$$S(\mathbf{k}) = \frac{F^2(K)}{F(K) + 2B_2\, cF^2(K) - B_2 c G(K,\kappa D)} \quad (15)$$

where the single rod structure factor is

$$F(K) = K^{-1}\, \mathrm{Si}(2K) - \left\{\frac{\sin K}{K}\right\}^2 \quad (16)$$

and G is a very small correction term slightly dependent on κD. Eq. (15) was also derived by Maeda[8] using the principle developed in ref. 9. In conclusion, to a good approximation the effective diameter D_{eff} (eq. (4)) is also a parameter worthwhile in assessing the static scattering. For a numerical evaluation of eq. (12) see ref. 8 in which the depolarized scattering is also discussed.

5. DYNAMICS

In the semidilute region we need a model describing the entanglement among polyions. Even for uncharged rods a complete theory is lacking but recent studies of the Brownian dynamics[10,11] of rod suspensions conclude that Fixman's theory[12] predicts the rotational diffusion fairly well. I will first give a qualitative picture of Fixman's "kinetic cage" for ideal rods.

A test rod is buffeted by the surrounding rods. We postulate that the relevant volume is of order L^3 *i.e.* cL^3 rods interact with the test rod if we average over a sufficient length of time. We want to calculate the mean-square torque exerted by the rods on the test rod which is in turn related to the mean-square orientation of the test rod with respect to the "dynamic cage"

$$<\tau^2> \approx \frac{(k_B T)^2}{<\gamma^2>} \tag{17}$$

Here, the average is over a time shorter than the reptation time of the test rod. Eq.(17) is the analogue of the mean-square force/mean-square extension relation pertaining to a harmonic spring in a heat bath.

Most of the forces exerted on the test rod by the surrounding cL^3 rods cancel because the collisions are random. The mean-square torque is determined by a Gaussian random process so that the square root of cL^3 is the effective number of collisions. Now one rod exerts a torque k_BT on the test rod on average: the (ideal) pressure on the test rod is k_BT/L^3 acting on an area L^2 times the length L. Accordingly we have

$$<\tau^2>^{1/2} \approx (cL^3)^{1/2} k_B T \tag{18}$$

After a reptation time, the test rod reorients $<\gamma^2>^{1/2} \approx (cL^3)^{-1/2}$ so that the rotational diffusion constant becomes

$$D_{rot} \approx \frac{D_{ro}}{cL^3} \tag{19}$$

To derive this, note that $<\gamma^2>^{-1}$ steps are needed for a test rod to reorient completely by diffusion (also a Gaussion random process). For cL^3 equal to unity, one regains the dilute solution value D_{ro}.

A more formal derivation[12] starts from the ideal pressure $\pi_p = k_BTc$ which is modified to eq.(5) for polyions. Hence, we have

$$D_{rot} \approx \frac{D_{ro}}{L^3 c\ (1 + B_2 c)} \tag{20}$$

which shows that the effective diameter (or second virial coefficient) is still the quantity of interest.

In the above argument electrohydrodynamic effects have been neglected. In analogy with the usual hydrodynamics[13] one expects the electrohydrodynamic friction to be determined by a factor $\ln d_e/L$ where d_e is a small length scale depending in a complicated way on other microscales like Q, κ^{-1}, D, ... For slender rods at low ionic strength eq.(20) is expected to hold to a good approximation.

Dynamical Structure Factor

The kinetic theory leading to the dynamical structure factor S(k,t) is involved so here I sketch the analysis. For uncharged rods Doi *et al.*[6] wrote a preaveraged equation

$$\frac{\partial f}{\partial t} = Lf \tag{21}$$

where L is a complicated nonlinear operator containing the rotational diffusion coefficient D_{rot}, the external field ϕ, the parallel and perpendicular diffusion coefficients $D_{\parallel}$ and $D_{\perp}$, and the kernel B_2 (sin γ). In the polyelectrolyte case D_{rot} is given by eq.(20) and B_2 (sin γ) by eq. (2). Again, we look at small deviations from equilibrium $f = f_{eq} + \delta f$ so that we get a linear equation which can be Fourier transformed

$$\frac{\partial\, \delta f(\mathbf{k},\mathbf{u},t)}{\partial t} = \Omega\, \delta f(\mathbf{k},\mathbf{u},t) \tag{22}$$

where again Ω is a complicated operator.

There exists a general theorem from kinetic theory stating that the autocorrelation function satisfies the same equation[7]

$$\frac{\partial}{\partial t} \langle \delta f(-\mathbf{k},\mathbf{u},0)\, \delta f(\mathbf{k},\mathbf{u},t) \rangle = \Omega \langle \delta f(-\mathbf{k},\mathbf{u},0)\, \delta f(\mathbf{k},\mathbf{u},t) \rangle \tag{23}$$

In this way we attain a formal expression for the structure factor $S(\mathbf{k}, t) = \langle c(-\mathbf{k},0)\, c(\mathbf{k},t) \rangle / c^2$ in view of the time dependent analogue of eq. (10).

An explicit analytic expression for S(**k**,t) is unknown although two limiting forms have been derived for uncharged rods[6]. To discuss polyions we need merely replace the uncharged second virial coefficient (A_2 in the notation of ref. 6) by B_2 as defined by eq. (3). In other words the expressions do not have a functional dependence on the kernel B_2 (sin γ).

(i) The $k \to 0$ limit (translational co-operative diffusion)

$$S(\mathbf{k},t) \approx \frac{\exp\{-(1 + 2B_2c)\}\, D_G k^2 t}{1 + 2B_2 c} \tag{24}$$

$D_G \equiv {}^1/_3\,(D_{\parallel} + 2D_{\perp})$

(ii) The t→0 limit with initial decay rate Γ

$$\Gamma(\mathbf{k}) \equiv -\lim_{t\to\infty} \frac{\partial\{\ln S(\mathbf{k},t)\}}{\partial t}$$

$$= -\frac{\dot{S}(\mathbf{k},0)}{S(\mathbf{k},0)}$$

In the small k limit this has the form[6]

$$\Gamma(\mathbf{k}) = (1 + 2B_2c)\, D_G\, k^2\, [1 + E(kL)^2] \tag{25}$$

where the coefficient E is given by

$$E = \frac{1}{1080}\,\frac{D_{rot}L^2}{D_G} - \frac{(D_{\parallel} - D_{\perp})}{135\, D_G} - \frac{B_2c}{18\,(1 + 2B_2c)} \tag{26}$$

and B_2 again depends on the ionic strength. As we diminish the salt concentration, the effective diameter D_{eff} and the second virial coefficient B_2 both increase whereas the rotational diffusion coefficient D_{rot} (eq. (20)) decreases: there is a distinct possibility of a change in sign of the coefficient E.

In both limits it so happens that B_2 is the sole electrostatic parameter of experimental relevance. It is reasonable to surmise that this holds to a good approximation for the entire wave vector and time dependence of S(**k**,t).

6. CONCLUDING REMARKS

I have put forth arguments that both the static and dynamic structure factors of semidilute rodlike polyelectrolyte solutions should depend on the electrostatics *via* the effective diameter only. This conclusion is not exact but ought to hold to a high degree subject to the following conditions.

(i) The polyions must be sufficiently slender: they must be much longer than the effective diameter D_{eff} (eq. (4)).

(ii) The polyions must be fairly rigid: they should not exceed one or two persistence lengths.

(iii) Electrohydrodynamic interactions are presumed negligible which is a good assumption at fairly low ionic strength. In this range they cannot compete with the strong ionic strength dependence of the usual eletrostatic interaction as expressed in the effective diameter. Electrohydrodynamic corrections are of logarithmic order.

(iv) The diffusion coefficients have been preaveraged in the kinetic theory. This is a standard assumption in the dynamics of semidilute and concentrated polymer solutions. Its effect is unknown.

One might be tempted to conclude that our treatment should be valid for any interaction. For instance, eqs. (3) and (5) seem to be reasonable for interactions of the general form B_2 (sin γ). However, for attractive dispersion interactions those configurations for which the rods are oriented almost parallel are weighted heavily - by contrast the same configurations have essentially zero weight for repulsive electrostatic interactions - so approximation schemes like eq. (15) break down. Attractive forces require a different analysis[14].

REFERENCES

1. A. Stroobants, H.N.W. Lekkerkerker and T. Odijk, *Macromolecules*, 1986, *19*, 2232.
2. I.A. Nyrkova and A.R. Khokhlov, *Biofizika*, 1986, *31*, 771.
3. L. Onsager, *Ann. N.Y. Acad.Sci.*, 1949, *51*, 627.
4. For a recent review see T. Odijk, *Macromolecules*, 1986, *19*, 2313.
5. T. Nicolai and M. Mandel, *Macromolecules*, 1989, *22*, 438.
6. M. Doi, T. Shimada and K. Okano, *J. Chem. Phys.*, 1988, *88*, 4070.
7. P. van der Schoot and T. Odijk, *J. Chem. Phys.*, 1990, *93*, 3580; *erratum*: *J. Chem. Phys.*, 1991, *94*, 2377.
8. T. Maeda, *Macromolecules*, 1991, *24*, 2740.
9. P. van der Schoot and T. Odijk, *Macromolecules*, 1990, *23*, 4181.
10. I. Bitsanis, H.T. Davis and M. Tirrell, *Macromolecules*, 1988, *21*, 2824.

11. I. Bitsanis, H.T. Davis and M. Tirrell, *Macromolecules*, 1990, *23*, 1157.
12. M. Fixman, *Phys.Rev.Lett.*, 1985, *55*, 2429.
13. T. Odijk, *Macromolecules*, 1986, *19*, 2073.
14. P. van der Schoot and T. Odijk, *J. Chem. Phys.*, submitted.

13

Quasi-elastic Light Scattering from an Evanescent Wave to Probe Particle/Wall Interactions

By N. Ostrowsky and N. Garnier

LABORATOIRE DE PHYSIQUE DE LA MATIERES CONDENSEE, CNRS URA 190, UNIVERSITE DE NICE SOPHIA ANTIPOLIS, F-06034 NICE CEDEX, FRANCE.

1. INTRODUCTION

The dynamics of colloidal particles in the immediate vicinity of a liquid/solid interface is an important fundamental process that is an essential for the understanding of a number of biologically relevant phenomena, such as particle sedimentation and adhesion on a substrate.

A number of theoretical[1] and numerical[2] studies have dealt with that problem, and, more recently, some molecular dynamics simulations[3,4] have helped in the understanding of the role of particle/surface hydrodynamic and finite range static interactions.

As far as experiments go, the problem is far less advanced. Macroscopic experiments have monitored the fall of suspended balls onto a solid surface, thus measuring the friction coefficient $\Lambda(z)$ as a function of the distance z between the particle surface and the solid surface[5]. On a more microscopic scale (particles $\approx$ 10 μm in diameter) static experiments have studied the height distribution of the suspended particles above a given transparent plate, from which the static particle/wall interaction potential can be deduced[6]. Static fluorescence techniques have also been used to measure the concentration profile of particles labelled with fluorescent probes in the vicinity of a transparent wall[7,8].

We have taken a different approach and used Quasi-Elastic

Light Scattering (QLS) techniques to measure the Brownian dynamics of particles very close to a wall, using an evanescent wave as the incident beam[9]. In section 2, we will first describe our experimental set-up and derive the form of the correlation function. We will then examine in section 3 the influence of particle/wall interactions on the correlation function and summarize our results obtained with different salt concentrations in the suspension. Finally, in section 4 we will discuss the molecular dynamic calculations which enable us to directly relate the measured correlation function to the static and hydrodynamic particle/wall interactions.

2. EVANESCENT QLS EXPERIMENTAL SET-UP AND CORRELATION FUNCTION

Our experimental set-up is shown in Fig. 1. The fairly concentrated latex suspension (particle diameter = 0.09 μm, concentration, $c \approx 3 \times 10^{-4}$ g/cm^3, - equivalent to a mean inter-particle distance of 1 μm) is contained in a semi-cylindrical cell, sealed by the flat surface of a larger semi-cylindrical glass prism. The sample holder is placed on a precision turntable, so as to facilitate the easy change of the incident angle θ_i of the vertically polarized Argon laser (300 mW at λ =514.5 nm). The critical angle for total internal reflection is given by the usual relation: $\sin\theta_c = n_{liquid}/n_{glass}$ which yields $\theta_c = 64.3°$.

For $\theta_i > \theta_c$, the incident wave vector in the medium has a real component $k_i = 2\pi n_{glass}/\lambda$ parallel to the flat surface of the prism, and an imaginary component equal to the inverse of the penetration depth ξ given by:

$$\xi = \frac{\lambda}{2\pi\, n_{glass}} \{\sin^2\theta_i - \sin^2\theta_c\}^{-1/2} \qquad (1)$$

which means that we are conducting a light scattering experiment with an incident beam always parallel to the flat glass/liquid interface and whose intensity decreases exponentially as $\exp(-2z/\xi)$ with the distance z from the wall. The light scattered in the liquid suspension at an angle θ from the incident wave vector k_i is collected *via* an optical fibre onto a photomultiplier whose output is analyzed using standard correlation techniques. To ensure 100% heterodyning, an adjustable fraction of the reflected laser beam was made to coincide with the path of the scattered light and thus mixed on the photomultiplier surface.

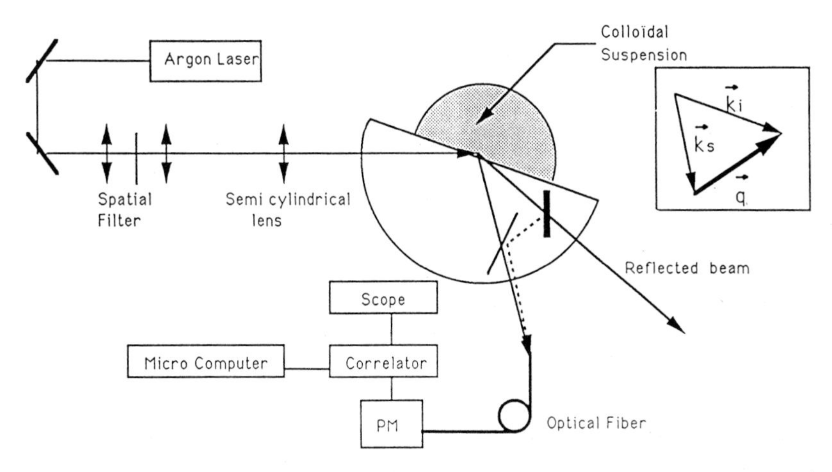

Figure 1. Experimental set-up of the evanescent wave QLS. The inset shows the incident, scattered and scattering wave vectors.

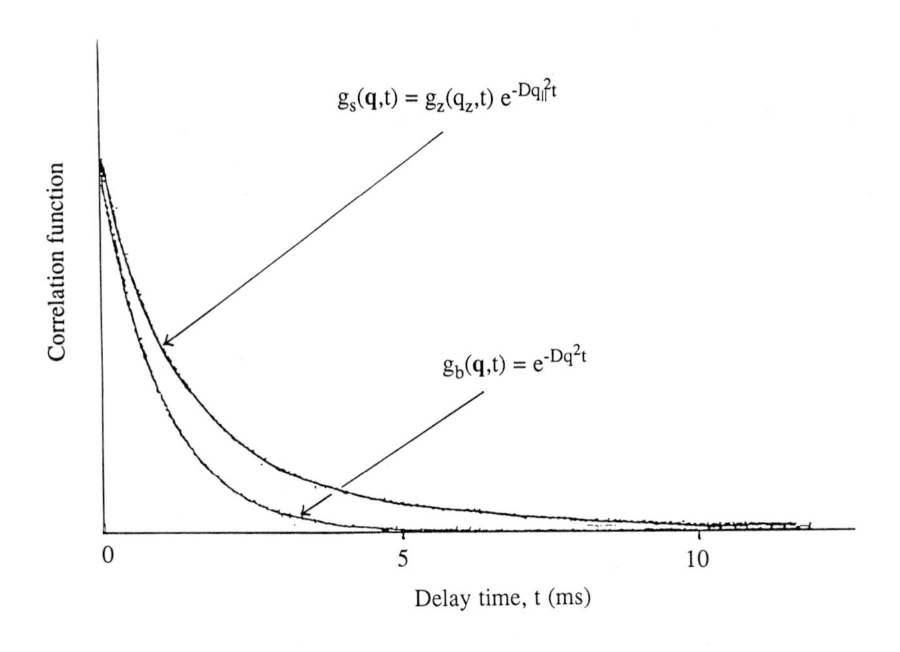

Figure 2. Typical correlation functions recorded in the bulk and surface geometries

Typical correlation functions recorded for θ_i just below (ξ = 0.8 μm) and just above ($\xi = \infty$) the critical angle θ_c, but at the same scattering angle θ, are shown in Fig. 2, illustrating the so-called "surface" and "bulk" correlation functions whose theoretical expression we will now discuss.

In heterodyne experiments, the intensity correlation function is a linear function of the first order correlation function of the scattered electric field, which, with homogeneous illumination of the scattering volume and in its normalized form, can be written as:

$$g^{(1)}(t) = \mathrm{Re} \langle e^{i\mathbf{q}.\mathbf{r}(t)} \rangle \tag{2}$$

where the scattering wave vector $\mathbf{q} = \mathbf{k}_i - \mathbf{k}_s$ is the difference between the incident and the scattering wave vectors, and $\mathbf{r}(t)$ is the vector position of the particle which was at the origin at time zero. The average is to be taken for a great number of independent Brownian particles.

In the case of the usual bulk geometry, the probability density for a particle to be at time t at a distance r from its starting point at time zero is the well-known Gaussian probability

$$P\,(r,t) = (4\pi Dt)^{-3/2} \exp(-r^2/4Dt) \tag{3}$$

where $D = kT/6\pi\eta R$ is the bulk diffusion coefficient, related to the particle radius R and the suspension's viscosity η. Using this probability density to compute the average in eq. (2) leads to the usual expression for the bulk correlation function:

$$g_b^{(1)}(t) = \int e^{-i\mathbf{q}.\mathbf{r}(t)}\, P(r,t)\, d^3r \tag{4}$$

$$= \exp(-Dq^2t) \tag{5}$$

In the presence of a wall, it is useful to decompose the three-dimensional Brownian motion into two independent motions, one parallel and one perpendicular to the wall, along the direction Oz. The first motion obeys the usual two-dimensional Brownian motion statistics, and the second one includes the "mirror" effect of the wall. In the evanescent wave geometry, one must further take into account the fact that the scattering volume is not uniformly illuminated, which requires that the average in eq. (2) be properly weighted by the electric field intensity at the particle's position at time 0 and time t.

The theoretical form of the "surface" correlation function $g^{(1)}{}_s(t)$ has been found[9] to be equal to

$$g^{(1)}{}_s(t) = \exp(-Dq_{\parallel}{}^2 t)\, g_z(q_z,\xi,D,t) \tag{6}$$

where $q_{\parallel}$ and q_z are the components of the scattering vector parallel and perpendicular to the wall, and $g_z(q_z,\xi,D,t)$ is an analytical function whose limited expansion can easily be implemented on a microcomputer. The theoretical correlation functions, eqs. (5) and (6), have been used to draw the solid lines on Fig. 2 leaving as an adjustable parameter the diffusion coefficient, D. The value found for the best adjustment of the surface correlation function was roughly 10% lower than the value found for the bulk measurement, which is due to the repulsive hydrodynamic and electrostatic interactions between the particles and the wall, as we will now discuss.

3. PARTICLE/WALL INTERACTIONS

As mentioned in the introduction, two kinds of interactions must be considered.

First, the *static particle/wall* interactions, represented by an interaction potential, U(z): these result in a non-uniform distribution of particles in the vicinity of the wall. As we studied latex particles suspended in salt solution in the vicinity of a glass wall, both negatively charged, the interaction is repulsive

$$U(z) = B \exp(-\kappa z) \tag{7}$$

where κ^{-1} is the Debye screening length

$$\kappa^{-1} = \left\{\frac{\varepsilon kT}{2e^2 N_A c_{salt}}\right\}^{1/2} \tag{8}$$

(with N_A Avogadro's number, ε the dielectric constant of the medium, e the electronic charge), and B is a numerical coefficient which is determined by the surface charge densities on the particle and on the wall. Simulations of the particle concentration profiles are given in Fig. 3b for different values of NaCl concentration.

The effect of the *hydrodynamic interactions* between the particles and the wall (mediated by the water molecules) are included through a position-dependent friction tensor Λ, which, when multiplied by the velocity vector **v** of the particle, yields the friction force, **F** experienced by the particle

$$\begin{pmatrix} F_{\parallel} \\ F_z \end{pmatrix} = \begin{pmatrix} \Lambda_{\parallel} & 0 \\ 0 & \Lambda_z \end{pmatrix} \begin{pmatrix} v_{\parallel} \\ v_z \end{pmatrix} \tag{9}$$

Using the Einstein-Smoluchowsky relation, a position dependent diffusion tensor can be generated, whose components $D_{\parallel}(z)$ and $D_z(z)$ have been calculated in the literature[4,10] (see, Fig. 3a).

To account for this position dependence of the diffusion coefficient in the computation of the correlation function $g^{(1)}(t)$ is not a trivial matter, except for short times compared to the correlation function relaxation time. In this limit, a limited expansion of eq. 6 yields:

$$g^{(1)}{}_s(t) \cong \{1 - D_{\parallel} q_{\parallel}^2 t - D_z (q_z^2 + 1/\xi^2)\, t\} \tag{10}$$

For such short times, one can assume that a given Brownian scatterer is confined to a volume small enough so that its diffusion coefficients $D_{\parallel}(z)$ and $D_z(z)$ can be considered as constant. The observed correlation function is an average of eq. (10) for all the Brownian particles contained in this scattering volume. Taking into account the fact that the concentration c(z) of particles near the wall may be position dependent (note c(z) as a function of h={(z-R)/R} in Fig. 3b), and that particles closer to the wall receive and thus scatter a higher intensity according to the exponential law $\exp(-2z/\xi)$, leads to the approximation

$$g^{(1)}{}_s(t) \cong \frac{\int_0^{\infty} \{1 - D_{\parallel}(z) q_{\parallel}^2 t - D_z(z)(q_z^2 + 1/\xi^2)t\}\, c(z) \exp(-2z/\xi)\, dz}{\int_0^{\infty} c(z) \exp(-2z/\xi)\, dz} \tag{11}$$

$$\cong \{1 - \bar{D}(\xi)\,(q^2 + 1/\xi^2)\} \tag{12}$$

thus defining the weighted average $\bar{D}(\xi)$ which has been computed numerically for different concentration profiles c(z), *i.e.* different static particle/wall interactions (see dotted lines in Fig. 3c).

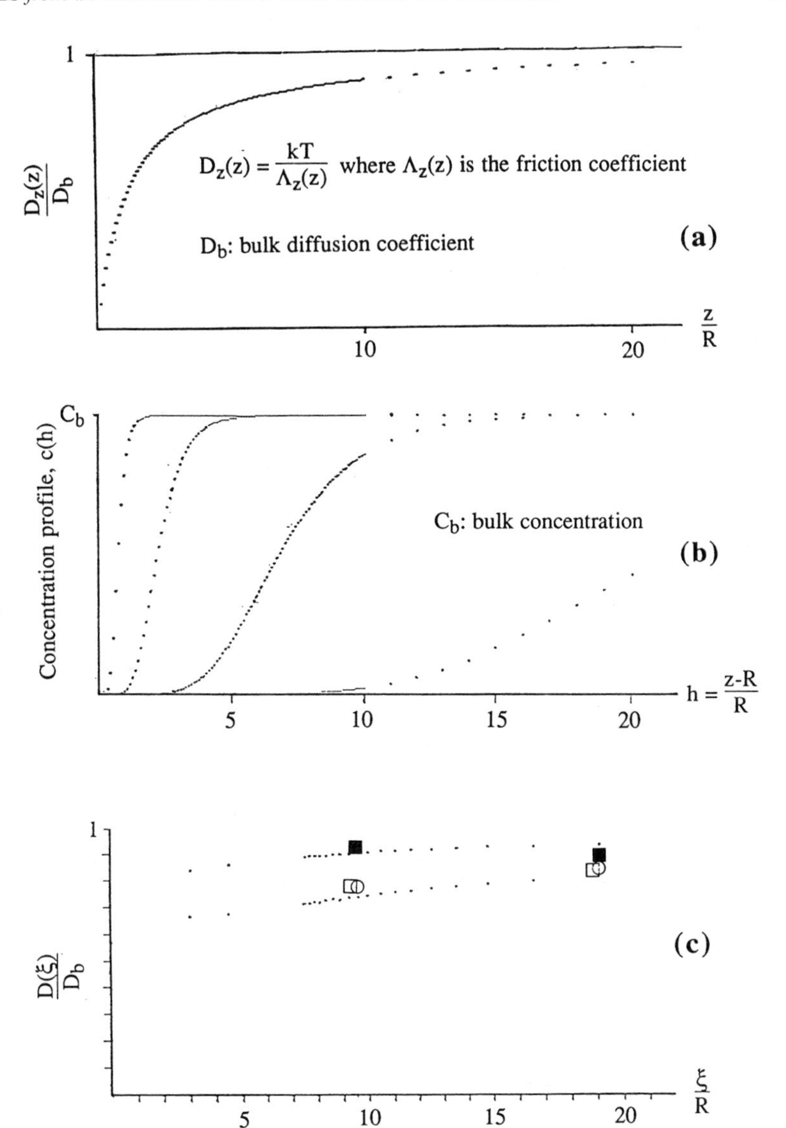

Figure 3. (a) Position dependence of the diffusion coefficient D_z as computed from a limited expansion of its theoretical expression[4,10]. (b) Particle concentration profiles for different salt concentrations. (c) Average diffusion coefficient for different penetration depths and salt concentrations. The experimental points are shown as black squares (no salt added) and open circles ([NaCl] = 10^{-2} mol/l). The dotted lines are computed according to eqs. (11) and (12) for zero and 10^{-2} mol/l salt concentrations. The molecular simulation predictions are shown as open squares for [NaCl] = 10^{-2} mol/l for two different penetration depths.

It must be remembered, however, that this approach is only valid for short times, and the above result should be compared with the very beginning of the experimentally measured correlation function, *i.e.* its slope at the origin. Unfortunately, this comparison cannot be very precise, as the surface correlation function is far from an exponential and its slope at the origin can only be poorly defined. We thus looked for a better way of analyzing our experimental data, and this led us to the simulations we now describe.

4. MOLECULAR DYNAMICS SIMULATIONS

Starting from the Langevin equation describing the motion of a Brownian particle submitted to a position-dependent friction coefficient and to a static position-dependent force, Ermak and McCammon[3] derived the expression for the position-dependent step such a Brownian particle accomplishes. In our problem, the random displacement L(z) of a walker along the Oz axis during a time interval τ is the sum of three terms:

$$L(z) = \pm\,(2D_z(z)\tau)^{1/2} + \frac{dD_z}{dz}\tau + \frac{D_z(z)}{kT}F_z\tau \qquad (13)$$

The first term is the usual random step, corrected by the fact that the diffusion coefficient is z-dependent. The second one reflects the fact that the hydrodynamic interaction acts as a repulsive force and the third is simply the drift of the walker away from the wall, due to the static force F_z. We have thus implemented on a Mac II microcomputer a programme which computes the position z(t) of a walker starting at time zero from a randomly chosen position z(0), repeating the procedures for a great number of independent walkers, and building progressively the function

$$g_z(q_z,\xi,t) = \mathrm{Re} \langle \exp\{q_z[z(t)-z(0)]\}\rangle \qquad (14)$$

the average being properly weighted by the illumination profile (penetration depth ξ) and the concentration profile. This "simulated" correlation function was then fitted with the theoretical expression found for g_z and the results are summarized in Table 1 for two different penetration depths and different kinds of Brownian motions.

Table 1. Values of $D_z(\xi)/D_b$ found by fitting the molecular dynamics simulations of the correlation function $g_z(t)$ to the theoretical expresion for $g_z(t)$. The differences between walkers A, B, C and D are explained in the text.

Penetration depth	0.43 μm	0.86μm
Walker A	1.001 ± 0.003	0.996 ± 0.005
Walker B	0.692 ± 0.004	0.797 ± 0.003
Walker C	0.771 ± 0.003	0.827 ± 0.002
Walker D	0.782 ± 0.002	0.831 ± 0.003
Computed $\bar{D}(\xi)/D_b$	0.693	0.788

Walker A is a simple Brownian particle submitted to the mirror effect of the wall and the illumination profile, but ignoring the hydrodynamic and static repulsions; it has a constant, random step: $\pm(2D_b\tau)^{1/2}$, determined by the bulk diffusion coefficient, D_b. Walker B is almost identical, except that its random step is position dependent: $\pm(2D_z(z)\tau)^{1/2}$. The step of Walker C includes the first two terms on the right hand side of eq. 13 and finally Walker D uses all three terms for L(z) with [NaCl]=10^{-2} mol/l. For the sake of comparison, we have added $\bar{D}(\xi)/D_b$ computed directly from eq. 12.

The experimental results we have obtained for two different penetration depths and two different salt concentrations are summarized in Table 2, and plotted in Fig. 3c.

Table 2. Experimentally determined diffusion coefficients for two different penetration depths at zero and 10^{-2} mol/l [NaCl] concentration.

[NaCl] (mol/l)	Penetration depth, ξ (μm)	10^{-3} x D_b (μm^2/ms)	10^{-3} x D_s (μm^2/ms)	D_s/D_b
0	0.86	4.64 ± 0.05	4.15 ± 0.10	0.89 ± 0.03
0	0.43	4.67 ± 0.02	4.35 ± 0.13	0.93 ± 0.02
0.01	0.86	4.70 ± 0.05	3.93 ± 0.13	0.84 ± 0.03
0.01	0.43	4.83 ± 0.02	3.76 ± 0.08	0.78 ± 0.02

5. CONCLUSIONS

To summarize, we have shown that the "Evanescent Quasi-elastic Light Scattering" technique is a sensitive tool for measuring Brownian dynamics in the immediate vicinity of a rigid surface. A net decrease of the measured diffusion coefficient is observed, due to the hydrodynamic retardation of the particles very close to the wall. This effect is all the more observable when the particles are allowed to get close to the wall, *i.e.* when the range of the static wall/particle repulsive interaction decreases.

This method could also prove to be very sensitive to the onset of the aggregation of particles on a surface, introducing into the computation a "residence time" during which a given particle would remain stuck on the wall before "drifting off" again into the suspension. Simulations are currently being performed to test this possibility.

ACKNOWLEDGEMENTS

The authors acknowledge stimulating discussions with Bruce Ackerson which led to the approximate form of the correlation function at short times and wish to thank Pierre Bezot for his valuable help with the experiments.

REFERENCES

1. G.K. Batchelor, *J. Fluid Mech.*, 1976, *74*, 1.
2. Z. Adamczyk and T.G.M. van de Ven, *J. Coll. Int. Sci.*, 1981, *84*, 497.
3. D.L. Ermak and J.A. McCammon, *J. Chem. Phys.*, 1978, *69*, 1352.
4. A.T. Clark, M. Lal and G.M. Watson, *Faraday Disc. Chem. Soc.*, 1987, *83*, 179.
5. Z. and M. Adamczyk and T.G.M. van de Ven, *J. Coll. Int. Sci.*, 1983, *96*, 204.
6. D.C. Prieve, F.Luo and Lanni, F., *Faraday Discuss. Chem. Soc.*, 1987, *83*, paper 22.
7. N.L. Thomson, T.P. Burghart and D. Axelrod, *Biophys. J.*, 1981, *33*, 435.
8. D. Aussere, H. Hervet and F. Rondelez, *Phys. Rev. Lett.*, 1985, *54*, 1948.

9. K.H. Lan, N. Ostrowsky and D. Sornette, *Phys. Rev. Lett.*, 1986, *57*, 17.
10. M. Stimson and G.B. Jeffery, *Proc. Roy. Soc.*, 1926, *A111*, 110.

Part II: Macromolecules

14

Compactness of Protein Molecules in Native and Denatured States as Revealed by Laser Light Scattering and X-Ray Scattering

By K. Gast, G.Damaschun, H. Damaschun, R. Misselwitz, D. Zirwer and V.A. Bychkova[1]

INSTITUTE OF MOLECULAR BIOLOGY, O-1115 BERLIN-BUCH, ROBERT-ROESSLE-STRASSE 10, GERMANY.

[1]INSTITUTE OF PROTEIN RESEARCH, ACADEMY OF SCIENCES OF THE USSR, 142292 PUSHCHINO, MOSCOW REGION, USSR

1. INTRODUCTION

Though most of the biochemical and biophysical studies on proteins are directed towards the native state, there is an increasing interest in denatured and unfolded states. This is mainly due to the following reasons: (i) the elucidation of the pathway of protein folding has become a very exiting field in protein research, which necessitates investigations of conformational states other than the native one; (ii) it became evident that non-native states appear in the translocation process of proteins in cells, especially during the transport of protein molecules through membranes; (iii) partly unfolded or misfolded protein molecules have been observed during the expression of recombinant proteins.

Conformational states of proteins can only be described in a satisfactory manner by the combination of the results of several biophysical and biochemical methods. By the late 1960's Tanford[1] had already given a thorough review of the state of the knowledge and the results of the methods used in denaturation experiments. The experimental techniques currently used include calorimetric, spectroscopic (UV, CD, fluorescence, IR, NMR), hydrodynamic and scattering methods. Whilst spectroscopic techniques provide very sensitive probes for monitoring changes of the local structure, hydrodynamic and scattering methods can yield information about the gross conformation and the dimensions of the protein as a whole.

Most proteins, in their native states, are folded into well-defined, usually essentially rigid, three-dimensional structures. As a rule this state is most compact and denaturation leads to an unfolding of the polypeptide chain. Denatured proteins can adopt a great variety of conformations from native-like states, such as the so-called 'molten globule' state[25,26], to highly unfolded states at high concentrations of strong denaturants. Therefore, the compactness of a protein molecule is an essential parameter.

We consider here only proteins consisting of a single polypeptide chain. The best measure of compactness can be achieved by measuring the 'radius of gyration' of the protein molecules. But, as we will show later on, the determination by direct methods such as small-angle X-ray scattering (SAXS) is difficult under certain conditions. The hydrodynamic Stokes' radius R_S, which can be measured precisely by dynamic light scattering (DLS), is also a very useful parameter for following changes of the compactness, but it depends on the physical dimensions in a more complicated manner.

For quantitative estimations of the compactness a reference point is necessary. In principle, the native as well as the totally unfolded and disordered state can be taken into consideration. The use of each of them however offers slight problems: for example, the compactness of the native state differs for various proteins; on the other hand, the dimensions of the protein in the totally unfolded state depend on the solvent. This means that one has to find conditions to estimate the unperturbed dimensions, the so-called 'theta-conditions'. Problems of this kind will be discussed later on. In our work we compare the dimensions of the protein in any conformational state with its dimensions in the native state. In the following we will firstly describe the methodological background and the most important experimental details. The experimental results cover the following topics: (i) the influence of different denaturing conditions on the compactness of various proteins; (ii) the dependence of the dimensions of a disordered protein chain on the solvent conditions (studied on apo-cytochrome c at acid pH); (iii) the compactness of the 'molten globule' state of alpha-lactalbumin.

2. PROBING THE MOLECULAR DIMENSIONS BY DLS AND SAXS

Physical quantities describing the molecular conformation

The molecular dimensions of a protein chain can be best described by the weighted root mean-square distance of all atoms from the centre of gravity, which is the root mean-square 'radius of gyration', R_g. The radius of gyration can be determined directly by total intensity light scattering or X-ray scattering. However, the angular dependence of the light scattering intensity of small proteins is too weak for a precise estimate of R_g. This problem does not exist with SAXS. But, some other difficulties can make measurements of R_g hardly feasible, *e.g.* high protein concentrations are needed and furthermore, the scattering contrast is very low in the presence of high concentrations of salt. As a consequence, SAXS investigations at 6 M guanidine hydrochloride (GuHCl) or 8 M urea are extremely difficult.

Thus, precise measurements of the translational diffusion coefficient D of the protein molecules by DLS offer a great advantage. Extrapolating the values of D to zero concentration one can calculate the hydrodynamic Stokes' radius *via* the Stokes-Einstein-equation

$$R_s = \frac{kT}{6\pi\eta D} \tag{1}$$

where k is Boltzmann's constant, T is the temperature in Kelvin and η is the solvent viscosity. Now we ask for the relation between the hydrodynamic radius R_s and the radius of gyration, R_g. Hydrodynamic theories[2,3,4] based on the pioneering work of Kirkwood and Riseman[5] have lead to equations for the frictional coefficient $f = kT/D$ of the type

$$f = P\sqrt{6}\,\eta\, R_g \tag{2}$$

The factor P depends on the structural model, *e.g.* $P = \pi\sqrt{10}$ for a compact sphere and $P = (3\pi)^{3/2}/2^{5/2} = 5.11$ for a linear disordered chain[2]. Hence, the connection between R_g and R_s is given by

$$R_g = (\pi\sqrt{6}/P)\, R_s \tag{3}$$

Burchard[5] has introduced the factor $\rho = R_g/R_s$. P, ρ or the factor ξ ($= 1/\rho$) used by Tanford[6] are equivalent and have typical values for particular molecular conformations. By measuring R_g by SAXS and R_s by DLS one has strong evidence for the structural type of the conformational state of a protein. On the other hand, it is possible to estimate the radius of gyration from DLS measurements if a reasonable structural model can be anticipated. In the following we will use the factor ρ. ρ-factors for some model structures typical for globular proteins are shown in Table 1. The native state of globular proteins is generally well fitted by an assembly of smaller spheres[7] or by ellipsoids[8]. The ρ-factor of an ideal sphere is 0.775. Usually, the axial ratio of typical globular proteins does not exceed 3:1 and the ρ-factor of such ellipsoids is close to that of a sphere (hydration, especially the outer hydration shell, leads to a slight decrease of ρ). According to Eq. 3 and using a value for P of 5.11, the ρ-factor for a randomly coiled chain is 1.51. As we will show later on, proteins in this conformational state may be more or less compact depending on the stiffness of the chain and on the solvent. It is still an open question on how to deal with intermediate conformations, for example, models of rigid blocks joined by flexible connections may be reasonable.

Table 1. ρ-factors for some structural models.

Structural Model	ρ-Factor
Sphere	0.77
Two-axial ellipsoid (prolate) 2:1	0.83
Two-axial ellipsoid (prolate) 3:1	0.92
Two-axial ellipsoid (oblate) 3:1	0.85
Linear disordered chain	1.51

DLS investigations

The DLS measurements were performed by using a laboratory built apparatus, equipped with an argon-laser (ILA-120, Carl-Zeiss-Jena) mostly operating at $\lambda = 514.5$ nm and 400 mW. The optical part of the spectrometer was mounted on a modified X-ray scattering goniometer (HZG 4, Praezisionmechanik Freiberg (Sachs.)). The homodyne time-autocorrelation functions were calculated at 128 delay times in the single clipped mode. Data acquisition and normalization were performed on an 8 bit microcomputer. A second on-line microcomputer was used for data storage and preliminary data analysis performed either in parallel or immediately after the running experiment (the latter is very important to ensure that data of sufficient precision has been collected for the problem under study).

Total intensity light scattering measurements (TILS) were performed with the same apparatus. Simultaneous measurements of TILS and DLS are not only important for estimating the molar mass and the second virial coefficient, A_2, and additionally for D and R_S but also to check whether the observed change in R_S is possibly due to aggregation or degradation of protein molecules. We employed 7 mm diameter flow cells to protect the solutions filtered through 100 nm Nuclepore filters from any contact with air. Final data analysis was performed using the inverse Laplace-type of transformation by the constrained regularization method of the program 'CONTIN'[9]. In our case the Laplace integrals

$$g^1(\tau) = \int_0^\infty A(D) \exp(-q^2 D\tau)\, dD \tag{4}$$

did not contain broad distributions of the diffusion coefficient A(D) but rather delta-function-like members $A(D_i)$, where i stands either for the fraction of protein or salt, respectively. $g^1(\tau)$ is the first order correlation function, τ is the delay time, and q is the scattering vector. At high concentrations of denaturants and protein concentrations of the order of 1 mg/ml, the scattering contribution of the salt becomes comparable to that of the protein solution. Sometimes a small background term due to particles which have not been removed by our filtering procedure had to be taken into account. Without a separation of these components of $g^1(\tau)$, the diffusion coefficient of the protein molecules cannot be evaluated with sufficient precision. The use of 'CONTIN' with a weak regularizor was found to be more reliable than the fitting of $g^1(\tau)$ by a sum of exponentials.

A peculiarity sometimes involved in investigations of denatured proteins is connected with their poor solubility in aqueous media. The exposure of hydrophobic regions of the molecule may lead to an aggregation, especially in the transition region between native and denaturing conditions. Furthermore, reliable results of unfolding and refolding experiments can only be obtained if the system is fully reversible.

SAXS measurements

SAXS was measured on a microcomputer-controlled small-angle / wide-angle (SAXS-WAXS) diffractometer. The CuK_{α}-radiation was collimated by 5 slits and Soller slits. The methods of data processing and de-smearing have been described by Damaschun[10] and Mueller[11]. The molar mass and the second virial coefficient A_2 were evaluated from the absolute scattering intensity using a calibrated Lupolen standard as a reference.

3. EXPERIMENTAL DATA OF THE COMPACTNESS OF PROTEINS IN DIFFERENT CONFORMATIONAL STATES

Comparative consideration of different denaturing conditions

In the following we will represent both our own and some other data from the literature to obtain an idea of the extent of unfolding under various denaturing conditions, namely: high concentrations of guanidine hydrochloride (GuHCl), acid pH and rise in temperature. The particular proteins for these experiments were chosen either because of their model character or because of their biochemical and pharmacological importance.

Unfolding in 6M GuHCl. It is well known[1] that most proteins with an ordered native structure undergo a marked transition upon the addition of GuHCl. Most of the hydrodynamic investigations of the degree of unfolding have been performed by measuring the intrinsic viscosity $[\eta]$[1] whereas measurements of the Stokes' radius R_s by DLS are scarcely known[12,18]. A first experiment of this kind had been reported by Dubin *et al.*[12] for lysozyme. These data are shown together with our experimental results for human α-lactalbumin (HLA) and streptokinase (SK) in Table 2. The relative increase in size S is of the same order of magnitude for all proteins under study. The apparent size increase is smaller in the case of intact disulfide bridges. Because directly measured values of the radius of gyration are not available, we have calculated R_g from R_s *via* Eq. 3 assuming the protein to have the conformation of a disordered chain and hence a value of $P = 5.11$.

It is now interesting to compare the experimental values of R_g (see Table 2.) with the unperturbed dimensions of a linear disordered chain $R_{g,0}$ on the basis of the expansion factor

Table 2. Dimensions in the native (N) and denatured state at 6 M GuHCl (D) at 20°C.

Protein	R_S^N (nm)	R_S^D (nm)	S	R_G^D (nm)	α	
					Eq. 5	Eq. 6
Lysozyme* (0.1 M Na-acetate; pH 4.2)	2.02	2.93[a]	1.45	4.42	1.00	1.14
		3.75[b]	1.86	5.66	1.29	1.46
H.α-lactalbumin (20 mM Tris-HCl; pH 7.5)	1.77	2.47[a]	1.40	3.37	0.87	0.98
Streptokinase (10 mM Na-phosphate; pH 7.5)	3.58	5.54	1.55	8.37	1.06	1.20

R_G^D was calculated from R_S^D using eq. 3 and P = 5.11.
Experimental error in all radii is ± 2%.

([a]) Oxidised and ([b]) reduced disulfide bridges.
* Data from Dubin *et al.*[12].

$$\alpha = R_g/R_{g,0}$$

Real chain dimensions differ from the unperturbed dimensions due to excluded volume effects. The unperturbed dimensions depend on the number of monomer units, the number of amino acids n_A in our case, and the composition of the polypeptide chain. Theoretical calculations[13,14,15] for long polypeptide chains containing 10% glycine and no proline yield for the mean-square end to end distance $\langle R^2 \rangle_0$ (which is related to R_g for disordered chains by $\langle R^2 \rangle_0 = 6R^2_{g,0}$) the following relation (R in Å)

$$\langle R^2 \rangle_0 = 6R^2_{g,0} = 90\ n_A \tag{5}$$

Tanford *et al.*[16] have derived the following quantitative relation on the basis of measurements of the intrinsic viscosity

$$\langle R^2 \rangle_0 = 6R^2_{g,0} = (70 \pm 15)\ n_A \tag{6}$$

We have calculated $R_{g,0}$ and α on the basis of both equations. The values of α_{70} and α_{90} are shown in Table 2. McDonnell and Jamieson[18] have published relations between the diffusion coefficient D and the number of amino acids and the molar mass for several proteins in 6 M GuHCl.

Conformational states at acid pH. One of the oldest known methods of denaturing proteins is the addition of acids. The results obtained for streptokinase, apo-cytochrome c and human α-lactalbumin at acid pH are shown in Table 3. As can be seen at first sight, rather different degrees of unfolding have been measured. Apo-cytochrome c molecules adopt according to the experimental ρ-factor the conformation of a disordered chain at low and high ionic strength. In the case of streptokinase we find a very extended conformation. The ρ-factor of 1.7 exceeds the value for random chains. At the present time it is not clear whether long rod-like or helical structures or excluded volume effects are responsible for this behaviour. In contrast to this, the expansion of α-lactalbumin is rather weak. According to the ρ-factor the structure remains compact. This conformational state is discussed later in more detail. It is noteworthy, that lysozyme[17] is unaffected by a reduction in pH at room temperature. Hence, no general rules can be formulated with respect to conformational transitions at low pH.

Table 3. Dimensions of proteins at acid pH compared to that of the native state at 20°C.

Protein	R_S^N (nm)	R_S^D (nm)	S	ρ^D
Streptokinase (10 mM HCl; pH 2.2)	3.58	6.65	1.86	1.71
Apo-cytochrome c				
Solvent I (pH 2.3)	1.90[a]	3.02	1.59	1.55
Solvent II (pH 2.3)		2.54	1.34	1.60
H.α-lactalbumin (50mM KCl-HCl; pH 2.0)	1.77	1.99	1.12	0.79
Lysozyme (acid pH)	No unfolding at 20°C[b]			

Experimental error: Stokes' radii: ± 2%, S and ρ: ± 5%.
[a] Data from 'Handbook of Biochemistry' [28].
[b] According to Nicoli and Benedek[17].
Solvent I: 20 mM Na-phosphate; Sovent II: solvent I plus 0.25 NaCl.

Thermal denaturation of proteins The first DLS studies of the thermal unfolding of proteins were reported by Nicoli and Benedek[17] for lysozyme and other proteins. The thermal transition of streptokinase measured by us at pH 7.5 is shown as an example in Fig. 1. As can be seen clearly, the transition proceeds in two steps. This two step transition was also verified by calorimetric data (K. Welfle and W. Pfeil, personal communication). The total increase in R_S is about 15%. The lack of change in the Stokes' radius after cooling down to room temperature at the end of the experiment points to reversibility of the unfolding /refolding process. The results for lysozyme (data from ref. 17), streptokinase and α-lactalbumin are summarized in Table 4. Obviously, the effect of heat on the measured Stokes' radius is only moderate and similar in all three cases. But, a surprising conclusion can be drawn by comparing the ρ-factors of streptokinase and alpha-lactalbumin. Although the relative size increase S is only slightly larger for streptokinase, the protein molecules adopt a disordered chain conformation, whereas alpha-lactalbumin still remains in a native-like conformation. This may be due to the fact that streptokinase is already less compact in the native state compared with lactalbumin (the ratios of frictional

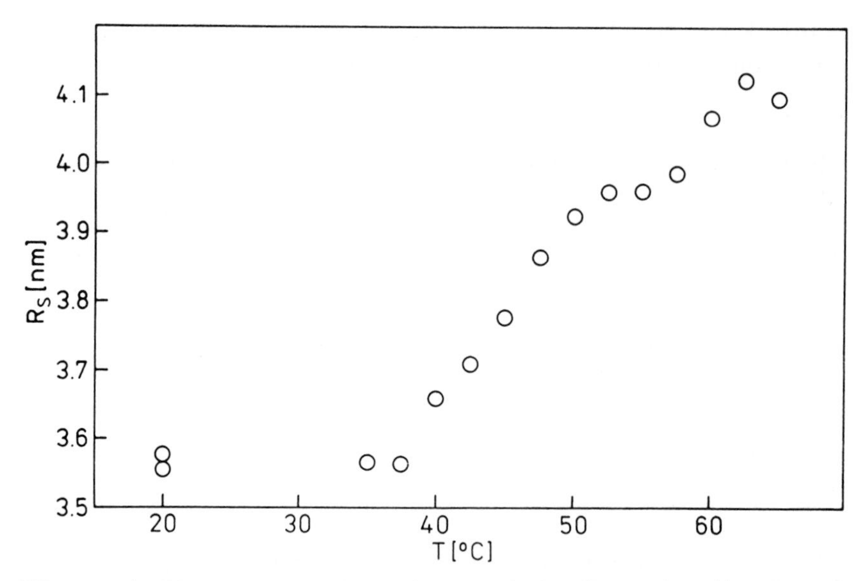

Figure 1. Temperature dependence of the Stokes' radius R_S of streptokinase in 10 mM Na-phosphate, pH 7.5, c = 0.74 mg ml^{-1}. The second, slightly larger, value of R_S at 20°C was measured at the end of the experiment.

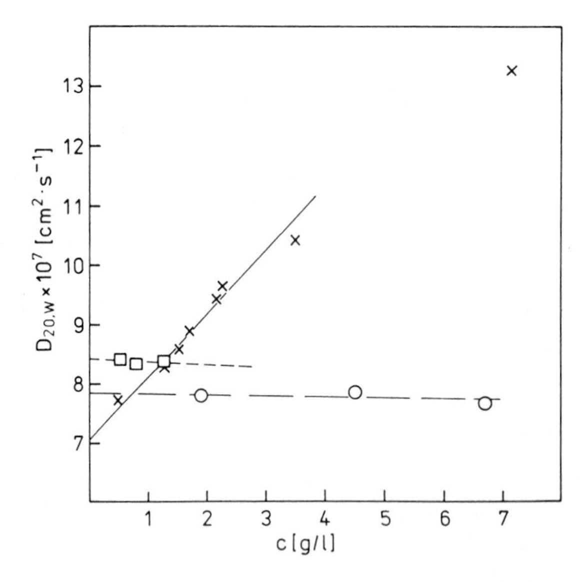

Figure 2. Concentration dependence of the diffusion coefficient $D_{20,w}$ for apo-cytochrome c at 20°C in three different solvents:

X 20 mM Na-phosphate/HCl, pH 2.3,

□ 20 mM Na-phosphate/HCl, pH 2.3, 0.25 M NaCl,

○ 20 mM Na-phosphate/HCl, pH 2.3, 6.65 M GuHCl.

coefficients are 1.5 and 1.1, respectively). But nevertheless, these findings are a warning to interpret the size increase S carefully.

Table 4. Results for thermal unfolding of three selected proteins. The thermal transition was complete at the indicated temperatures.

Protein	T(°C)	R_S^N (nm)	R_S^T (nm)	S	ρ^T
H.α-lactalbumin (20 mM Tris-HCl; pH 7.5; 10 mM EDTA)	50	1.77	1.96	1.11	0.79
Streptokinase (10 mM; Na-phosphate; pH 7.5)	65	3.58	4.12	1.15	1.57
Lysozyme (0.2 M KCl; pH 1.45)	60	1.85	2.18	1.18	-

The values of R_g^T used to calculate ρ are not shown.

The dependence of the dimensions of the disordered protein chain on solvent conditions

The theoretical relation between the radius of gyration and the Stokes' radius for linear disordered chains ($R_g = 1.51\ R_S$), has not been verified experimentally for protein molecules until now. Deviations from this relation could be expected for protein chains with n_A of the order of 100. To check this we have chosen apo-cytochrome c (ACC) because this protein is known to be completely unfolded at acid pH: it has a high intrinsic viscosity[19], its CD-[19] and NMR-spectra[20] are typical for unfolded proteins and it does not melt upon heating[20]. The preparation of holo- and apo-cytochrome c from horse heart is described elsewhere[19,21].

The dimensions of ACC have been evaluated at pH 2.3 in three solvents listed in Table 5, namely in water at low and high ionic strength and in the presence of 6 M GuHCl. The concentration dependences of the translational diffusion coefficient D for all three solvent conditions are shown in Fig. 2. The Stokes' radii from the extrapolation to zero concentration and further results are given in Table 5.

Table 5. Solvent dependence of the dimensions of apo-cytochrome c at pH 2.3 and 20°C.

	Solvent	R_S (nm)	R_g (nm)	ρ_{expt}	α	
					A	B
I	20 mM Na-phosphate, 10 mM mercaptoethanol	3.02	4.68	1.55	1.15	1.24
II	Solvent I + 0.25 M NaCl	2.54	4.07	1.60	1	1.08
III	Solvent I + 6 M GuHCl	2.72	(4.22)*	-	1.04	1.12

* Calculated using ρ = 1.55.
The values of α have been calculated for: (A) $R_{g,0}$ = 4.07 nm; and (B) $R_{g,0}$ = 3.77 nm, respectively.

Apparent (*i.e.* not extrapolated to zero concentration) Stokes' radii and radii of gyration are shown in Fig. 3 for ACC at low ionic strength. In accordance with the common representation of the concentration dependence of the translational diffusion coefficient, we have drawn the reciprocals of both. We have scaled R_S^{-1} in such a way that the values fit R_g^{-1} from SAXS. From this we get R_g/R_S = 1.55, which corresponds to the experimental ρ-factor. This is very close to the theoretical value 1.51.

We have calculated the molar mass and the second virial coefficient A_2 from TILS and SAXS data. The results are M = 12,800 g mol^{-1} and A_2 = 8.6x10^{-3} mol ml g^{-2} from light scattering and M = 11,200 g mol^{-1} and A_2 = 7.8x10^{-3} mol ml g^{-2} from SAXS for ACC at low ionic strength. The large positive second virial coefficient and the pronounced concentration dependence of R_S and R_g (Fig. 3) reflect strong intermolecular interactions.

At high ionic strength the dimensions of the protein chain are considerably reduced but, according to the experimental ρ-factor of 1.60 the protein chain has still the conformation of a linear disordered polymer. A_2 measured by light scattering was rather small so that its precise value has not been evaluated.

At 6 M GuHCl only R_S has been determined by DLS. Its value is between that for water at low and high ionic strength. A ρ value of 1.55 was used to calculate R_g.

We will now try to find a quantitative relation between the experimentally determined and the unperturbed dimensions. From the second virial coefficient and the concentration dependence of D (Fig. 2) it is reasonable to expect theta-point conditions, and therefore $\alpha = 1$, in the high salt concentration region. We will now consider two particular cases.

At first we assume that our experimental conditions at the ionic strength $\mu = 0.27$ (solvent I + 250 mM NaCl) fit the theta-conditions. Accordingly, we get for the unperturbed radius of gyration $R_{g,0} = 4.07$ nm. This value of $R_{g,0}$ can be related to the number of amino acids $n_A = 104$ by the equation

$$\langle R^2 \rangle = 6R^2_{g,0} = 96\, n_A$$

With this assumption we get the lower values of α in Table 5. A lower limit of R_g is probably the value from an extrapolation to infinite ionic strength $R_{g,ex}$. This can be obtained using the formula[22,23]

$$R^3_g = R^3_{g,ex} + B/\sqrt{\mu} \qquad (7)$$

where μ is the ionic strength and B is a constant. Taking $R_{g,ex}$ as $R_{g,0}$ we find $R_{g,0} = 3.77$ nm which fits the relation

$$\langle R^2 \rangle = 6R^2_{g,0} = 82\, n_A$$

The corresponding values of α for our experimental solvent conditions are also shown in Table 5. Note that in both cases we find an equation which is very similar to the theoretical relation (Eq. 5).

The compactness of the 'molten globule' state of human alpha-lactalbumin

In the results presented in the preceding sections data for human α-lactalbumin have been included. Note particularly the states found at acid pH and at high temperatures (see Table 3 and Table 4 or refs. 24,29,30). For a more complete representation Fig. 4 shows the temperature dependence of the Stokes' radius R_S and of the relative size increase S. Fig. 4 illustrates the close similarity of the acid and

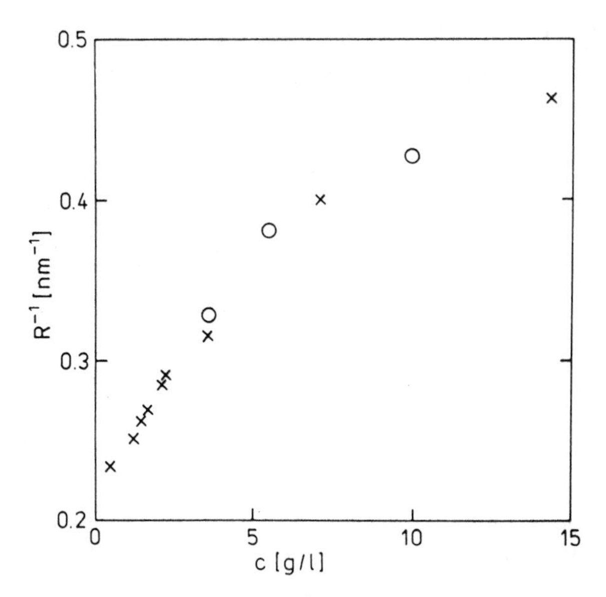

Figure 3. Dependence of R_g^{-1} (O) and scaled values of R_S^{-1} (X) on concentration for apo-cytochrome c at low ionic strength (μ= 0.025) and 20°C. The scaling factor corresponds to a ρ-factor of 1.55.

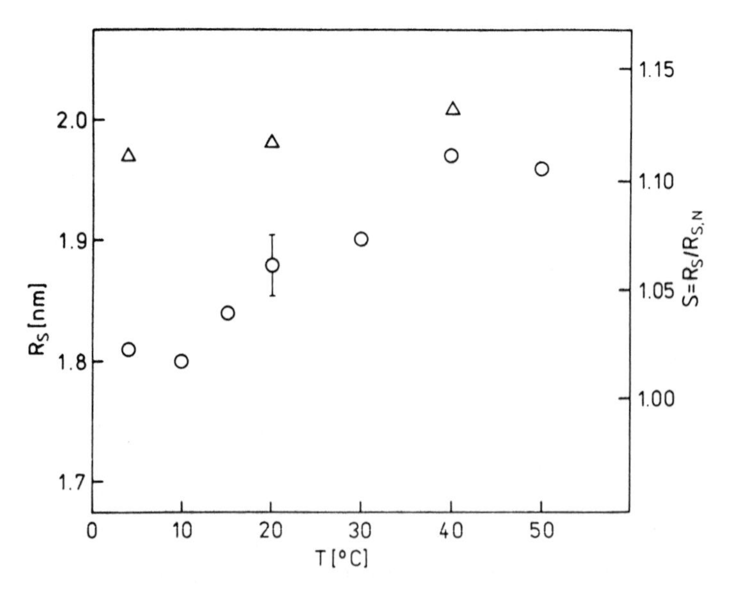

Figure 4. Temperature dependence of R_S and the relative size increase $S = R_S/R_{S,N}$ for the Ca-free (O) and the acid (Δ) forms of human α-lactalbumin. The Ca-free form was used in order to decrease the thermostability of the native protein.

temperature denatured forms. According to this the conclusion can be drawn that both forms belong to the same conformational state which has a Stokes' radius about 10 % larger than the native state: this was indeed confirmed by various other physical methods. This state was termed the 'molten globule' state. First of all it was observed during unfolding experiments of lactalbumin but more and more results have been published giving evidence that the 'molten globule' state is a general intermediate in protein folding. Reviews concerning this subject have been published by Ptitsyn[25] and Kuwajima[26]. On the basis of the circular dichroism spectra in the far and near UV regions[27] general features of the 'molten globule' can be described very well. The native-like CD spectrum in the far uv region hints at a nearly intact secondary structure. In contrast, the small CD amplitude in the near UV region compares favourably with that of the totally unfolded protein. This implies that the tight packing of aromatic side groups is lost reflecting the absence of a specific tertiary structure.

The 'molten globule' state is considered as an specific intermediate state on the non-random pathway of protein folding. The detection and the investigations of such intermediate states are very important for a better understanding and the formulation of particular models for the pathway of protein folding.

ACKNOWLEDGEMENT

The authors are very grateful to Prof. O.B. Ptitsyn for many stimulating suggestions and for fruitful discussions.

REFERENCES

1. C. Tanford, *Advan. Protein Chem.*, 1968, *23*, 121.
2. P.J. Flory, 'Principles of Polymer Chemistry', Cornell Univ. Press, Ithaca, N. Y., 1953.
3. H. Yamakawa, 'Modern Theory of Polymer Solutions', Harper and Row, New York, 1971.
4. J.G. Kirkwood and J. Riseman, *J. Chem. Phys.*, 1948, *16*, 565.
5. W. Burchard, *Advan. Polymer Science*, 1983, *48*, 1.
6. C. Tanford, 'Physical Chemistry of Macromolecules', Wiley and Sons, New York, 1961.
7. V.A. Bloomfield, W.D. Dalton and K.E. Van Holde, *Biopolymers,* 1967, *5*, 135.

8. S.E. Harding in: 'Dynamic Properties of Biomolecular Assemblies' (S.E. Harding and A.J. Rowe, eds.) Chap. 2, Royal Society of Chemistry, Cambridge, 1989.
9. S.W. Provencher, *Comp. Phys. Commun.*, 1982, *27*, 213.
10. G. Damaschun and L.A. Feigin, *studia biophysica*, 1986, *112*, 110.
11. J.J. Mueller, T.N. Zalkova, G. Damaschun, R. Misselwitz, I.N. Serdyuk and H. Welfle, *studia biophysica*, 1986, *112*, 151.
12. S.B. Dubin, G. Feher and G.B. Benedek, *Biochemistry*, 1973, *12*, 714.
13. D.A. Brandt, W.G. Miller and P.J. Flory, *J. Mol. Biol.*, 1967, *23*, 47.
14. W.G. Miller, D.A. Brandt and P.J. Flory, *J. Mol. Biol.*, 1967, *23*, 67.
15. W.G. Miller and C.V. Goebel, *Biochemistry*, 1968, *7*, 3925.
16. C. Tanford, K. Kawahara and S. Lapanje, *J. Biol. Chem.*, 1966, *241*, 1921.
17. D.F. Nicoli and G. Benedek, *Biopolymers*, 1976, *15*, 2421.
18. M.E. McDonnell and A.M. Jamieson, *Biopolymers*, 1976, *15*, 1283.
19. W.R. Fisher, H. Taniuchi and C.B. Anfinsen, *J. Biol. Chem.*, 1973, *248*, 3188.
20. J.S. Cohen, R.I. Shrager, M. McNeel and A.N. Schechter, *Nature*, 1970, *228*, 642.
21. E. Margoliash and O.F. Walasek, *Methods in Enzymol.*, 1967, *10*, 339.
22. A.M. Elyashevich and O.B. Ptitsyn, *Dokl. Akad. Nauk SSSR*, 1964, *156*, 1154.
23. M. Tricot, *Macromolecules*, 1984, *17*, 1698.
24. K. Gast, D. Zirwer, H. Welfle, V.E. Bychkova and O.B. Ptitsyn, *Int. J. Biol. Macromol.,* 1986, *8*, 231.
25. O.B. Ptitsyn, *J. Protein Chem.*, 1987, *6*, 273.
26. K. Kuwajima, *Proteins,* 1989, *6*, 87.
27. D.A. Dolgikh, R.I. Gilmanshin, E.V. Brazhnikov, V.E. Bychkova, G.V. Semisotnov, S.Yu. Venyaminov and O. B. Ptitsyn, *FEBS Lett.*, 1981, *136*, 311.
28. 'Handbook of Biochemistry', Selected data for Mol. Biology, (ed. H. A. Sober), The Chem. Rubber Co., 1970.
29. G. Damaschun, Ch. Gernat, H. Damaschun, V.E. Bychkova and O.B. Ptitsyn, *Int. J. Biol. Macromol.*, 1987, *8*, 226.
30. A.A. Timchenko, D.A. Dolgikh, H. Damaschun, G. Damaschun, *studia biophysica*, 1986, *112*, 201.

15

Dynamic Light Scattering Studies of Concentrated Casein Micelle Suspensions

By David S. Horne

HANNAH RESEARCH INSTITUTE, AYR. KA6 5HL SCOTLAND

1. INTRODUCTION

Quasi-elastic light scattering (QLS) is extensively used in colloid and polymer science - as well as increasingly in biology - as a powerful probe of the dynamics of particle motion. In the study of dilute suspensions QLS is probably the simplest and quickest method of measuring sub-micron particle sizes. This simplicity demands, however, that QLS experiments be performed in the strictly single-scattering limit. As the concentration of scatterers increases, a multiple-scattering regime rapidly obtains and the system becomes effectively opaque. Significant practical and theoretical limitations have prevented meaningful interpretation of QLS data taken in such circumstances. On those occasions where concentrated systems have been studied, experimental constraints such as short pathlength cells[1] or contrast-matching of refractive indices to reduce the scattering to that of a few tracer particles[2] have been employed. Similarly, available theories to tackle multiple scattering have only been applicable to situations where relatively weak multiple scattering contributions existed so that effectively single scattering with minor corrections was being considered. Recently, however, it has been demonstrated that QLS measurements taken in optically thick media can be analysed to yield particle size values by exploiting the diffusive nature of light transport in concentrated dispersions[3-5]. The relatively trivial problem of conveying light through the sample between source and detector we have solved by the use of fibre optic light guides operating in a back-scattering geometry[5].

The technique has come to be known as diffusing wave spectroscopy (DWS). In the following sections we describe briefly its underlying theoretical background and the verification of its applicability to our experimental configuration using monodisperse polystyrene latices as a model colloid system. The main theme of the paper is, however, its application in a practical polydisperse suspension: the casein micelle system of bovine milk. We first of all consider the employment of DWS as a particle size measuring technique in undiluted systems, its sensitivity to the interaction properties of the casein micelles as evidenced by the concentration dependence of the light scattering behaviour and finally, its use as as a monitor of curd formation through the observation of hindered motion and decreased mobility as the gel network is fashioned.

2. THEORY

As in all QLS experiments, the motion of the particle is probed by monitoring the time dependence of the fluctuations of the scattered light. Using autocorrelation techniques these fluctuations are translated into an intensity correlation function whose temporal decay is non-exponential[3]. To calculate this function, the light, multiply-scattered by a random distribution of particles, is assumed to execute a random walk through the sample, the detected light intensity being the incoherent sum of contributions scattered through all possible paths[6]. This approximation is valid for the transport of light over distances which are large compared to the diffusive mean free path, l^*, of the light. The non-exponential character of the correlation function betrays the existence of multiple time constants, indicative of the differing decay rates of the paths of various lengths. The longest paths with the greatest number of scattering events decay most rapidly, while the slowest decay rate is that of single scattering characterised by the time, τ, required for a scattering particle to move by one wavelength. The summation of intensity over all possible light diffusion paths is highly influenced by the geometrical configuration of the measuring system. Size, shape and positioning of source, sample and detector all play a part. The overall problem then reduces to the solution of a diffusion equation subject to certain boundary conditions.

All our light scattering measurements on latices and milks were carried out in backscattering mode. A more extensive description of the apparatus has been given previously[5]. The instrument is based on a Malvern digital autocorrelator and incorporates a bifurcated optical fibre bundle as light guide. Half the fibres, distributed randomly over the face of the common leg, carry light from a randomly polarised He-Ne laser. When dipped into the scattering suspension, the other half of the fibres carry backscattered light to the photomultiplier. Their masking by slit and pinhole ensures that light from only a small area impinges on the detector.

This experimental configuration approximates to an extended plane wavelight source incident on the front face of a slab of thickness L (the depth of the solution) and effectively infinite extent with the scattered light collected at a point on the same front face. Pine *et al.*[4] have shown that under such experimental geometry constraints the electric field correlation function for the multiply scattered light can be written as

$$g^{(1)}(t) = \frac{1}{1 - (\gamma l^*/L)} \left\{ \frac{\sinh \{(L/l^*)\,(6t/\tau)^{1/2}\,(1 - \gamma l^*/L)\}}{\sinh \{(L/l^*)\,(6t/\tau)^{1/2}\}} \right\} \quad (1a)$$

$$\rightarrow \exp \{-\gamma\,(6t/\tau)^{1/2}\} \text{ for } L \gg l^* \quad (1b)$$

where l^* is the diffusive mean free path for light in the suspension. The parameter γ is introduced as a multiplicative factor to locate the conversion point within the sample from which the passage of light can be considered diffusive, providing a physically meaningful boundary condition for the analytical solution of the diffusion equation. A more rigorous treatment of the transition from ballistic to diffusive light transport predicts the value of γ to depend on both the polarization and anisotropy of the scattering[6]. Experiment confirms this polarization memory effect together with a particle size dependence reflecting single scattering anisotropy[7]. Parallel polarization shows γ increasing with size whilst perpendicular polarization shows the converse, both asymptotically approaching ~2.0 as particle size increases above ~100 nm. Because our apparatus employs a bundle of multimode fibres as light guide, each fibre presents to the sample light at a random polarization angle. Parallel and perpendicular effects cancel and hence a constant value of γ equal to the average of ~2.0 is obtained.

When infinite depth of solution is employed, the derived equation predicts that a linear plot should be obtained when the logarithm of the correlation function is plotted against the square root of the delay time. Because of fortuitous cancellation when the sinh functions are expanded, the slope of this plot is independent of the diffusion length, l^*, yielding the single scattering delay time, τ, when γ is known. Reducing the depth of solution introduces curvature into the plot at shorter delay times due to the loss of the longer light paths. Measurements of the intensity correlation function made on dispersions of monodisperse polystyrene latices and using our fibre optic apparatus confirm these predictions[5]. The thicker or more concentrated samples gave linear plots (Fig. 1), the thinner or less concentrated ones showed curvature at low values of $\sqrt{t}$, their long time tails running parallel to the straight line behaviour of their thicker counterparts. Using relaxation times calculated from the straight line behaviour and values of l^* and γ obtained independently from transmission measurements, it is found that eq. 1a gives an excellent fit to all curves (Fig. 1).

3. CASEIN MICELLE SUSPENSIONS

It was suggested above that the relaxation time τ characterised the slowest decay rate observed, corresponding to single scattering and was the time required for a particle to move by a single wavelength. It is therefore equivalent to a diffusional lifetime. The questions we now ask are, "Does it possess the characteristics expected of such a lifetime? How does it vary with particle size? Does it obey the Stokes-Einstein relation in its variation with medium viscosity? How is it influenced by particle-particle interactions?" These are some of the questions considered in the application of this diffusing wave spectroscopy technique to the practical example of the casein micelle system of skim milk. Milk is often described as an emulsion which owes its whiteness to the scattering power of the fat droplets. While it is undoubtedly true that these do scatter light, it is equally true that they are not the only light-scattering entities in milk. This is most forcefully demonstrated by the observation that skim milk, milk from which the fat has been removed by centrifugation, is also densely white. It is noticeably a bluer white indicating that the scattering particles, for it is a particle dispersion, are smaller than the average in the original whole milk. These scattering particles are casein micelles, aggregates of the casein family of phospho-proteins with calcium phosphate. They are very polydisperse, ranging in diameter from 50 nm to 500 nm, with an average diameter of about 200 nm [8].

A preliminary study of the light transmission of fresh, undiluted skim milk in a variable pathlength apparatus yielded a value of 0.54 ± 0.01 mm for the diffusion pathlength, l^*. This value

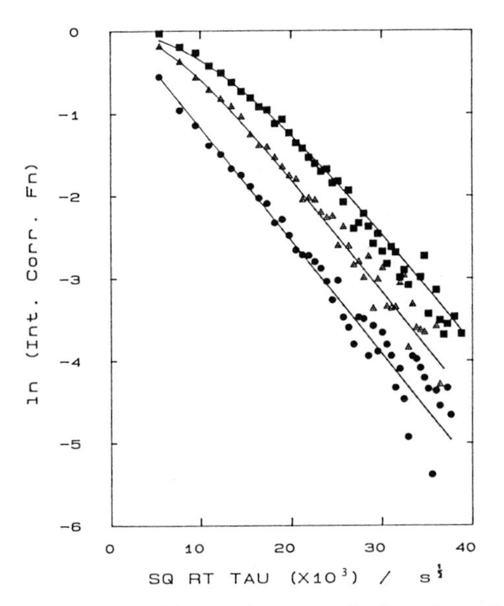

Figure 1. Variation of intensity correlation functions obtained from light scattering of 330 nm polystyrene latex (ϕ~0.01) with depth, L, of suspension. Depths were: ● 2 mm; Δ 0.5 mm; ■ 0.25 mm. Full curves are fits of Eq. 1a to the data using l^*=0.217, γ = 0.9 and τ = 9.614 x 10^{-4} throughout.

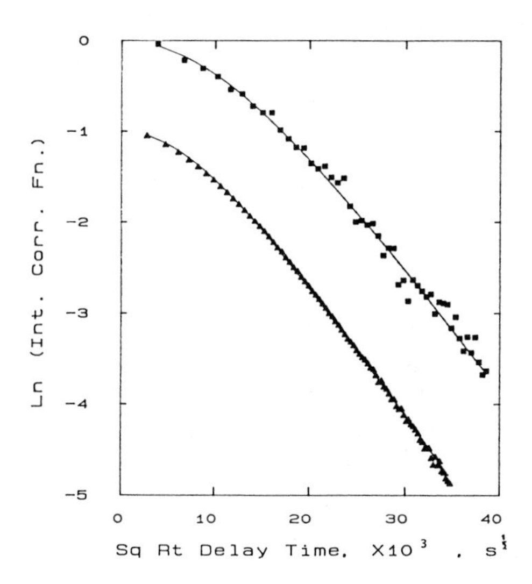

Figure 2. Semi-log plot of intensity correlation function from undiluted skim milk against square root of delay time. Depths were 1 mm (■) and 100 mm.(▲). Full curves are fits of Eq. 1a to the data using l^*/L = 0.6 (1 mm), and 0.51 (100 mm) γ = 0.9 and τ = 1 x 10^{-3} s for both.

of l^* was confirmed by fitting the autocorrelation functions obtained using slabs of milk up to 1 mm thick to Eq. 1a. Such correlation functions, shown in Fig. 2, behaved similarly to those obtained with similar concentrations of polystyrene latex and rolled over at short delay times owing to the finite thickness of the slab cutting out the longest paths of the diffusing light, the paths which decay fastest and on the shortest timescales. For still thicker slabs up to 100 mm the correlation functions obtained for skim milk continued to roll over at short delay times, unlike the polystyrene latex scattering behaviour where linear plots were obtained with these thicknesses. The failure to achieve straight-line behaviour at depths of milk two orders of magnitude greater than the light diffusional pathlength suggested that some other factor in our experimental configuration was limiting access to the longest pathways for diffusing light. Allowing the ratio l^*/L to float in the fitting procedure for these functions from thicker slabs of milk produced a best fit value of 0.5 for the ratio. Incorporating the value of l^* obtained independently gave a value of 1.1 mm for the critical length L. This is close to the 1.5 mm radius of the common leg of the optical fibre bundle employed, suggesting that a finite source radius is a plausible cause of the truncated diffusional pathways. Using reconstituted milk solutions at concentrations ranging from well below to up to four times that of normal milk, the derived ratio, l^*/L, decreased linearly with reciprocal concentration. With L fixed by the geometry of the apparatus, the diffusional pathway was thus inversely proportional to the particle number concentration, as expected theoretically and as confirmed previously for polystyrene latices[5].

Relaxation Time and Viscosity

At long delay times, the autocorrelation function from undiluted milk behaves asymptotically as the single stretched exponential form given by Eq.1b. From this segment of the plot, assuming the value of γ to remain unchanged at 2.0, we can derive a value of the relaxation time, τ, for the milk sample. Setting the time constant equal to $(Dk^2)^{-1}$ and $k = 2\pi/\lambda$, a diffusion coefficient, D, would be calculable. The Stokes-Einstein relation predicts this diffusion constant to be inversely proportional to both the viscosity of the suspending medium and to the particle radius. The viscosity of skim milk is close to that of water and for the purposes of this exercise is taken as equal to it. Tabulated values of water viscosity as a function of temperature are available[9]. Measuring the autocorrelation functions of skim milk at various temperatures between 5°C and 70°C allows a diffusional mobility to be calculated and as demonstrated in Fig. 3 this shows the expected linear relation to reciprocal viscosity.

Relaxation Time and Particle Size

The average casein micelle size measured in dilute suspensions using conventional QLS is not a constant quantity but varies between milk samples. This is particularly true for the milks of individual cows, where variations of up to a factor of two can be found, depending on feeding regime, season and stage of lactation[10]. With no particular size selection process in mind, we have measured the average hydrodynamic diameters of casein micelles in milks from a group of individual cows in the Institute herd. The measurements were made at a scattering angle of 90° only, diluting 5µl of each milk into 3ml of a buffer solution in which micelle integrity was known to be preserved. Average diffusion coefficients were determined from the dynamic scattering data by the cumulants method[11] and from these particle diameters were calculated using the Stokes-Einstein relation for spheres. Measurements of the dynamic light scattering behaviour of the undiluted milks were also made using our fibre-optic backscattering apparatus. The relaxation times calculated from the long-time tails of these functions are plotted against the QLS average sizes in Fig. 4: the relaxation time obviously increases with particle size, as predicted. The correlation coefficient of a straight-line fit to the data points is, however, only 0.85. This may be partly due to the small size range covered by the measurments under consideration, this tending to amplify the short-comings of the technique. Over a much wider range of monodisperse polystyrene latices, an excellent linear relation between relaxation time and particle size was observed[5].

Relaxation Time and Polydispersity

In the case of casein micelles, as in many systems of practical interest the particles are not all of the same size or optical composition. As we have demonstrated above this does not prevent us from gaining average size information from DWS measurements, but what average are we measuring or how is that size value influenced by the particle size distribution? In the case of conventional QLS in dilute solutions, polydispersity of size leads to a non-exponential decay of the intensity correlation function. In the presence of strong multiple

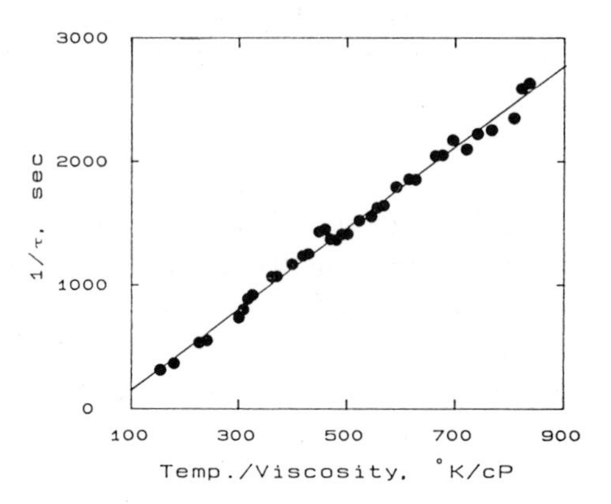

Figure 3. Variation of micellar diffusional mobility (τ^{-1}) with reciprocal viscosity of aqueous suspending medium. Changes in viscosity were achieved by adjusting sample temperature between 0°C and 70°C. Relaxation times were calculated from the slopes of the tail regions of plots similar to those of Fig. 2, using Eq. 1b.

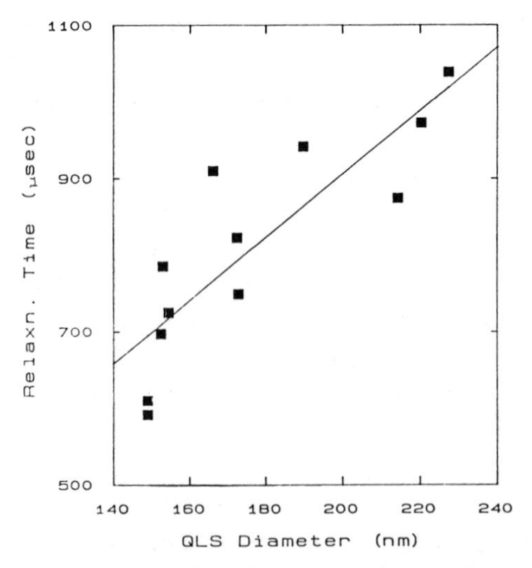

Figure 4. Diffusional relaxation times for casein micelles of individual cow milk samples plotted against the average hydrodynamic diameters measured in dilute solution using conventional QLS equipment at a scattering angle of 90°. The straight line is a linear regression fit to the data points.

scattering, the decay of the correlation function in a monodisperse system is already non-exponential and geometry dependent. This makes the precise consequences of polydispersity more difficult to discern from the data than in QLS.

Pine *et al.* have demonstrated that the form of the correlation function can be calculated if the temporal dephasing of the scattering over a pathlength of n steps can be formulated[12]. This dephasing occurs by a random walk of phase shifts as a result of the motion of individual particles along a particular scattering path. For a polydisperse system where the average magnitude of phase shift depends on particle size, this results in a random walk with different step sizes. The central limit theorem tells us that such a random walk of n different length steps is no different in result from a walk of n steps with a single average step size. In consequence, the shape of the autocorrelation function from a polydisperse system should be the same as for a monodisperse one. In DWS, there can be no information about polydispersity in the measurements. Only average properties are measurable.

Using bimodal mixtures of polystyrene latices of sufficient depth and concentration, Horne[5] confirmed the linearity of the correlation function plots obtained in backscattering mode and derived an empirical form for the appropriate weighting factor to be used in calculating the effective slope from the individual contributions. This weighting factor, $W_i = C_i r_i^2$, or concentration times area, is identical in form to that derived by Pine *et al.*[12], from consideration of the diffusional model for light scattering. They showed the weighting factor for each individual species to be the inverse transport mean free path *i.e.* number density times scattering cross-section. In their theory, however, this weighting factor was applied to individual particle diffusion coefficicients in calculating an average effective particle diffusion coefficient whereas Horne's empirical approach[5] considered the slope of the semi-log plot of the correlation function against the square root of delay time, which is proportional to the square root of D. Data of higher quality are required to differentiate between these results.

It is emphasised that, unlike QLS, DWS cannot provide additional information on degree of polydispersity or size distribution in a sample. Average quantities can be calculated from a known or assumed distribution function, allowing the size measured by DWS to be related to values obtained by other techniques.

Relaxation Time and Particle Interactions

The relaxation time, τ, characterizes the time for a single particle to move by a distance of one wavelength. In a concentrated dispersion, it is therefore to be anticipated that this motion will be influenced by the particle's interactions with its neighbours. It was therefore perhaps somewhat surprising that Maret and Wolf[3] found this diffusional relaxation time for polystyrene latex particles to be equal to the infinite dilution free particle value. Up to volume fractions of 0.1 they and other groups[4,5] found τ independent of concentration for these monodisperse systems. This lack of sensitivity to the interaction effects expected to influence particle diffusive motion was explained by Maret and Wolf as a counterbalancing of hydrodynamic and structure factor effects. For milk solutions, however, we find relaxation time to increase as the micellar concentration is increased, indicating decreased micellar mobility. When these data are normalised to their extrapolated infinite dilution value, the decay of the normalised reciprocal relaxation time with micellar volume fraction superimposes on the behaviour of the long-time self-diffusion coefficient observed by several other groups[13-16] in studies using tracer techniques (Fig. 5). Diffusing wave spectroscopy is measuring dynamic light scattering behaviour sensitive to the local motions of single particles which at higher volume fractions require the accommodating movements of their neighbours to diffuse an appreciable distance.

Relaxation Time and Gel Formation

As noted above, in more and more concentrated micellar suspensions, relaxation time is lengthened as particle motion is hindered and slowed. A logical progression from here would be to entrapment and immobility when the particle becomes part of a gel network. At this stage, we should distinguish between true entrapment where, say, a latex particle is caught up in a biopolymeric gel of, say, gelatin and the case where the particle itself becomes part of the network. The latter scenario is closer to the situation believed to exist when gelation of casein micelles is induced either by acidification of milk or by proteolysis with chymosin, or by a combination of both methods.

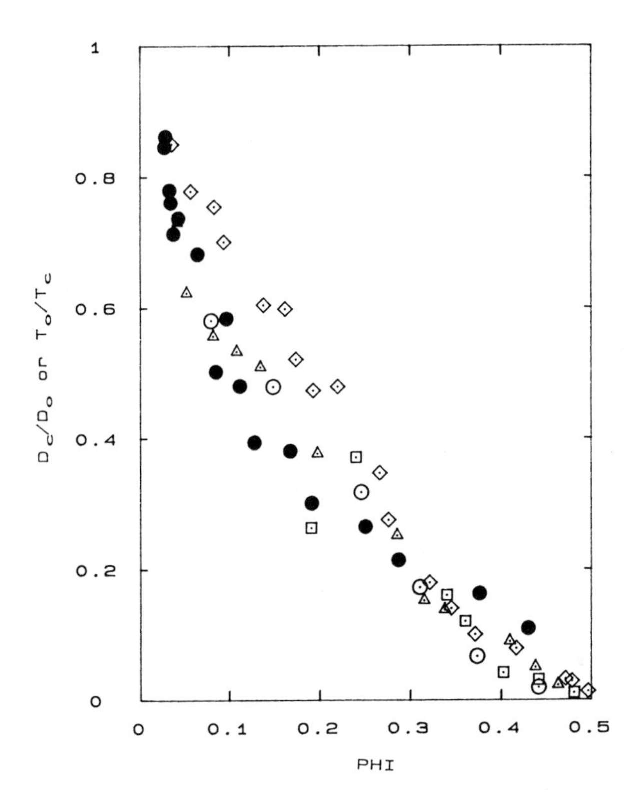

Figure 5. Normalized reciprocal relaxation time as a function of casein micelle volume fraction (●) measured in reconstituted milk powder solutions. Tracer diffusion measurements of the long time self-diffusion coefficient (D_c/D_0) from references 14(◊), 15(□), 16(Δ) and 17(O) are also shown for comparison.

A complete discussion of the milk clotting process is beyond the scope of this article. Briefly enzymatic clotting occurs in a series of overlapping stages designated primary, secondary and tertiary. During the primary stage, the κ-casein of the micelle is attacked at a specific bond, Phe_{105}-Met_{106} to give two peptides. The loss of the hydrophilic carboxyl peptide which then diffuses away from the micelle leads to a progressive destabilization of the micelle as the proteolysis of its κ-casein content continues. The subsequent aggregation of the destabilized micelles through to curd or gel formation then constitutes the secondary stage of the process. In the overall clotting reaction, these stages are not, strictly speaking, separable since aggregation is initiated before the conversion of κ-casein substrate is completed by the enzyme. It has, however, been demonstrated that normally a conversion of between 85% and 90% is required before significant aggregation is observed[17].

Following the addition of enzyme to the milk, the changes in relaxation time as the reaction proceeds are illustrated in Fig. 6. The plot shows a lag phase where τ remains constant at its untreated milk value. At the end of this period, designated a clotting time, the relaxation time increases rapidly by several orders of magnitude before levelling off. This clotting time coincides with the onset of visible clot formation in a separate test tube assay. As a critical time in the progress of the reaction, this coagulation time can be usefully related to many reaction parameters and its variation studied as a function of enzyme and substrate concentrations, system pH or temperature, for example. This light scattering technique is intrinsically more useful than this rudimentary analysis suggests, in that as a non-destructive measurement it provides a continuity of output from which may be derived fundamental kinetic and mechanistic information starting from the earliest stages of the gelation process.

Further interesting relaxation time behaviour is encountered during combined enzyme/acid induced gelation of skim milk, as exemplified in Fig. 7. For this reaction starter bacteria and a low level of chymosin were added at time zero. As shown relaxation time exhibits a maximum. The rates of rise and fall, the final value of τ, are all variable depending on reaction conditions. If particle motion is considered to be governed by the resultant of viscous and elastic forces, then the relaxation time can be expressed simply as the ratio of viscosity to elastic modulus *i.e.* $\tau=\eta/E$. As aggregation begins, the

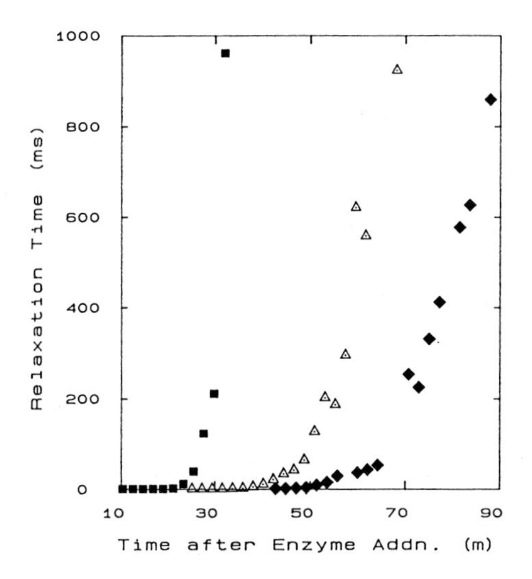

Figure 6. Changes in relaxation time, τ, due to curd formation following the addition of chymosin to milk at time zero. Data are shown for milks at pH 6.7(♦) and adjusted to pH 6.5(Δ) and 6.3(■).

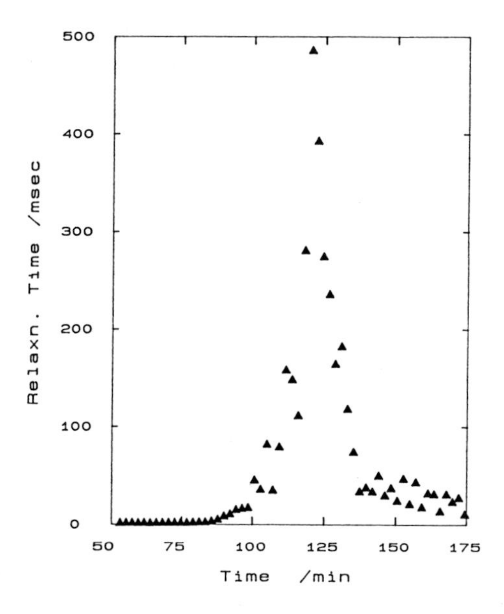

Figure 7. Variation of relaxation time during combined rennet and acid-induced gelation of skim-milk. Reagents again added to milk at time zero.

aggregate meets growing frictional resistance to its motion and an increasing viscous component becomes apparent as the relaxation time diverges from the fluid value. Later in the reaction, as gel network formation dominates, the elastic forces come to the fore and relaxation time decreases. It therefore appears that observation of DWS behaviour in these gelling systems, by virtue of its measurement of relaxation in the system, must eventually reflect changes in viscoelastic properties. Should this approach prove fruitful, light scattering possesses the major advantage over traditional rheological techniques in that it is non-perturbing. No mechanical action is employed and only the results of spontaneous thermal fluctuations are measured, so that the viscoelastic properties of the weakest gels should be accessible with hitherto unparallelled sensitivity.

4. CONCLUSIONS

Diffusing wave spectroscopy is opening up the application of dynamic light scattering techniques for the analysis of optically dense suspensions. Our interest in the technique obviously stems from its possible uses with milk systems where doubts concerning intrinsic stability of casein micelles to dilution have cast shadows over conventional QLS measurements. The ability of this approach to provide calibrated "average" micelle sizes, to monitor micelle size during processing and to probe the transition from fluid to gel clearly demonstrates its practical benefits in this and many other fields where like systems are encountered.

REFERENCES

1. P.G. Cummins and E.J. Staples, *J. Phys. E: Sci. Instrum.*, 1981, *14*, 1171.
2. P.N. Pusey and W. Van Megen, *J. Physique*, 1983, *44*, 285.
3. G. Maret and P.E. Wolf, *Z. Phys. B*, 1987, *65*, 409.
4. D.J. Pine, D.A. Weitz, P.M. Chaikin and E. Herbolzheimer, *Phys. Rev. Lett.*, 1988, *60*, 1134.
5. D.S. Horne, *J. Phys. D: Appl. Phys.*, 1989, *22*, 1257.
6. F.C. MacKintosh and S. John, *Phys. Rev. B*, 1989, *40*, 2383.
7. F.C. MacKintosh, J.X. Zhu, D.J. Pine and D.A. Weitz, *Phys. Rev. B*, 1989, *40*, 9342.
8. T.C.A. McGann, W.J. Donnelly, R.D. Kearney and W. Buchheim, *Biochim. Biophys. Acta*, 1980, *630*, 261.

9. 'CRC Handbook of Chemistry and Physics', CRC Press, Boca Raton, Florida, USA, 60th Edn., 1979, p. F-51.
10. C. Holt and L. Baird, *J. Dairy Res.*, 1978, *45*, 339.
11. D.E. Koppel, *J. Chem. Phys.*, 1972, *58*, 4814.
12. D.J. Pine, D.A. Weitz, G. Maret, P.E. Wolf, E. Herbolzheimer and P.M. Chaikin, 'Scattering and Localization of Classical Waves in Random Media', (Ping Sheng ed.), World Scientific, Singapore, 1990, p.312.
13. W. Van Megen and S.M. Underwood, *J. Chem. Phys.*, 1989, *91*, 552.
14. M.M. Kops-Werkhoven and H.M. Fijnaut, *J. Chem. Phys.*, 1982, *77*, 2242.
15. R.H. Ottewill and N.St.J. Williams, *Nature*, 1987, *325*, 232.
16. A. Van Veluwen and H.N.W. Lekkerkerker, *Phys. Rev. A*, 1988, *38*, 3758.
17. D.G. Dalgleish, *J. Dairy Res.*, 1979, *46*, 653.

16

Aggregation of Proteins

By J.G. Rarity, P.C.M. Owens, T. Atkinson[1], R.N. Seabrook[1] and R.J.G. Carr[1]

ROYAL SIGNALS AND RADAR ESTABLISHMENT, ST. ANDREWS ROAD, MALVERN, WORCESTERSHIRE. WR14 3PS U.K.

[1]DIVISION OF BIOTECHNOLOGY, PHLS CENTRE FOR APPLIED MICROBIOLOGY AND RESEARCH, PORTON DOWN, WILTSHIRE. SP4 0JG U.K.

1. INTRODUCTION

The attraction of fractal theories of nature is that the application of large scale geometrical rules can lead to simple universal properties in a variety of complex random growth processes. One such process that has been popular with the colloid scientist since the days of the Faraday gold sol is the phenomenon of colloidal aggregation[1-4]. In recent years there has been an upsurge of interest in the study of aggregation due to the discovery of the fractal nature of the aggregates produced in these reactions[5-7]. Colloids, being suspensions of small particles in a fluid, are thermodynamically unstable and only survive because of strong repulsive forces between the particles. Aggregation is thus inevitable when repulsive barriers are reduced, by a variety of means specific to the constituents of the colloid. The large scale rule which leads to fractal properties of the aggregates is simply the geometrical limitations of close diffusive approach of two irregularly shaped objects. If particles stick on first contact and the bonds are stiff we expect quite tenuous structures to evolve. If there is a low sticking probability per contact the aggregates will have time to explore the contact surface and possibly interpenetrate somewhat before sticking leading to less tenuous structures. The fractal nature of the resulting structure arises because each successful collision leads to a drop in the mean density of the aggregate and, on average, we can relate the mass contained in the aggregate M to a measure of its radius R by a power law

$$M \propto R^{d_f} \qquad (1)$$

where d_f is the fractal dimension. d_f can range in value between 3 for space filling objects (or aggregates whose density does not drop with mass) to 1 in the case of extremely tenuous (rod like) structures.

Here we study the aggregation of a variety of antibodies in the presence of their respective antigens. For systems where large aggregates form we find that the aggregate structure is a random fractal and measure the fractal dimension d_f by multi-angle static light scattering.

2. THEORY

A fractal object is characterised by a non-integer power law form for the pair correlation function g(r) which measures the probability of finding material at a point r given material at r = 0. For an object embedded in three dimensional space

$$g(r) - 1 \propto r^{d_f - 3}\, \Psi(-r/R) \qquad (2)$$

where d_f is the correlation fractal dimension. Random clusters have a finite extent hence the equation includes a cut-off function dependent on the cluster radius. An integration of this function over all space gives us the relation in eq. 1. Evidence for the fractal nature of random aggregates was originally obtained[5,6] from the power law dependence of scattered intensity J(q) on the modulus of the scattering vector **q** measured when average cluster size is large.

$$\frac{I(q)}{P_0(q)} \propto q^{-d_f} \qquad (3)$$

where $P_0(q)$ is the form factor for the seed particles ($q = 4\pi n/\lambda \sin \theta/2$, varied by changing scattering angle θ, λ is the light wavelength and n is the medium refractive index). This relationship arises from the power law form of the pair correlation function in the limit $qR \gg 1$[8] and has been shown to hold despite the broad distribution of cluster sizes present in a typical aggregation[9].

3. EXPERIMENTAL METHOD

Three aggregation systems were investigated: Antisera/IgG, Protein A/IgG and Pertussis Toxin/Fetuin.

Antisera (antihuman IgG Sigma Chemical Co.) was diluted in phosphate buffered saline (PBS) as recommended (to make a 5mg/ml stock) and filtered through a 0.22 μm low protein binding filter. 42 μl of this solution was added to 600 μl of a solution of 1% polyethylene glycol (PEG) in PBS and filtered into a sample cuvette. PEG enhances the rate of aggregation but does not appear to affect aggregate structure strongly[10]. The solutions were all degassed before preparing the sample, either by heating or by vacuum. The sample was centrifuged (3000 r.p.m. in a bench-top centrifuge) for a few minutes to expel air bubbles and to force any large dust particles to the bottom of the sample cell. It is essential to have a clean sample as any dust scatters a lot of light and markedly increases the estimated particle size.

42 μl of a 200 μg/ml filtered solution of purified human IgG in PBS was added to commence the aggregation. A suitable mixing technique was developed which involved injecting the antigen (human IgG here) into the sample and reciprocating an equivalent amount of sample rapidly in and out of the pipette two or three times to effect mixing. This whole process took less than 5 seconds.

The aggregation of Protein A with Immunoglobin G was investigated using the same method as above but using 1 ml of 90 μg/ml human IgG in PBS plus 1% PEG. 130 μl of 0.1 mg/ml protein A in PBS was added to commence the aggregation.

The aggregation of Pertussis Toxin with Fetuin, a glycoprotein cell surface receptor analogue, was investigated using the same general method but with 0.2 mg/ml pertussis toxin in 2 M urea + 0.5% PEG all in PBS. 100μl of 2 mg/ml fetuin was added to aggregate. Urea was used as a solvent to ensure that the pertussis was monodisperse. If the pertussis is monodisperse and the system aggregates then we have proved that pertussis is at least divalent in its reaction with Fetuin.

A Malvern K7032 correlator operating in linear mode was used to size the protein. Two PCS systems were used: one with a krypton ion laser operating at a wavelength of 647.1 nm and the other used an argon ion laser of wavelength 488 nm. Typical measured radii before reaction were below 20 nm. Some time after initiating the reaction the aggregates grow to a size greater than the inverse of the scattering vector. This point is characterised by a saturation of the scattered intensity. After this time the intensity was measured as a function of the scattering angle, in 5° steps from θ = 20° to θ = 140°. In addition to correcting for background counts and dead time, the intensity was also multiplied by sin θ to allow for the variation in scattering volume at different angles.

4. RESULTS

A logarithmic plot showing the scattered intensity *versus* scattering vector for several experiments is shown in Fig. 1. Measured gradients were all were near 2.65 (solid line in figure) indicating fractal dimension $d_f \sim 2.65$ for all three systems.

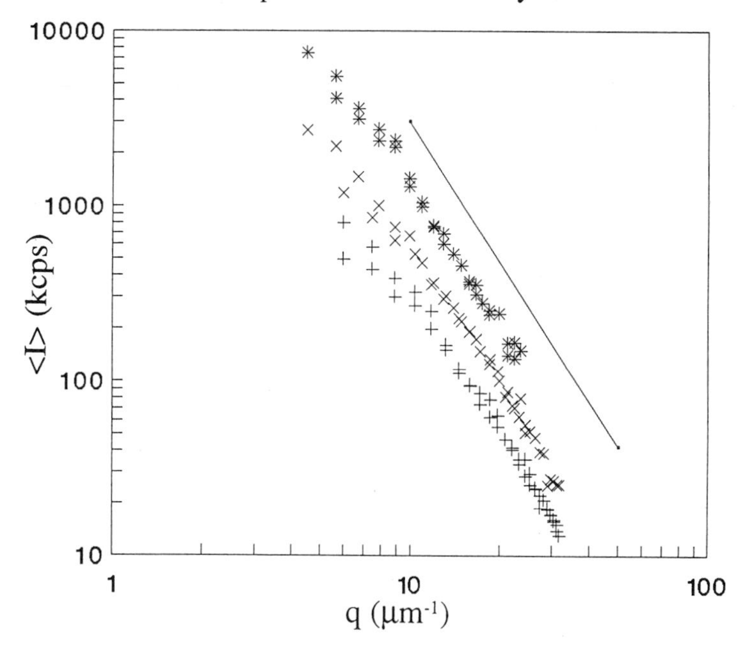

Figure 1. Log-log plot of mean scattered intensity I *versus* scattering vector q for the three reactions: * Alpha sera/IgG, X protein A/IgG and + Pertussis toxin/fetuin (arbitrarily scaled to avoid overlap). Solid line shows a gradient of 2.65. Measurements were made after particle radius reached 500 nm ensuring saturation of scattered intensity. Drooping at low q indicates that saturation has not been reached for scattering in the forward direction.

5. CONCLUSIONS

We have gathered more evidence for a common fractal dimension for aggregates formed in antibody-antigen aggregation. Using multi-angle static light scattering we have measured $d_f \approx 2.65$ in three different systems. This is higher than the range of d_f values obtained in the aggregation of model colloids[7] and may be a result of restructuring due to the reversibility of the individual bonds, non-specific binding within the aggregates or more probably denaturation (internal melting) of the proteins on binding. Other studies of protein association have led to similar high estimates for the fractal dimension[10-12]. Further studies of the growth kinetics using dynamic light scattering are underway to investigate this.

REFERENCES

1. M. Von Smoluchowski, *Phys. Z.*, 1916,*17*, 585.
2. E.J.W. Verwey and J.Th.G. Overbeek, 'Theory of the Stability of Lyophobic Colloids', Elsevier, Amsterdam, 1948.
3. H. Reerink and J.Th.G. Overbeek, *Disc. Faraday Soc.*, 1954, *18*, 74 (and papers therein).
4. R.H. Ottewill and J.N. Shaw, *Disc. Faraday Soc.*, 1966, *42*, 143, (and papers cited therein).
5. D. Schaeffer, J.E. Martin, P. Wiltzius and D.S. Cannel, *Phys. Rev. Letts.* 1984, *52*, 2371.
6. D.A. Weitz, and M. Oliveira, *Phys. Rev. Letts.*, 1983, *52*, 1433.
7. M.Y. Lin, H.M. Lindsay, D.A. Weitz, R.C. Ball, R. Klein and P. Meakin, *Nature*, 1989,*339*, 360.
8. J. Texeira, 'On Growth and Form', H.E. Stanley and N. Ostrowski, eds, Martinus Nijhoff, Boston, 1986, p. 145.
9. J.E. Martin, *J Appl. Crystallogr.*, 1986, *19*, 25.
10. J.G. Rarity, R.J.G. Carr and R.N. Seabrook , *Proc. Roy. Soc.*, 1988, *A423*, 89,.
11. J. Feder, T. Jossgang and E. Rosenqvist, *Phys. Rev. Letts.*, 1984, *53*, 1403.
12. S. Magazu, G. Maisano, F. Mallamace and N. Micali, *Phys. Rev. A*, 1989, *39*, 4195.

17

Hydrodynamic Behaviour of Myosin Filaments Studied by Laser Light Scattering

By Satoru Fujime and Noriko Mochizuki-Oda[1]

GRADUATE SCHOOL OF INTEGRATED SCIENCE, YOKOHAMA CITY UNIVERSITY, 22-2 SETO, KANAZAWA-KU, YOKOHAMA 236, JAPAN.

[1]OSAKA BIOSCIENCE INSTITUTE, URUEDAI, SUITA 565, JAPAN.

1. INTRODUCTION

A method of laser light scattering has been extensively applied to the study of the dynamic behaviour of macromolecules in solution. For suspensions of long filaments, laser light scattering provides us with information about translational, rotational and bending motions of the filament. As an example of laser light scattering in biochemistry, we present our study on synthetic myosin filaments of rabbit skeletal muscle.

Theoretical Background

We first summarize the theoretical background in order to facilitate discussion and to introduce relevant terminology. For free Brownian motion of a semiflexible rod, the first cumulant of the time correlation function $g^1(\tau)$ of the electric field of the scattered light can be written as[1,2]

$$\frac{\bar{\Gamma}}{K^2} = D_0 + \frac{L^2}{12}\Theta f_1(k) + (D_3 - D_1)\{f_2(k) - {}^1/_3\} + \sum_m D_{[m]} a_m(k) \tag{1}$$

where $D_0 = (2D_1 + D_3)/3$ is the overall translational diffusion coefficient, D_3 and D_1 are, respectively, the translational diffusion coefficients parallel and perpendicular to the mean axis of the rod undergoing the bending motions, Θ is the end-over-end rotational diffusion coefficient, K is the length of the scattering vector, and L is the contour length of the rod. $f_i(k)$ (i = 1 and 2) and $a_m(k)$ ($m \geq 2$) depend only on the flexibility parameter σ (or the inverse of the Kuhn length) and k = KL/2. $D_{[m]}$ is the diffusion coefficient of the mth bending mode of motion[2]. On the right-hand-side of eq. (1), the first term is due to overall translation, the second term to rotation, the third term to anisotropy in translation, and the fourth term(s) to bending of the semiflexible filament. When $\sigma = 0$ (rigid rod), the last term in eq. (1) should be ignored. Eq. (1) predicts that for a given condition, the more flexible the rod is, the larger the $\bar{\Gamma}/K^2$ value is.

For a smooth cylinder (rigid rod) with length L and diameter d, the diffusion coefficients are given by[3]

$$D_1 = \frac{k_B T}{4\pi\eta L} \{Lp - 0.19 + 4.2(Lp^{-1} - 0.39)^2\} \tag{2a}$$

$$D_3 = \frac{k_B T}{2\pi\eta L} \{Lp - 1.27 + 7.4(Lp^{-1} - 0.34)^2\} \tag{2b}$$

$$\Theta = \frac{3k_B T}{\pi\eta L^3} \{Lp - 1.45 + 7.5(Lp^{-1} - 0.27)^2\} \tag{2c}$$

$$D_{[m]} = \frac{k_B T}{4\pi\eta L} [1 + f_m] \qquad (m \geq 2) \tag{2d}$$

where k_B is the Boltzmann constant, T is the absolute temperature, η is the solvent viscosity, Lp = ln(2L/d), and an explicit expression of f_m is found elsewhere[2]. For a semiflexible rod, these diffusion coefficients are also functions of the flexibility parameter σ, but their dependence on σ is negligibly small in this particular case of synthetic myosin filaments.

Structure of Myosin Filament

Figure 1(a) schematically shows the structure of an isolated myosin molecule, where arrows indicate the universal joint-like portions. Under an appropriate condition, myosin molecules polymerize into bipolar filaments. Figure 1(b) schematically shows the structure of a myosin filament, where the two "S1"s of each myosin molecule are represented by a single circle. In intact filaments, the number of myosin molecules per 14.3-nm axial repeat is three instead of two in this figure.

Aim of This Study

From the structure of the myosin filament, there arise such questions as:

(1) Do the projections produce an extra frictional drag?

(2) Is each myosin filament made up of subfilaments? If yes, is there any dependence of their structural stability on ionic strength?

In this study, we examine dependence on ionic strength of morphology by electron microscopy, and of hydrodynamic behaviours by laser light scattering and analytical ultracentrifugation, of synthetic myosin filaments[4].

2. MATERIALS AND METHODS

Synthetic Myosin Filaments

Rabbit skeletal myosin and synthetic filaments were prepared according to the method of Persechini and Rowe[5]. Note that pH values of Tris buffers were those at 10°C. Extracted and solubilized myosin was diluted into an equal volume of cold water. The resultant suspension was dialyzed exhaustively against a buffer containing 120 mM KCl (not 150 mM KCl as buffer D in ref. 5), 10 mM Tris-HCl, 1 mM $MgSO_4$, and 0.2 mM dithiothreitol (DTT); final pH was 8.3. The suspension was centrifuged at 70,000 g for 2 h in a swinging bucket rotor. The protein concentration was adjusted to 1 mg/mL. The ionic strength of the solution was changed by dialysis for 2 h at 4°C against buffers containing various concentrations of KCl, 10 mM Tris-HCl, 1 mM $MgSO_4$, and 0.2 mM DTT (pH 8.3). Since the exact value of ionic strength of the solution containing Tris was not available, we used apparent ionic strength calculated on the assumption that Tris was fully ionized.

Laser Light Scattering

A 488.0 nm beam from an Ar^+ laser (Lexel Corp.) was used as the light source. The spectrometer we used has been detailed elsewhere[6]. An (8 x N)-bit correlator with 128 channels (K7032-CE, Malvern Instruments) was used to measure the intensity correlation functions $G^2(\tau)$. All the measurements were made at (10 ± 0.1)°C. The net correlation functions, $G^2(\tau)/B - 1 = \beta[g^1(\tau)]^2$, were least squares fitted with eq. (3)

$$g^1(\tau) = \exp(-\bar{\Gamma}\tau)\,[1 + (\mu_2/2!)\tau^2 - (\mu_3/3!)\tau^3] \qquad (3)$$

where B is the baseline (known), β is the instrument constant, $\bar{\Gamma}$ is the first cumulant, and μ_i is the ith moment around $\bar{\Gamma}$ of the decay rate distribution.

3. RESULTS

Electron Microscopy

For a quantitative analysis of light scattering data, we measured the length distribution of the synthetic myosin filaments by electron microscopy (data not shown, see ref. 4). At any ionic strength (I = 134, 74 and 44 mM) in this study, the number N(z) of filaments with length L can be approximated by a Schulz-Zimm distribution[7]

$$N(z)\,dz = C_m z^m \exp[-(m + 1)\,z]\,dz \qquad (4)$$

with L_n (the number-average length) = 470 nm and m = 27 [4], where $z = L/L_n$ and C_m is a constant depending only on m. The lengths of our filaments were about one third of the intact thick filament, and the length distribution was very sharp; $L_w/L_n = (m + 2)/(m + 1) = 1.04$ for m = 27 where L_w is the weight-average length.

At I = 134 mM, the filament diameter d^i at the bare zone was about 15 nm for the major fraction, and about 10 nm for the very minor fraction, where the superscript "i" attached to d means the d value from electron microscopic images. After the ionic strength was lowered from 134 to 74 mM by the 2-h dialysis, the number of the 15-nm (10-nm) filaments drastically decreased (increased), keeping L_n and m unchanged. At I = 44 mM, almost all filaments had a diameter of about 10 nm (Table 1). Although two sizes in diameter are quite clear on micrographs, these numbers (10 and 15 nm) are not exact and probably overestimated, and so the filaments with d^i of about 15 nm and 10 nm are henceforth called M- and m-filaments, respectively.

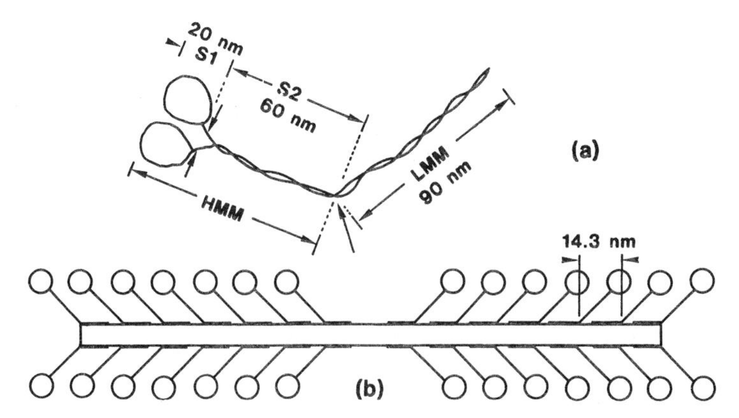

Figure 1. Schematic illustrations of the structures of a myosin molecule (a) and a myosin filament (b).

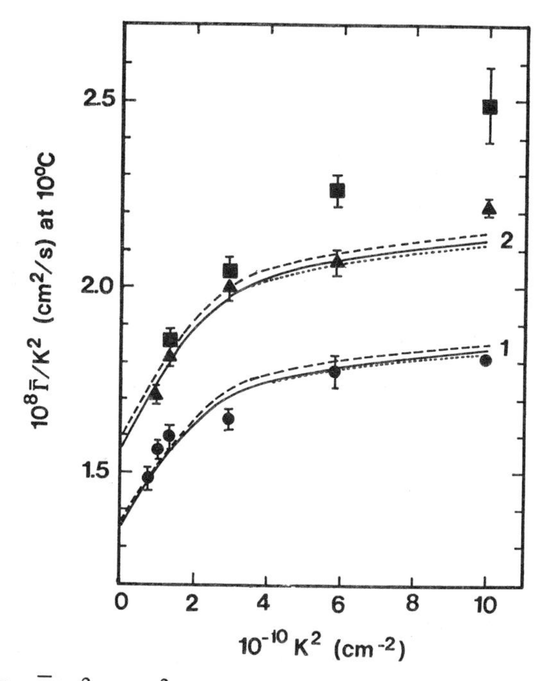

Figure 2. $\bar{\Gamma}/K^2$ *vs.* K^2 relationships of synthetic myosin filaments at I = 134 (circles), at 74 (triangles) and at 44 mM (squares). The solid lines denote the rigid-rod values of eq. (1) averaged over N(z) in eq. (4) with L_n = 470 nm and m = 27. The dashed and dotted lines denote the rigid-rod values of eq. (1) for a monodisperse distribution with L = L_n and L = L_w, respectively. Curves with label 1 were computed for the rod diameter d^h = 60 nm, and those with label 2 for d^h = 43 nm. The viscosity η of water at 10°C was assumed. Reproduced with permission from ref. 4.

Table 1. Filament diameters estimated by various methods.

Ionic strength (mM)	EM image[a] d^i (nm)	Hydrodynamic[b] d^h (nm)	Sedimentation[c] d^s (nm)	Renormalized[d] d^* (nm)
134	15 (M-filament)	60	13	13
74	10 (m-filament)	43	9.3[e]	9
44	10 (m-filament)	40[e]	8.7[f]	

[a] Major filaments on electron micrographs (EM).
[b] Hydrodynamic diameter for the η value of water.
[c] By use of eq. (5).
[d] Hydrodynamic diameter renormalized against $\eta^*/\eta = 1.8$ (1.7) at I = 134 (74) mM.
[e] Indirectly estimated values such as $d^s = 13(43/60)$ and $d^h = 60(8.7/13)$, listed for reference only.
[f] For the assumed value of D_0 (see Table 2). Computed D_0 for $d^h = 40$ nm averaged over N(z), 1.62×10^{-8} cm^2/s, gave $d^s = 8.5$ nm.

Laser Light Scattering

Figure 2 shows the $\bar{\Gamma}/K^2$ *vs.* K^2 relationships for the synthetic myosin filaments at three ionic strengths. Measurements were made at only one concentration, 1 mg/mL myosin. It is noteworthy that the change in the $\bar{\Gamma}/K^2$ value was *reversible* against ionic strength, which was changed by the two hour dialysis.

The light scattering results were first examined in the following way: If the myosin filament was assumed to be a rigid rod ($\sigma L = 0$), only one adjustable parameter, the diameter d, was left in eqs.(1) and (2). The dependence on K^2 of $\bar{\Gamma}/K^2$ at I = 134 mM was then found to follow the theoretical curve (the solid line 1 in Fig. 2) for rigid rods with $d^h = 60$ nm, where the superscript "h" attached to d means the hydrodynamic d value, and $L_n = 470$ nm and m = 27 taken from the electron microscopic result. An anomalously large value of $d^h = 60$ nm compared with $d^i = 15$ nm can be attributed to extra friction due to myosin heads projecting from the shaft of the filament (Fig. 1b): this is discussed in a later section. To see the effect of sample polydispersity on the theoretical results, we assumed a monodisperse distribution with $d^h = 60$ nm and $L = L_n$ (the dashed line 1 in Fig. 2), or with $L = L_w$ (the dotted line 1 in Fig. 2). Because of sharp distribution of our sample, the effect of sample polydispersity on Γ/K^2 was very small.

When the ionic strength was lowered, the $\bar{\Gamma}/K^2$ *vs.* K^2 relationship changed greatly. At I = 74 mM, $\bar{\Gamma}/K^2$ increased at every K^2, and the theoretical curves for rigid rods with $d^h = 43$ nm were found to reproduce the observed values (the lines 2 in Fig. 2), except at the highest K^2. The hydrodynamic value of $d^h = 43$ nm is again much larger than both d^i values of 15 and 10 nm. Although D_0 (= $\bar{\Gamma}/K^2$ at $K^2 = 0$) at I = 44 mM was only slightly larger than that at I = 74 mM, $\bar{\Gamma}/K^2$ at 44 mM increased more rapidly with K^2 than that at 74 mM. No theoretical curve for rigid rods reproduced the observed data. A possible interpretation on high-frequency modes at I = 44 mM is given in a later section.

In order to estimate the D_0 values, the theoretical curves (solid lines in Fig. 2) can be used for extrapolation of $\bar{\Gamma}/K^2$ to $K^2 = 0$. The estimated d^h and D_0 values are listed in Tables 1 and 2, respectively.

Sedimentation Velocity

Sedimentation velocity measurements at 1 mg/mL myosin filaments and at the rotor speed of 7,000 rpm gave $s_{20,w}$ = 106 S at I = 134 mM, and $s_{20,w}$ = 54.4 S at I = 44 mM. At both ionic strengths, almost the same but very small amounts of monomers/dimers were detected after the major component had sedimented at the rotor speeds of 30,000 r.p.m. -50,000 r.p.m.; no appreciable increase was observed in the concentration of monomers/dimers on lowering ionic strength.

It is expected that the anomalous friction affects both s and D_0 values equally. If this is the case, a correct molecular weight M of the filament will be estimated from the experimental values of s and D_0. Since $\pi(d/2)^2L$ is the volume of a cylinder in which the mass distribution is uniform, we have a relationship

$$M/N = \left\{\frac{k_BT}{(1 - \bar{v}\rho)}\right\}\left\{\frac{s}{D_0}\right\} = \pi\ (d/2)\ 2\left\{\frac{L}{\bar{v}}\right\} \tag{5}$$

where N is Avogadro's number, $\bar{v}$ is the partial specific volume (assumed to be 0.73), and ρ is the solvent density. Eq. (5) with L = 470 nm, D_0 values corrected to 20°C in water, and $s_{20,w}$ values gives d^s = 13 nm for I = 134 mM and d^s = 8.7 nm for I = 44 mM, where the superscript "s" attached to d means the d value from sedimentation. These d^s values are close to but a little smaller than d^i values. These d^s values may be the lower bounds of the diameter (not at the bare zone but overall) of the filament, because the d *vs.* M relationship in eq. (5) is for a "compact" rod, although the M *vs.* (s/D_0) relationship is highly correct irrespective of the compactness of the rod. The estimated d^s values by use of eq. (5) are listed in Table 1, and the measured $s_{20,w}$ values as well as estimated M values are in Table 2.

4. DISCUSSION

Anomalous Friction of Myosin Filaments

If the d^s values were assumed for the filament diameter, the theoretical curves based on eqs.(1) and (2) gave about twice larger Γ/K^2 values than the experimental ones. From eq. (2) we see that the

Table 2. Observed values of diffusion coefficient D_0, sedimentation coefficient s, and molecular weight M, for synthetic myosin filaments at the concentration of 1 mg/mL myosin.

Ionic Strength (mM)	$10^8 \times D_0$ (cm^2/s) at 10°C	$s_{20,w}$ (S)	$10^{-7} \times M$ (Da)
134	1.3_5	106	5.2
74	1.5_7		
44	1.5_7[a]	54.4	2.3

[a] Assumed value. Note that D_0 at I = 44 mM seems to be a little larger than that at I = 74 mM (see Fig. 2). This is inferred to be due to the absence of the M-filaments at I = 44 mM. Computed D_0 for d^h = 40 nm averaged over N(z), 1.62×10^{-8} cm^2/s, gave $10^{-7}M = 2.2$ at I = 44 mM.

diffusion coefficients, D_i and Θ, are proportional to $(1/\eta)$ [ln(2L/d) + minor terms]. Therefore, if we assumed an effective viscosity $\eta^* = 1.8\eta$ for I = 134 mM and 1.7η for I = 74 mM, we could obtain small values of d^* = 13 and 9 nm for I = 134 mM and 74 mM, respectively, as shown in Fig. 3. The d^* values renormalized against η^* are also listed in Table 1 for a comparison. The size of the friction ratio η^*/η depends on the number of projections per unit length of the filament, and the assumption of two values in our case is reasonable as shown later.

It is worth noting here the following. The importance of the third term in eq. (1), the term due to anisotropy in translation, is observed in this sample too, as has also been observed in suspensions of tobacco mosaic virus[8]. If this term were ignored in the analysis in Figs. 2 and 3, we would have to assume a diameter (d^h or d^*) for Θ which is different from the diameter for D_0. This difficulty has been encountered in a previous study[9].

Substituting D_0(observed) = 1.3_5 and D_{shaft} (for d^* = 13 nm) = 2.4_3 (note that $D_{shaft}/D_0 = 1.8$) into

$$(1/D_0) = (1/D_{shaft}) + (1/D_{proj}) \tag{6}$$

we have $D_{proj} = 3.0_4$ in units of 10^{-8} cm^2/s, where D_{proj} symbolically represents the effect of the projections on the friction ratio η^*/η. Theoretically speaking, we may have

$$D_{proj} = (D_{S1}/n)[1 + f(n)] \tag{7}$$

where D_{S1} is the diffusion coefficient of myosin subfragment-1 (S1 in Fig. 1a), n is the number of "S1"s per filament, and f(n) is the hydrodynamic interaction factor. A myosin filament can be modelled as something in between the following two limits:

(a) a shaft plus many "S1"s connected with each other in a random-coil form, and

(b) a shaft plus many "S1"s connected with each other in a rod form.

According to Kirkwood[10], we have for n = 200 and $D_{S1} = 4.5 \times 10^{-7}$ cm^2/s at 10°C

$$D_{coil} = (D_{S1}/n)\,[1 + 1.84n^{1/2}] = 6.0 \times 10^{-8} \text{ cm}^2/\text{s} \tag{8a}$$

and

$$D_{rod} = (D_{S1}/n)\,[1 + \{\ln(n) - 1\}] = 1.2 \times 10^{-8}\,cm^2/s \qquad (8b)$$

An inequality relation $D_{rod} < D_{proj} < D_{coil}$ qualitatively estimates the size of D_{proj}. Since, however, it is very difficult to quantitatively discuss the size of f(n) in eq. (7), we can only say at the moment that the very large values of d^h = 60 nm or 43 nm mentioned before might correspond to a "hydrodynamic envelope" of the mass distribution of a myosin filament.

Bending Motion of Myosin Filaments

The dramatic increase in $\bar{\Gamma}/K^2$ with K^2 at I = 44 mM may be due to the bending motion of m-filaments as a whole as a result of loosening of the structure of the m-filament shaft at low ionic strength. To examine this possibility, the field correlation function $g^1(\tau)$ was computed for $L = L_n$, $d^* = 9$ nm, $\eta^*/\eta = 1.7$, and various values of K and σL. The intensity correlation function $g^2(n\delta\tau)$ was constructed according to the formula $g^2(n\delta\tau) = 1 + [g^1(n\delta\tau)]^2$, where n = 1, 2, ..., 128, and $\delta\tau$ (sampling time) was set to the value at the experiment. The simulated correlation functions were analyzed in the same way as that in the analysis of experimental correlation functions. The details of the numerical method are found elsewhere[11]. The simulated results of the $\bar{\Gamma}/K^2$ *vs.* K^2 relationships are shown in Fig. 3. The experimental $\bar{\Gamma}/K^2$ values for I = 74 mM can be simulated by the curve for σL = 0 to 0.025. Although most of the filaments at I = 74 mM are m-filaments, the $\bar{\Gamma}/K^2$ values are very close to the rigid-rod values. On the other hand, the experimental $\bar{\Gamma}/K^2$ values at I = 44 mM can be simulated by the curve for σL = 0.1, or σ = 0.2 μm^{-1}; the flexibility of the m-filaments increased on lowering ionic strength from 74 mM to 44 mM. (See ref. 4 for more details on the analysis of light scattering data.)

Dynamic Equilibrium of Myosin Subfilaments

The ratio (5.2/2.3) of the molecular weights roughly suggests that each M-filament is converted into two m-filaments on lowering ionic strength from 134 mM to 44 mM, and *vice versa*. This conversion *via* monomers/dimers is not likely, because

(a) filaments are stable at both ionic strengths, and the

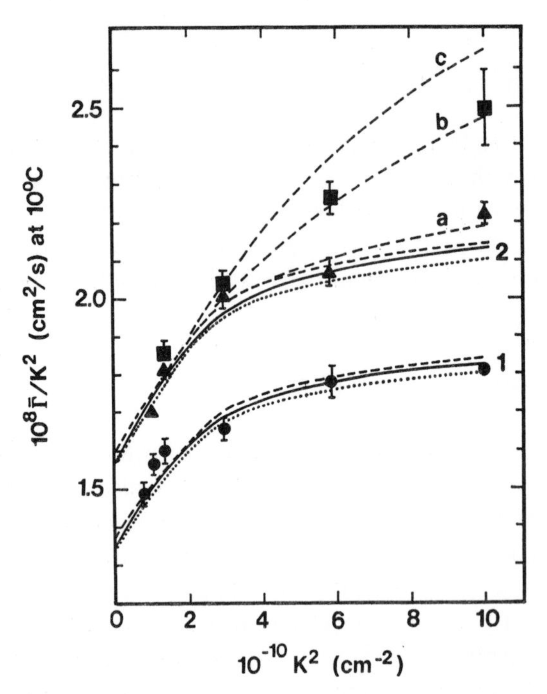

Figure 3. $\bar{\Gamma}/K^2$ *vs.* K^2 relationships. For symbols, see caption of Fig. 2. Curves with label 1 were computed for $d^* = 13$ nm and $\eta^* = 1.8\eta$ (see text), and those with label 2 for $d^* = 9$ nm and $\eta^* = 1.7\eta$. The dashed curves with labels a-c were obtained by constructing the theoretical "$g^2(\tau)$"s for $L = L_n$, $d^* = 9$ nm, $\eta^*/\eta = 1.7$, and various values of the filament flexibility; $\sigma L = 0.025$ for a, 0.1 for b and 0.2 for c. Only when slight flexibility of the filament is assumed, does $\bar{\Gamma}/K^2$ start from the same value at $K^2 = 0$ but increase more rapidly with K^2 than the rigid-rod values. This is qualitatively seen from eq. (1): the increase of the fourth term with σL. Reproduced with permission from ref. 4.

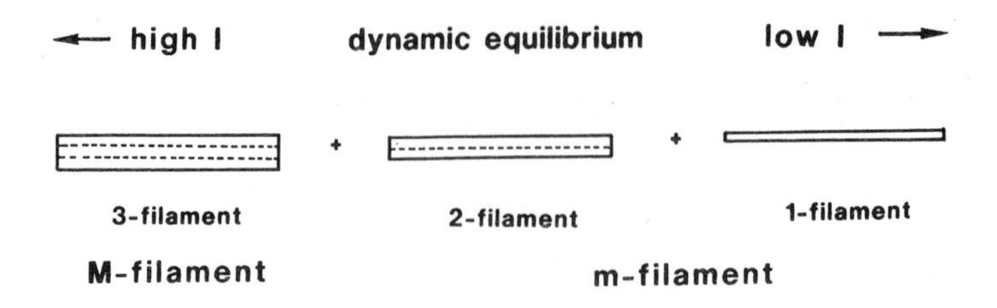

Figure 4. Schematic illustration of a dynamic association-dissociation equilibrium of the μ-filaments.

depolymerization of filaments within 2 h dialysis at 4°C is inferred to be too fast, and

(b) slow polymerization of myosin monomers/dimers by dialysis results in very long filaments, but no changes in L_n and N(z) were observed after up and down cycles of ionic strength.

The reversible change in $\bar{\Gamma}/K^2$ against the ionic strength suggests that the M- and m-filaments are in a dynamic association-dissociation equilibrium, keeping a very low critical concentration of monomers/dimers coexisting with filaments.

Combination of the length and molecular weight of the myosin filament yields an estimate of the number N of myosin molecules per 14.3 nm axial repeat (see Fig. 1b). For this number, literature values are N = 6.3 for L_w = 810 nm - 840 nm [5], N = 6 for L_w = 750 nm [9], N = 3 - 6 for L_w = 600 nm - 700 nm [12], and similar values to those (quoted in refs. 5, 9, and 12). Our value for the M-filament is N = 3 - 3.2 for L_n = 470 nm. If the M-filament is assumed to be composed of three subfilaments, the association-dissociation equilibrium mentioned above will be visualized as shown in Fig. 4, and the diameter of the "μ-filament" composed of μ subfilaments is given by

$$d^s(\mu) = 13(\mu/3)^{1/2} \text{ (nm)} \tag{9}$$

Table 3 lists the computed values of $d^s(\mu)$, D_0 and η^*/η.

Table 3. Estimated Values of d^s, D_0 and η^*/η for μ-filaments with $L = L_n$.

μ-filament	$d^s(\mu)$ (nm)	$10^8 D_0$ (cm²/s)	η^*/η
3-(M-)filament	13	1.3_8	1.8
2-(m-)filament	$10._6$	1.5_5	1.7
1-(m-)filament	7.5	1.6_9	1.7

The static scattering intensity I_s is given by the square root of B in an arbitrary scale, where B is the baseline of $G^2(\tau)$. Our light scattering data gave

$$IR = I_s\,(134 \text{ mM})/I_s\,(44 \text{ mM}) = 1.9 \pm 0.1 \tag{10}$$

For given L_n and N(z), the static scattering intensity is also given by

$$I_s = i_0 \Sigma\, n_\mu M_\mu^{\,2} = j_0 \Sigma\, n_\mu d^s(\mu)^4 \tag{11}$$

where i_0 and j_0 are proportionality constants, and n_μ and M_μ are, respectively, the number and molecular weight of the μ-filaments. From electron microscope observations, we assume for simplicity that all filaments are the M-filaments at I = 134 mM and the m-filaments at 44 mM. For splitting of one M- (3-)filaments into n_1 1-filaments and n_2 2-filaments, we have

$$IR = d^s(3)^4/[n_1 d^s(1)^4 + n_2 d^s(2)^4] \tag{12}$$

with a constraint for the conservation of the number of the subfilaments

$$n_1 + 2n_2 = 3 \qquad \text{(per one M-filament)} \tag{13}$$

and the molecular weight (the square of the diameter) of the m-filament in terms of those of 1- and 2-filaments

$$d^2 = [n_1 d^s(1)^2 + n_2 d^s(2)^2]/(n_1 + n_2) \tag{14}$$

Table 4. Estimation of population of the 1- and 2-filaments at I = 44 mM.

n_1	n_2	IR	d^s (nm)	population $n_1 : n_2$
3	0	3.0_0	$d^s(1)$	
1.4_6	0.77	1.9_9	8.7	2 : 1
1.2_4	0.88	1.9	8.9_2	3 : 2
0.98	1.0_1	1.8	9.2_0	1 : 1
0	0.5	1.5_1	$d^s(2)$	

Observed values are d_s = 8.7 nm (I = 44 mM) and IR = 1.9 ± 0.1.

From numerals in Table 4, combination of eqs.(12-14) suggests that the m-filaments at I = 44 mM is a mixture of 1- and 2-filaments in a number ratio n_1/n_2 of 2:1 to 3:2. Then, we can estimate the weight-average diffusion coefficent of the m-filament at I = 44 mM as

$$\langle D_0 \rangle = \frac{\Sigma\, n_\mu d^s(\mu)^4 D_0(\mu)}{\Sigma\, n_\mu d^s(\mu)^4} = 1.60 - 1.63 \tag{15}$$

in units of 10^{-8} cm^2/s, which should be compared with the $\bar{\Gamma}/K^2$ value at $K^2 = 0$ in Fig. 3.

5. CONCLUDING REMARKS

Our result may have a deep connection with the fact that the intact myosin filament frays into three subfilaments by washing with distilled water[13]. There are many previous studies on dynamics of myosin filaments by various physical methods, but only a few by the laser light-scattering method[4,9,14-16]. Laser light scattering is expected to provide us with valuable information, in aspects different from previous ones, about the architecture of myosin filaments. This article is entirely based on our first work on myosin filaments[4].

REFERENCES

1. T. Maeda and S. Fujime, *Macromolecules*, 1984, *17*, 2381.
2. S. Fujime and T. Maeda, *Macromolecules*, 1985, *18*, 191.
3. S.J. Broersma, *J. Chem. Phys.*, 1969, *32*, 1626, 1632.
4. N. Mochizuki-Oda and S. Fujime, *Biopolymers*, 1988, *27*, 1389.
5. A. Persechini and A. Rowe, *J. Mol. Biol.*, 1984, *172*, 23.
6. S. Fujime, S. Ishiwata and T. Maeda, *Biophys. Chem.*, 1984, *20*, 1.
7. G. V. Schulz, *Z. Phys. Chem.*, 1939, *B43*, 25.
8. K. Kubota, H. Urabe, Y. Tominaga and S. Fujime, *Macromolecules*, 1984, *17*, 2096.
9. N. Suzuki and A. Wada, *Biochim. Biophys. Acta*, 1981, *670*, 408.
10. J. G. Kirkwood, 'Macromolecules, Documents on Modern Physics', Gordon & Breach, New York, 1967, p.23.
11. T. Maeda and S. Fujime, *Macromolecules*, 1985, *18*, 430.
12. C.H. Emes and A. Rowe, *Biochim. Biophys. Acta*, 1978, *537*, 125.
13. M. Maw and A. Rowe, *Nature*, 1980, *286*, 412.
14. K. Kubota, B. Chu, S.-F. Fan, M.M. Dewey, P. Brink and D.E. Colflesh, *J. Mol. Biol.*, 1983, *166*, 329.
15. S. Fujime and K. Kubota, *Macromolecules*, 1984, *17*, 441.
16. S.-F. Fan, M.M. Dewey, D. Colflesh, B. Gaylinn, R. Greguski and B. Chu, *Biophys. J.*, 1985, *47*, 809.

Quasielastic and Total Intensity Light Scattering Studies of Mucin Glycoproteins and Cartilage Proteoglycans

By Alex M. Jamieson, John Blackwell, David Zangrando and Audrey Demers

DEPARTMENT OF MACROMOLECULAR SCIENCE, CASE WESTERN RESERVE UNIVERSITY, CLEVELAND, OHIO 44106 U.S.A.

1. INTRODUCTION

Mucins are high-molecular weight O-glycosylated glycoproteins, which impart viscoelastic properties to mucous secretions[1,2]. They contain 60% - 90% w/w sugar, depending on sub-species, and have a common molecular architecture, consisting of a protein backbone with numerous oligosaccharide side-chains, attached *via* serine linkages. Submaxillary mucins, present in saliva, have least carbohydrate, with side-chains of 1-5 sugars; cervical mucins (1-9 sugars), gastrointestinal, and tracheobronchial mucins (1-20 sugars) have longer side-chains. Provided precautions are taken to preclude proteolytic, hydrolytic and mechanical degradation during isolation, high molecular weights (M_r »10^6) are obtained[3-11]. Such large biopolymers must be formed by covalently-linking subunits together. Large decreases in molecular weight of mucins after reduction and alkylation indicate that some of the inter-subunit linkages involve disulphide bonds[3,4,7,12]. The linked subunit mucin structure is necessary for the highly viscoelastic mucosal rheology[13,14]. Also, the carbohydrate side-chain lengths play an important role[13,14], since, at similar concentrations and molecular weights, submaxillary mucins show lower elasticity than cervical[13] or tracheobronchial mucins[15].

The chondroitin sulphate proteoglycans (CSPG) form a viscous hydrogel surrounding the collagen fibrils in cartilage[16].

Collagen imparts strength to cartilage, whereas proteoglycans contribute to the dissipative properties. The native structure is a giant aggregate (PGA; $M_r = 10^7 - 10^8$), with subunits (PGS; $M_r = 1 \times 10^6 - 4 \times 10^6$) non-covalently bound to a hyaluronic acid chain (HA; $M_r = 0.5 \times 10^6 - 2 \times 10^6$). Cartilage PGS has a protein core ($M_r = 2 \times 10^5 - 3 \times 10^5$) with side-chains of keratin sulphate ($M_r \sim 12{,}000$) and chondroitin sulphate ($M_r \sim 20{,}000$) covalently attached to serine residues. The PGS-HA interaction is stabilized by a small hydrophobic link protein ($M_r \sim 4 \times 10^4$) [16]. The native aggregate is dissociated by chaotropic solvents (4M guanidine hydrochloride (GuHCl)) and can be reconstituted in 0.5 M GuHCl. Electron microscopy[17] indicates that reconstituted aggregates differ from native PGA in having a lower subunit density. The intact aggregate is necessary for the enhanced viscoelasticity of PGA solutions[18-20].

2. MATERIALS AND METHODS

When isolating high molecular weight mucin glycoproteins and cartilage proteoglycans, suitable procedures must be utilized to minimize mechanical and proteolytic degradation[9,10].

Isolation of Human Tracheobronchial Mucins

Human Tracheobronchial Mucin (HTM) was prepared from freshly collected sputum of one patient suffering from Cystic Fibrosis (CF). Purification procedures used were based on the work of Chase *et al.*[21]. The freshly collected expectorate was immediately treated with aqueous sodium azide (Sigma Chem. Co.) (0.02%), an antibacterial growth factor (Sigma Chem. Co.); phenylmethylsulfonyl fluoride (PMSF) (0.1M), a serine protease inhibitor (Sigma Chem. Co.); and gentamycin sulphate (100 mg/ml), an antibiotic (Sigma Chem. Co.). The sample was then dialyzed against 0.02% sodium azide and distilled deionized water, lyophilized, and stored at 4°C.

Lyophilized sputum was solubilized in 0.1 M (hydroxymethyl)-aminomethane hydrochloride (Tris-HCl) buffer, pH 7.5, containing 0.22 M potassium thiocyanate and 0.02% aqueous sodium azide, then centrifuged at 27,000 g for four hours at 4°C to remove any insoluble debris. Portions (30 ml) of this solution were chromatographed on a 5 x 90 cm column containing Bio-Gel A-15 M equilibrated with Tris-HCl buffer, pH 7.5 and 0.02% sodium azide containing 0.22 M Potassium thiocyanate. Three peaks were found, referred to as I, II, and III, respectively. Compositional analyses showed that peak I contained high M_W glycoprotein plus minor amounts of low molecular weight proteins and DNA, whereas peaks II and III contain predominantly serum-type proteins and low molecular weight glycoproteins. The fractions in peak I were combined, dialyzed against distilled water and lyophilized.

To separate DNA contaminants in peak I, the lyophilized material was resolubilized in 0.1 M Tris-HCl buffer, pH 7.5, containing sodium azide (0.02% w/v) and 0.22 M potassium thiocyanate by mechanical shaking for four days at 4°C. DNA was digested for six hours by incubation of 20 ml solution with 0.2 ml DNase solution, (DNase, 18 mg/ml, 2000 units/mg protein) and 0.2 ml of a solution containing 2.2 mg/ml PMSF in dimethyl sulphonyl oxide (DMSO). Incubation for six hours at 37°C was followed by centrifugation at 27,000 g and 4°C for two hours. The PMSF prevents competing proteolytic activity during the DNA hydrolysis.

The centrifuged DNase digest was chromatographed on a Bio-Gel A-15M column, and showed two peaks, designated as peaks A and B. SDS gel electrophoresis studies showed that peak A contained high molecular weight mucin glycoprotein. Peak B contained degraded DNA and a minor amount of low molecular weight proteins. The pooled fractions were dialyzed against distilled water and lyophilized.

Finally, material in peak A was chromatographed on a 2.5 x 9 cm Hydroxylapatite column equilibrated with 0.01 M Potassium phosphate buffer, pH 6.8 (equilibration buffer), using a discontinuous gradient of 0.15 M, 0.30 M and 0.5 M potassium phosphate buffer, pH 6.8. Eluted fractions were monitored for optical density at 230 nm. Three peaks were observed, designated as fractions A1, A2 and A3, respectively. Peak A1, contained the native glycoprotein, which was pooled, dialyzed against distilled water, lyophilized, and stored in a freezer. Fractions A2 and A3 contain low molecular weight proteins[1].

The carbohydrate and amino acid compositions of the purified HTBM were determined in the laboratory of Dr. Neil Jentoft using methods described elsewhere[11,22,23]. The results are similar[26] to literature data for bronchial mucins[11].

Isolation of Bovine Nasal Septum Cartilage Proteoglycans

Proteoglycan subunit and aggregate specimens were isolated from bovine nasal septum using purification methods described in the literature[24]. The subunit was an A1D1 fraction representing material that had been subjected first to density gradient centrifugation under associative conditions (0.4 M GuHCl) followed by density gradient centrifugation under dissociative conditions (4 M GuHCl), and isolated as the included material *via* column chromatography on a Sepharose CL2B column in 4 M GuHCl. The aggregate was an A1 fraction[24], isolated as the void volume fraction by column chromatography on a Sepharose CL2B column in 4 M GuHCl.

Static and Dynamic Light Scattering Methods

Aqueous solutions of human tracheobronchial mucins (HTBM), proteoglycan subunit (PGS) and aggregate (PGA) were prepared for light scattering analysis in appropriate solvents. All solvent solutions were prepared using distilled water which was passed through a mixed bed deionizer/ultrafilter and clarified by filtration through 0.22 mm Millipore filters. GuHCl and NaCl solutions contained, in addition, 10 mM phosphate buffer, pH 7.0 ± 0.1 and 0.02% (w/v) sodium azide. HTBM solutions in $CaCl_2$ contained 50 mM Tris-HCl buffer, pH 7.0 ± 0.1, and 0.02% (w/v) sodium azide, since $CaCl_2$ is insoluble in phosphate buffer. Solutions were prepared by adding precision-weighed amounts of lyophilized material to buffered solvent and slowly shaking (2 Hz) for five days at 3°C. The resultant solutions were filtered through 8 mm Millipore filters in order to remove dust. Dilutions were made by adding known solvent volumes directly into the light scattering cell. The light scattering cells were prewashed with chromic acid for 24 hours, rinsed thoroughly with distilled water, steam rinsed, wrapped in aluminum foil and oven-dried for 24 hours prior to use. All solutions were centrifuged at 6000 r.p.m. for 30 minutes at 20°C before light scattering analyses. The refractive indices of the solvent and solutions were measured on a Bausch and Lomb Abbe Refractometer. The refractive index increments at constant chemical potential, $(dn/dc)_\mu$, were determined using a differential refractometer model RF-600 from C.N. Wood Co., Newtown, PA.

Light scattering intensities were interpreted using conventional Zimm plot analyses to yield estimates of weight-average molecular weight, $\bar{M}_w$, and z-average radius of gyration, $\bar{R}_{g,z}$. Dynamic light scattering autocorrelation functions were fitted to third-order cumulant expansions to yield the first cumulant $\bar{\Gamma}$, and the second cumulant $\mu_2 = (\bar{\Gamma}^2) - (\bar{\Gamma})^2$. From $\bar{\Gamma}$, we obtain the z-average translational diffusion coefficient since $\bar{\Gamma} = D_{t,z}q^2$, where q is the scattering vector, and hence we can determine the mean frictional radius $\bar{R}_f$ from the Stokes-Einstein equation, $D_{t,z}{}^0 = kT/6\pi\eta\bar{R}_f$, where $D_{t,z}$ is in the limit of zero concentration.

3. RESULTS

Light Scattering from Mucin Glycoproteins

Static and dynamic light scattering studies of mucins indicate that their configuration is a linear, semi-flexible chain. Thus, data on the radius of gyration, $\bar{R}_{g,z}$, and the frictional hydrodynamic radius, $\bar{R}_f$, for various mucin species, with M_r ranging from 10^6 to 20×10^6, are uniformly fit by[9-11]

$$R_{g,z}\ (\text{Å}) = 0.57\ M_{p,w}{}^{0.56} \qquad (1)$$

and

$$R_f\ (\text{Å}) = 0.27\ M_{p,w}{}^{0.58} \qquad (2)$$

where $M_{p,w}$ is the weight-average protein core molecular weight, calculated from M_w of the intact mucin by $M_{p,w} = cM_w$, where c is the weight fraction of protein. Equations (1) and (2) yield a ratio $R_g/R_f = 1.6 - 1.7$, consistent with theoretical expectation for linear semi-flexible chains. However, the radii of mucins (eq. (2)) are three-fold larger than for non-glycosylated protein random coils[25]. Studies of asialo ovine submaxillary mucin, which has a single sugar side-chain demonstrate[26] that two sugars are necessary for maximal expansion of the protein core. The $\bar{R}_{g,z}$ data for fractionated mucins, fitted to the wormlike model, indicate[9,26] the persistence length of the protein backbone is $\lambda^{-1} = 145$Å, compared to $\lambda^{-1} = 62$Å for the asialo form[26] and $\lambda^{-1} = 9.3$Å for a non-glycosylated protein coil. Comparison of $\bar{R}_{g,z}$ and $\bar{R}_f$ values for porcine submaxillary mucins[9,27] in 6M GuHCl and in 0.1M NaCl, indicates that a non-covalent self-association occurs in the latter solvent[9,27] producing a further large increase in hydrodynamic volume. More recently, a similar observation has been made for human tracheobronchial mucins.[28] These results are summarized in Table 1. Substantially larger values of M_w, $\bar{R}_{g,z}$, and $\bar{R}_f$ are observed for HTBM in aqueous NaCl and $CaCl_2$ when contrasted with 6M GuHCl. These associative interactions lead to increased viscosity of concentrated solutions and enhanced gel-forming capability in these solvents[14].

Table 1. Static and dynamic light scattering studies of human tracheobronchial mucins in various solvents.

Solvent	Ionic Strength (M)	M_w (g/mol)*	R_H (Å)**	R_g (Å)**
NaCl	0.1	5.2×10^6	733	1237
$CaCl_2$	0.1	5.2×10^6	684	1153
NaCl	0.75	5.1×10^6	777	1270
$CaCl_2$	0.75	5.1×10^6	702	1180
NaCl	1.5	4.6×10^6	800	1275
$CaCl_2$	1.5	4.3×10^6	718	1254
GuHCl	6	2.5×10^6	592	913

* Intercept from square root plot of light scattering intensities; error ±5%;

** Experimental error ±5%.

Light Scattering from Cartilage Proteoglycans

By contrast, light scattering shows that proteoglycan aggregates and subunits behave like impermeable ellipsoids of uniform segmental density. Thus, $\bar{R}_f$ and M_w values for PGS and PGA follow the universal equation[29]

$$\bar{R}_f \text{ (nm)} = 0.179\, M_w^{0.377} \quad (3)$$

for specimens ranging from $M_w = 10^6$ to 10^8. The corresponding molecular weight exponent for a uniform impermeable sphere is $^1/_3$.

In addition, for a purified PGS specimen from bovine nasal septum in 4M GuHCl, we determined[30] $M_w = 1.93 \times 10^6$, $\bar{R}_{g,z} = 608$ Å, $\bar{R}_f = 490$ Å, and the intrinsic viscosity $[\eta] = 120$ ml/g. From the Stokes-Einstein equation, $M_w[\eta] = (^{10}/_3\,\pi)\, N_A\, (\bar{R}_\eta)^3$, where N_A is Avogadro's number, we calculate the viscometric hydrodynamic radius $\bar{R}_\eta = 268$ Å [30]. These results lead to ratios $\bar{R}_{gz}/\bar{R}_f = 1.24$, and $\bar{R}_\eta/\bar{R}_f = 0.74$. For a comparable PGA preparation[30], we obtained $M_w = 6.6 \times 10^7$, $\bar{R}_{gz} = 2567$ Å, $\bar{R}_f = 1951$ Å, $[\eta] = 310$ ml/g, and $\bar{R}_\eta = 1658$ Å, hence $\bar{R}_{gz}/\bar{R}_f = 1.32$, and $\bar{R}_\eta/\bar{R}_f = 0.85$. Similar PGS values have been reported previously[31]. To our knowledge, numerical values of the ratio $\bar{R}_\eta/\bar{R}_f \ll 1.0$ have no theoretical precedence. However, experimental ratios $\bar{R}_\eta/\bar{R}_f$ slightly smaller than unity have been reported for globular proteins, and shown to be consistent with theoretical prediction[32] based on porous sphere hydrodynamics. It is possible that this anomalous result is due at least in part to the polydispersity of the PGS sample. We have characterized the size distribution by gel exclusion chromatography on Sepharose CL 2B[33] as shown in Fig. 1. This enables us to attempt polydispersity corrections to R_g, R_f and R_η, using eq. (3). This analysis is summarized in the Appendix. The chromatogram is divided into 33 fractions of equal elution volume. For simplicity, we assume the same molecular weight exponent, $\nu = 0.38$ applies to R_{gi}, R_{fi} and $R_{\eta i}$ where the subscript i refers to the ith fraction. To determine the molecular weight, M_i, of each fraction, we utilize a calibration relation[33] for the chromatographic distribution coefficient $K_d = (V_{e,i} - V_0)/(V_t - V_0)$, where $V_{e,i}$ is the elution volume of the ith segment, V_0 is the void volume, and V_t is the total volume. This relation, valid for the separation of PGS fractions on Sepharose CL-2B, with 4M GuHCl as solvent[33], is

$$\log \bar{M}_w = A - BK_d \tag{4}$$

with $A = 6.58 \pm 0.08$ and $B = 1.65 \pm 0.27$.

As shown in the Appendix, we obtain revised ratios $k' = R_{gi}/R_{fi} = 1.07$, and $k = R_{\eta i}/R_{fi} = 0.84$. An impermeable sphere of uniform density should have $R_g/R_f = 0.894$ and $R_\eta/R_f = 1.0$. Thus we see that the polydispersity correction leads to values closer to theoretical expectation. However, $R_{\eta i}/R_{fi}$ is still smaller than unity. This discrepancy may indicate inadequate polydispersity correction, as evident in Fig. 1, since a small portion elutes near the void volume[30]. For comparison[34] non-ionic synthetic star-branched polymers of high functionality (f = 16-18) have ratios of $\bar{R}_g/\bar{R}_f \cong 0.8$, and $\bar{R}_\eta/\bar{R}_f \cong 1.0$, comparable to the theoretical values for impermeable spheres.

One fortunate consequence of the dense quasi-spherical structure of PGS and PGA is the absence of QELS contributions from intramolecular motions at scattering angles which exceed the interference condition $qR_g \gg 1.0$. This enables us to apply QELS to determine size distributions of PGS and PGA specimens. We have used this approach to confirm that the hydrodynamic sizes of native

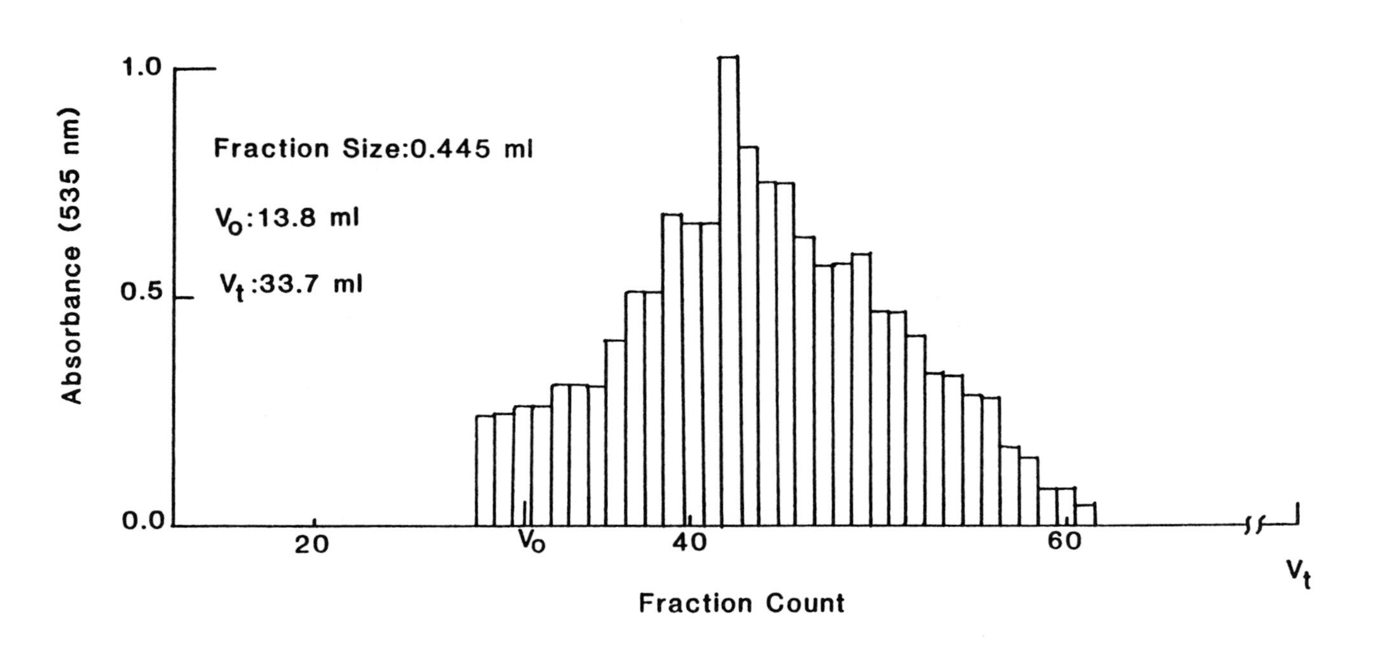

Figure 1. GPC profile of BNS proteoglycan subunit A1D1 on a Sepharose CL-2B column in 4 M GuHCl. Sample size: 0.445 ml.; Void volume: V_0 = 13.8 ml.; Total Volume: V_t = 33.7 ml.

PGA species are larger than reconstituted PGA[29], and to quantify the ability of link protein to enhance the thermal stability of PGA[35].

With regard to the latter question, we compare in Fig. 2 the experimental temperature dependence of the first cumulant $\bar{\Gamma}$ for dynamic light scattering from preparations of bovine nasal septum proteoglycan subunit (BNS PGS) and aggregate (BNS PGA) in 0.4M GuHCl. The data plotted are the values $\bar{\Gamma}_{norm}$, normalized to the viscosity of 0.4M GuHCl at 20°C:

$$\bar{\Gamma}_{norm} = \Gamma\,(T)\,\frac{293.16}{T}\,\frac{\eta\,(T)}{\eta\,(293.16°K)} \tag{5}$$

These values are shown at three scattering angles, $\theta = 30°$, $\theta = 55°$, and $\theta = 90°$. The corrected $\bar{\Gamma}$ values for BNS PGS are temperature-independent at small angles. At $\theta = 90°$ a small increase with temperature is observed. For BNS PGS, we find that, at each angle, $\bar{\Gamma}_{norm}$ increases with temperature and approaches the corresponding value for BNS PGS at T = 70°C. This is an indication of dissociation of the aggregate into subunit. The magnitude of the change is greatest at larger scattering angles because of the relatively higher contribution of subunit scattering[30].

Multi-exponential fitting of the correlation functions indicates that the PGA preparation contains a mixture of free subunits and multimers as well as aggregate species. Dissociation of the latter causes the increased $\bar{\Gamma}_{corr}$ with temperature. In Fig. 3, we plot the fraction of the total scattering amplitude of free subunit. At T ≅ 50°C, the scattering amplitude of free subunit suddenly increases, indicating onset of aggregate dissociation. On cooling, there is only partial recovery of aggregate suggesting that aggregate dissociation is due to irreversible denaturation of link protein and/or the subunit binding site.

4. DISCUSSION

Static and dynamic light scattering analysis indicate that the macromolecular structure of mucins is that of a linear semi-flexible coil whose large dimensions, expanded by steric interactions of the oligosaccharide side-chains, impart viscoelastic behaviour by forming an entanglement network at low concentrations. Further enhancement of the elasticity of mucin solutions is achieved by intermolecular association.

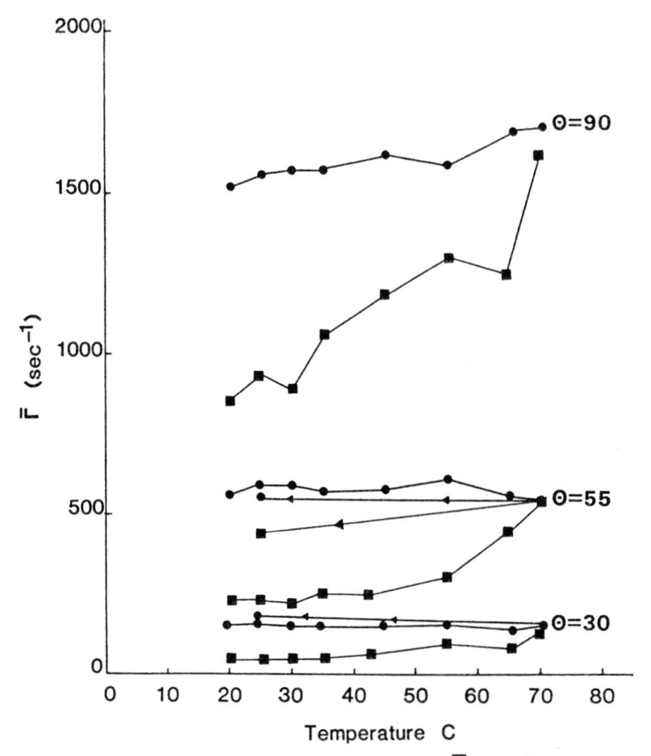

Figure 2. Mean relaxation frequency ($\bar{\Gamma}$ sec^{-1}) *vs*. T(°C) for BNS Proteoglycan Subunit (•) and Aggregate (square) at scattering angles θ = 30°, 55°, 90°. Solvent: 0.4 M GuHCl, 0.5 M NaOAc, 0.2% NaN_3, pH 7.0. Concentration: 3.0 mg/ml.

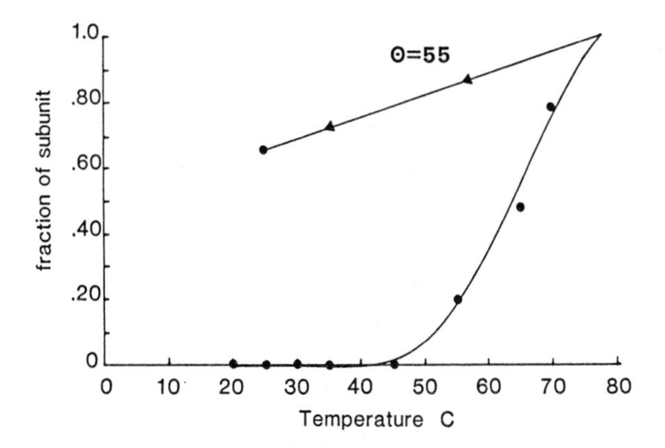

Figure 3. Fractional scattering amplitude of free BNS proteoglycan subunit in a solution of BNS proteoglycan aggregate as the sample is heated to 70°C. Solvent: 0.4 M GuHCl, 0.5 M NaOAc, 0.2% NaN_3, pH 7.0. Concentration: 3.0 mg/ml.

In contrast, dynamic light scattering studies demonstrate to a reasonable approximation that proteoglycan subunit and proteoglycan aggregate behave hydrodynamically like impermeable ellipsoids of uniform segmental density. Thus PGA resists interpenetration in concentrated solutions, and we propose that the viscoelastic response results from the formation of a pseudo-lattice (*cf.* ref. 36). The aggregate structure, which is crucial for maintaining the high elasticity of concentrated proteoglycan solutions, is stabilized by a specific binding site involving a carbohydrate-free region of the subunit, a link protein and a six-disaccharide sequence of hyaluronic acid. Thermal stability of the aggregate is maintained until 50°C when irreversible denaturation occurs.

ACKNOWLEDGEMENTS

This work was supported by NIH grants AG 02921, DK 33365 and AM 27651, and a grant from the National Cystic Fibrosis Foundation.

REFERENCES

1. I. Carlstedt, J.K. Sheehan, A.P. Corfield and J.T. Gallagher, *Essays Biochem.*, 1985, *20*, 40
2. A. Allen, *Trends Biochem. Sci.*, 1983, *8*, 169
3. R.L. Shogren, A.M. Jamieson, J. Blackwell, P.W. Chang, D.G. Dearborn and T.F. Boat, *Biopolymers*, 1983, *22*, 1657
4. I. Carlstedt, H. Lindgren and J.K. Sheehan, *Biochem. J.*, 1983, **213**, 427
5. F.A. Meyer, *Biochem. J.*, 1983, *215*, 701
6. I. Carlstedt and J. Sheehan, *Biochem. J.*, 1986, *217*, 93
7. R.L. Shogren, A.M. Jamieson, J. Blackwell, and N. Jentoft, *J. Biol. Chem.*, 1984, *259*, 14657
8. M.C. Rose, W.A. Voter, H. Sage, C.F. Brown, and B. Kaufman, *J. Biol. Chem.*, 1984, *259*, 3167
9. R.L. Shogren, A.M. Jamieson, J. Blackwell, and N. Jentoft, *Biopolymers,* 1986, *25*, 1505
10. R.L. Shogren, N. Jentoft, T.A. Gerken, A.M. Jamieson, and J. Blackwell, *Carbohydrate Res.*, 1987, *160*, 317
11. R. Gupta, N. Jentoft, A.M. Jamieson, and J. Blackwell, *Biopolymers*, 1990, *28*, 347
12. J.P. Pearson, A. Allen, and S. Darry, *Biochem. J.*, 1981, *197*, 155

13. A.E. Bell, A. Allen, E.R. Morris, and S.B. Ross-Murphy, *Int. J. Biol. Macromol.*, 1984, *6*, 309
14. L.M. Soby, A.M. Jamieson, J. Blackwell, and N. Jentoft, *Biopolymers*, 1990, *29*, 1359
15. C. McCullagh, A.M. Jamieson, J. Blackwell, and R. Gupta, 1990, unpublished results
16. V.C. Hascall, 'Biology of Carbohydrates' (V. Ginsburg, ed.) , Wiley, New York, 1981, *1*, p. 1
17. J.H. Kimura, P. Osdoby, A.I. Caplan, and V.C. Hascall, *J. Biol. Chem.*,1978, *258*, 6226
18. G.Matsumura, 'Solution Properties of Polysaccharides' (Brant, D.A., ed.), 1981, ACS Symp. Ser. Vol. *150*, p. 213
19. V.C. Mow, A.F. Mak, W.M. Lai, L.C. Rosenberg, L.C. and L.H. Tang, *J. Biomechanics,* 1984, *17*, 325
20. L.M. Soby, A.M. Jamieson, J. Blackwell, H.U. Choi, and L.C. Rosenberg, *Biopolymers*, 1990, *29*, 1587
21. K.V. Chase, M. Flux and G.P. Sachdev, *Biochemistry*, 1985, *28*, 7334
22. R.L. Heinrikson and S.C. Meredith, *Anal. Biochem.*, 1984, *136*, 65
23. N. Jentoft, *Anal. Biochem.*, 1985, *185*, 424
24. L. Rosenberg, W. Hellmann and A.K. Kleinschmidt, *J. Biol. Chem.*, 1975, *250*, 1877
25. M.E. McDonnell and A.M. Jamieson, *Biopolymers*, 1976, *15*, 1283
26. R.L. Shogren,T.A. Gerken, and N. Jentoft, *Biochemistry* , 1989, *28*, 5525
27. B.K. Varma, A.G. Demers, A.M. Jamieson, J. Blackwell, and N. Jentoft, *Biopolymers*, 1990, *29*, 441
28. Demers, A.G. (1990) Structural Studies of Glycoproteins in Solutions, Ph.D. Thesis, Case Western Reserve University, p. 127
29. H. Ohno, J. Blackwell, A.M. Jamieson, D.G. Pechak, D.A. Carrino, and A.I. Caplan, *Biopolymers,* 1986, *25*, 931
30. D. Zangrando, A.M. Jamieson and J. Blackwell, (1990) unpublished results
31. R.G. Kitchen, and R.L. Cleland, *Biopolymers*, 1978, *17*, 759
32. J.A. McCammon, J.M. Deutch, and V.A. Bloomfield, *Biopolymers,* 1975, *14*, 2479
33. H. Ohno, J. Blackwell, A.M. Jamieson, D.A. Carrino, and A.I. Caplan, *Biochem. J.*, 1986, *235*, 553

34. N. Khasat, R.W. Pennisi, N. Hadjichristidis and L.J. Fetters, *Macromolecules*, 1988, *21*, 1100
35. L.H. Tang, L.C. Rosenberg, H. Reihanian, A.M. Jamieson, and J. Blackwell, *Conn. Tiss. Res.*, 1989, *19*, 177-193
36. R.J. Ketz, R.K. Prud'homme and W.W. Graessley, *Rheol. Acta*, 1988, *27*, 531

APPENDIX: EFFECT OF POLYDISPERSITY

For the individual chromatography fractions, we have the "monodisperse" results:

$$R_{\eta i} = k\, R_{fi} \tag{A1}$$

where, from eq. (3),

$$R_{fi} = K\, M_i^{0.38} \tag{A2a}$$

and hence

$$R_{\eta i} = k\, K\, M_i^{0.38} \tag{A2b}$$

Now,

$$\bar{R}_\eta^{\,3} = \frac{\Sigma c_i\, M_i}{\Sigma c_i} \cdot \frac{\Sigma c_i\, [\eta]_i}{\Sigma c_i} = \frac{\Sigma c_i\, M_i}{\Sigma c_i} \cdot \frac{\Sigma c_i\, \{R_{\eta i}^{\,3}/M_i\}}{\Sigma c_i} \tag{A3}$$

and since R_f is computed from the z-average translational diffusion coefficient, D_t, *via* the Stokes-Einstein equation, we have

$$\bar{R}_f = \frac{\Sigma\, c_i\, M_i\, R_{fi}^{\,-1}}{\Sigma\, c_i\, M_i} \tag{A4}$$

Therefore:

$$(\bar{R}_\eta/\bar{R}_f)^3_{exp} = \frac{\Sigma c_i\, M_i}{\Sigma c_i} \cdot \frac{\Sigma c_i\, (kK)^3\, M_i^{0.14}}{\Sigma c_i} \cdot \left\{\frac{\Sigma c_i\, M_i^{0.62} K^{-1}}{\Sigma c_i\, M_i}\right\}^3$$

$$= k^3 \cdot \frac{\Sigma c_i\, M_i^{0.14}}{(\Sigma c_i)^2} \cdot \frac{(\Sigma c_i\, M_i^{0.62})^3}{(\Sigma c_i\, M_i)^2} \tag{A5}$$

Applying eq. (4) to determine M_i for each segment, we obtain

$$(\bar{R}_\eta/\bar{R}_f)^3_{exp} = k^3\, (0.76) \tag{A6}$$

Since

$$(\bar{R}_\eta/\bar{R}_f)^3_{exp} = 0.77 \quad (A7)$$

whence

$$k = \left\{\frac{(0.77)^3}{0.76}\right\}^{1/3} = 0.84 = R_{\eta i}/R_{fi} \quad (A8)$$

Likewise, for the z-average radius of gyration, we have

$$\bar{R}^2_{gz} = \left\{\frac{\Sigma c_i\, M_i\, R_{gi}{}^2}{\Sigma c_i\, M_i}\right\}^{1/2} \quad (A9)$$

Therefore,

$$(\bar{R}_{gz}/\bar{R}_f)_{exp} = \left\{\frac{\Sigma c_i\, M_i\, R_{gi}{}^2}{\Sigma c_i\, M_i}\right\}^{1/2} \cdot \frac{\Sigma c_i\, M_i\, R_{fi}{}^{-1}}{\Sigma c_i\, M_i} \quad (A10)$$

Again, for the "monodisperse" ratio we write

$$R_{gi} = k'\, R_{fi} \quad (A11)$$

and assuming

$$\bar{R}_f = K\, M^{0.38} \quad (A11a)$$

and

$$\bar{R}_g = k'\, KM^{0.38} \quad (A11b)$$

we obtain

$$(\bar{R}_{gz}/\bar{R}_f) = k' \left\{\frac{\Sigma c_i\, M_i{}^{1.76}}{\Sigma c_i\, M_i}\right\}^{1/2} \cdot \frac{\Sigma c_i\, M_i{}^{0.62}}{\Sigma c_i\, M_i} \quad (A12)$$

Applying eq. (4) to the chromatogram, we find

$$(\bar{R}_{gz}/\bar{R}_f)_{exp} = k'\ (1.15) \quad (A13)$$

Since

$$(\bar{R}_{gz}/\bar{R}_f)_{exp} = 1.24 \quad (A14)$$

we find

$$k' = \frac{1.24}{1.15} = 1.07 = \frac{R_{gi}}{R_{fi}} \quad (A15)$$

Note that from the chromatogram, we can also estimate the second moment of the correlation function of light scattered by our PGS sample:

$$\frac{\mu_2}{\bar{\Gamma}^2} + 1 = \frac{(\bar{\Gamma}^2)_z}{(\bar{\Gamma})_z^2} = \frac{\Sigma c_i\, M_i\, R_{fi}^{-2}}{\Sigma c_i\, M_i} \cdot \frac{(\Sigma c_i\, M_i)^2}{(\Sigma c_i\, M_i\, R_{fi}^{-1})^2} \qquad \text{(A16)}$$

where the symbol Γ refers to the decay rate $\Gamma_i = D_{ti}q^2$ with q the scattering vector.
We determine:

$$\left\{\frac{\mu_2}{\bar{\Gamma}^2}\right\}_{calc} = 0.10 \qquad \text{(A17)}$$

which is slightly smaller than our experimental result

$$\left\{\frac{\mu_2}{\bar{\Gamma}^2}\right\}_{exp} = 0.16 \pm 0.06 \qquad \text{(A18)}$$

19

Use of On-line Laser Light Scattering Coupled to Chromatographic Separations for the Determination of Molecular Weight, Branching, Size and Shape Distributions of Polysaccharides

By James E. Rollings

CHEMICAL ENGINEERING DEPARTMENT, WORCESTER POLYTECHNIC INSTITUTE, WORCESTER, MASSACHUSETTS, U.S.A.

1. INTRODUCTION

In this Chapter, there is little reason to review in detail the scientific background on laser light scattering: this subject has been adequately covered earlier in this volume. Instead I will focus primarily on the various types of information one can obtain from such devices (and experiments) and how this information relates to biochemistry by supplying additional (or alternative) data to the biochemical researcher. Here I will centre on the biochemical applications which my laboratory has had some experience. Our work has mainly dealt with examining (poly)saccharides. These materials are often orders of magnitude more difficult to assay when compared to (poly)amino acids [proteins] or (poly)nucleic acids [DNA or RNA]. This difficulty arises as each individual monomer unit of (poly)saccharides possesses multiple sizes which are capable of participating in polymer linkages (*i.e.* the hydroxyl sites). This fact allows (poly)saccharides to 'branch' and thereby exist in numerous physical structures that other biologically derived macromolecules cannot. Statistically, it can be shown that ten glucose monomer units can be arranged in approximately one million different structures due to these multiple bonding sites. Fortunately for us, nature has not chosen such a random architecture.

2. BACKGROUND

My interest in laser light scattering is more pragmatic than pure. Such topics have been reviewed earlier[1]. My laboratory has over the last ten or so years been interested in unraveling the interrelationships between biopolymer structure and these materials' functions. The functions of interest to us relate most often to industrial applications. These interests reflect my engineering bias and include such topics as enzymatic reaction processes of biopolymer systems, rheology or fluid flow phenomena, chemically reactive biopolymers, and bioactive polymers. Laser light scattering has been a means to an end. For me, the age-old adage 'necessity is the mother of invention' applies here, for had it not been for these interests, laser light scattering may not have been more than a scientific footnote.

Historically, it was my interest in biopolymer systems which led me into the field of chromatography. Here, like many biochemical investigators, I saw chromatography as a tool which offered the possibility of providing some information on biopolymer structure/function relationships. Again, this is a subject which has been reviewed[2]. As chromatography functions as a means of distributing a biopolymer sample, *via* some underlying thermodynamic laws, post column detection provides an avenue to draw conclusions about a biopolymer's chemical distributions. Here, however, we found that the 'normal' post column detection devices, such as differential refractometry, were, in and of themselves, inadequate for determining the information which we wished to gain about these biopolymer systems. This fact has led us into the use of laser light scattering devices (and other devices) as a means of solving this problem. Many of the reports from my laboratory group[3-9] have dealt exclusively with these analytical advances and not with biopolymer structure/function relationships *per se*. It has been necessary to make these advances in order to realize my larger strategic goals.

The 'global' aim of those of us in this field is to use biotechnology in creative ways to assist new product developments employing renewable, biologically derived materials. My personal 'global' aim is in using biotechnology for assisting economic development and stabilization of developing nations. I add this personal statement here as this latter interest developed while working in the former British colony of Kenya during the early 1970's and this is the first time which I have spoken professionally in the United Kingdom.

I will re-emphasize one of the primary themes of this volume - and the Conference on which it was based - this being biopolymers, or more generally, polymers. Polymers in some ways are more difficult to work with when compared to other materials; and require special (or at least several alternative) approaches for their examination. Historically, for many polymer systems of natural origin that possess sufficient industrial utility, numerous, very specific and often times 'strange' analytical devices have been invented. Devices such as the amylograph, the mixograph, and the farinograph are commonly employed industrially even today in the flour and dough related industries. In essence, the objective of theses analytical devices is to directly tie the 'end use' functionality of a material to modifications in its pretreatment or processing conditions. This is depicted in Fig. 1 as Path A; and here, it must be underscored; this route has clearly proved useful to those industries wishing to develop new products.

As basic scientists, we may be unsatisfied with this direct approach which attempts to tie functionality to past material pretreatments. Some of us may rather venture along Path B of Fig. 1 which may lead us through a more deeper level of understanding,the biopolymer system. Here the approach is to take a given biopolymer system, treat it in some fashion to alter its state, then by using some appropriate set of analytical tools seek to establish relationships between the biopolymer's physical chemical properties and the material's functional behaviour. By modifying the pretreatment processing conditions in a controlled and systematic manner, the analytical tools can be used to seek for corresponding changes in both the physical chemical properties and the functionality. We believe that this methodology will provide a greater insight and minimize trial and error methodologies more prevalent in Path A. This somewhat untraditional pathway is that which we shall explore here and that which is related to the main theme of this volume; the use of light scattering as a means of gaining such information. Before leaving Fig. 1, I must add that there are several other methodologies available to the biopolymer scientist who wishes to obtain physical chemical information on these systems. Methods such as viscometry or thermal gravimetric analysis are also available, but

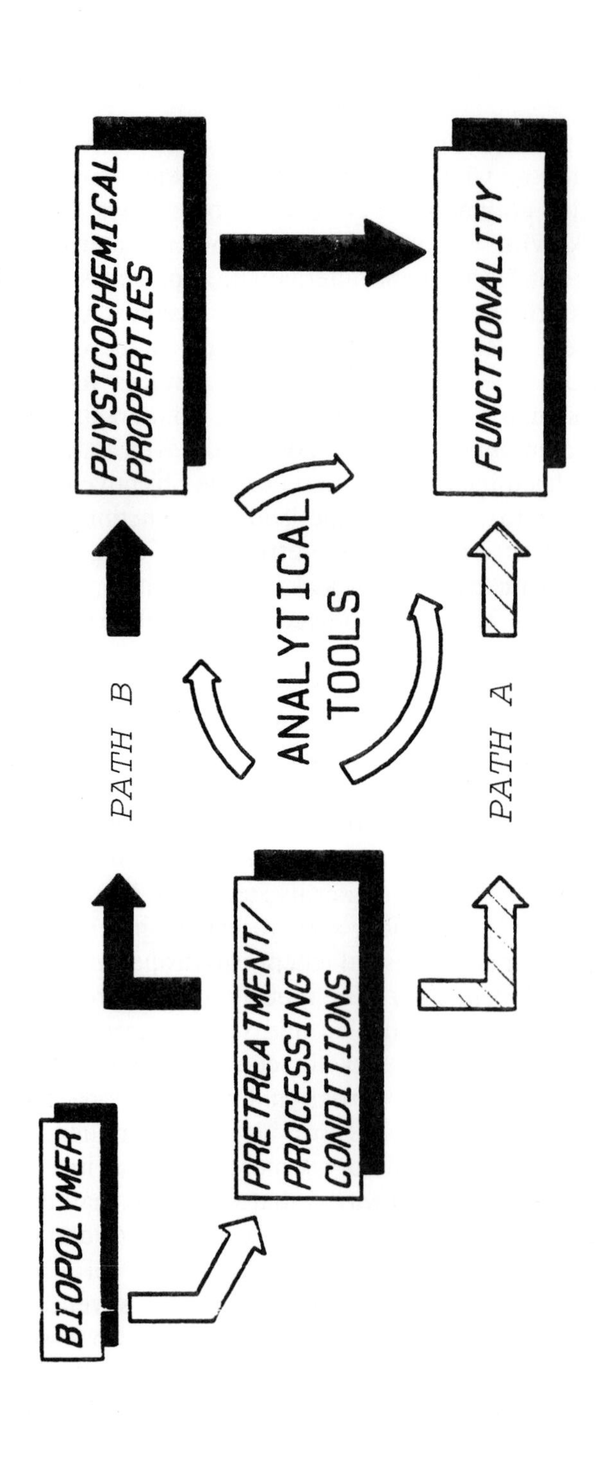

Figure 1. Alternative approaches for establishing biopolymer functionality or related pretreatment/processing conditions.

theses methods are not as advanced as light scattering methods particularly if one wished to combine the analytical device which provides the physical chemical data on the light scattering system to a separation device such as a chromatograph.

3. LASER LIGHT SCATTERING AND CHROMATOGRAPHY: BIOCHEMICAL TOOLS

Chromatography has become one of the main separation devices commonly used by life scientists. Column chromatography was, after all, invented by life scientists and thus its use by these investigators generally is met with little inertial barriers. In addition these devices are readily coupled to light scattering detection devices and will therefore (theoretically) provide the needed physical chemical data.

My reason for entering into the study of chromatography was as a means of gaining information on biopolymer systems. In many ways, I became fascinated with the separation technology itself, and much of my published work deals exclusively with chromatography and compatible down-stream detection devices and to a somewhat lesser extent with the study of biopolymer systems *per se*. But as I said at the outset, 'necessity is the mother of invention', and had it not been for my interest in the study of biopolymer structure and function relationships, I would never have ventured into the world of combined chromatography with novel down-stream detection devices such as light scattering. To this extent, I am a pragmatist. As a biochemical engineer; I was initially most interested in learning from these devices what they could tell me about the biologically derived polymer systems. But as a scientist, I have become intrigued with the analytical devices themselves.

Let me be self critical here for a moment: after all, I have taken no vows of loyalty to chromatography. Chromatography, notwithstanding its standing as one of the life bloods of the bioscientist's investigative tools, is only a means to an end. These scientists are, in general, not interested in the chromatography in and of itself, but rather in the information in which it can provide. For 'pure' size exclusion chromatography, certain published theories state that polymers will separate within such a device based upon their size in solution; big molecules emerging from these columns before their smaller counterparts. A typical SEC experiment supplies information relating polymer concentration to column elution. If the relationship between hydrodynamic size for such soluble polymers follows prescribed rules, then it is possible to relate this elution volume data through the expected hydrodynamic volume to molecular weight information. This is in essence the basis of all calibration procedures. Here I am talking about direct calibration and indirect calibration procedures including the so called 'universal calibration' procedure. The latter procedure purports to relate the product of the polymer's intrinsic viscosity and molecular weight to its column elution volume. In many systems (particularly in aqueous chromatography), other separation phenomena are present (*i.e.* other than pure size separation) and these complications or interactions tend to upset the theories. Moreover, for real polymer systems, such as that encountered in biological systems, the 'simple' polymer molecular size, molecular weight, and column elution volume relationships are further complicated by chemical heterogeneities such as branching. These compositional variations, in turn, effect the polymer's conformation. Other local laboratory environmental variables related to 'solvent' states (T, pH, *etc.*) further complicate the investigations. The investigator wishing to obtain molecular weight directly from elution volume information may simply be tilting at windmills unless he/she systematically explores all of these phenomena. These arguments are central to many of my publications.

The pragmatist, including the bioscientist, isn't interested in opening this Pandora's box for he/she is only interested in relating functional properties to molecular properties. As one of my practicing chemical engineering friends keeps reminding me, 'All this theory is fine, but I have to run the factory'.

We can summarize these limitations of conventional SEC by stating that 'traditional' application of this technique requires that the researcher determine some appropriate means of calibrating his chromatographic system. And he/she must do this while at the same time recognizing that no practical, appropriate means may be available. The chromatographic system may be plagued with incipient pit falls. Non-size separation effects may exist. The polymer system under study may fall beyond the range of separation limits of the column or be naturally present as a mixture of polymers. All bets are off here when using traditional techniques. Oh, yes, let us not forget the additional overriding constraint of making sure that the polymer under study is solvated in a column-compatible liquid medium.

Rather than making a career of chromatography, one must ask, 'Is there a more direct pathway to obtaining the macromolecular information on a biopolymer system?' On-line molecular probes, such as light scattering detectors, are not Maxwell demands; but they do lead us more directly to our objectives. Several such examples of the use these hybrid chromatographic techniques which avoid many of these pitfalls will be presented here.

Depending upon which property an investigator is interested in obtaining for a given biopolymer system, any of several techniques are available. A partial list of properties may include concentration, composition, molecular weight, molecular size and molecular conformation; and a partial list of available techniques include any of several wet chemical methods as well as refractometry, static and dynamic light scattering, dilute solution viscometry and sedimentation. Employing these techniques in 'batch' experiments is very tedious and, moreover, will only provide one averaged value; characteristic of a given polymer sample. If an investigator is interested in distributed biopolymer information, bulk techniques are inadequate.

Distributed biopolymer information can be obtained by employing the traditional methodologies advanced by polymer scientists three to four decades ago or by more modern methodologies. Polymer scientists demonstrated that solvent/ non-solvent fractionation schemes could be used to prepare subpopulations of a polymer sample and these fractions in turn could than be analyzed by the same bulk techniques listed here to construct distributed property information. Alternately, novel separation devices such as chromatography or field flow fractionation methods can be coupled directly to a selected few of these listed techniques compatible with the separation device and distributed biopolymer information obtained more easily. I shall emphasize the use of static light scattering techniques coupled to size exclusion chromatograph as such a means of obtaining the desired information.

Static light scattering experiments as we have seen will provide much of the data of interest to the biochemical scientist. These experiments typically generate measures of scattered light intensity due to the presence of a polymer in solution (presented as the excess Rayleigh ratio) as a function of observational angle. This type of data is illustrated in Fig. 2 using a Zimm Plot. When the intensities of scattered light are extrapolated to zero angle and zero concentration, the sample's weight average molecular weight is obtained. Moreover, by determining the initial angular dependent slope at zero concentration, the polymer sample's square mean radius is obtained. Conformational information is contained in the zero concentration line and second virial coefficient data is present at the incident beam *extremum*. This type of an experiment can be conveniently coupled to a size exclusion chromatograph to obtain molecular data at every slice of a chromatogram. In this way, it is possible to construct distributed data for a biopolymer system that includes distributions in the material's molecular weight, size and conformation. We have demonstrated both theoretically and experimentally that branching distributional data can be obtained by this method and the theoretical possibility exists for other compositional distributions as well. What I shall discuss in the remainder of this Chapter is several examples which will illustrate the use of this technique in providing fundamental information on biopolymer systems.

4. SELECTED CASE STUDIES

Table 1 lists molecular weight data for dextran standards typically used with aqueous SEC calibration schemes.[10] Data from both the manufacturer (Pharmacia Labs, Piscataway N.J., U.S.A) and a combined SEC/static light scattering photometer (Wyatt Technologies, Inc. Santa Barbara, CA, U.S.A.) are compared. Clearly these results illustrate the accuracy of this novel separation detection method which, according to Table 1, is reliable to within a few percent.

Table 1. Molecular weight data of dextran samples.

Sample Name	SEC/MALLS Data ± σ			Nominal mol.wt.[a]
	M_w	M_n	P.D.*	
T-500	498,000 ± 17,000	298,000 ± 5,000	1.67	500,000
T-70	70,500 ± 700	56,500 ± 4,950	1.25	70,000
T-40	40,000 ± 2,000	31,000 ± 4,600	1.29	40,000
T-10	10,000	7,500	1.33	10,000

* P.D.: Polydispersity index (M_w/M_n); [a] Data of Pharmacia

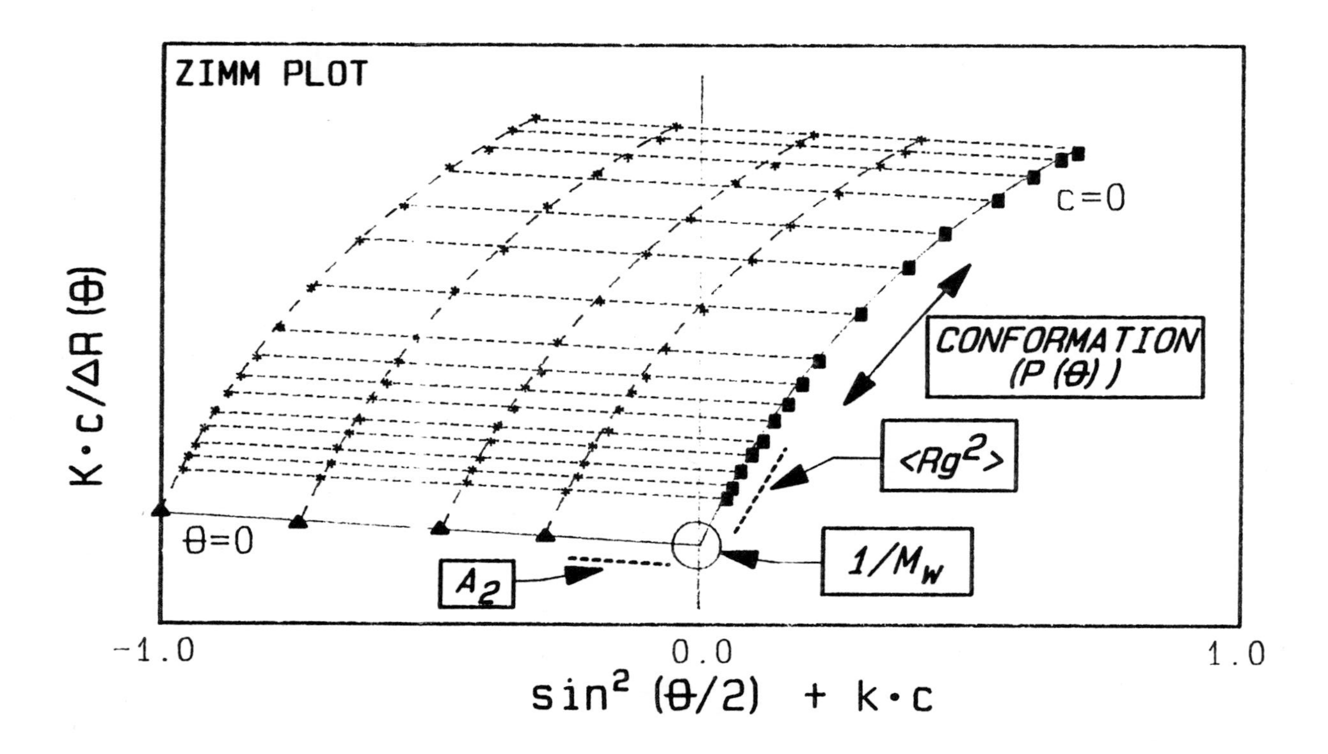

Figure 2. Typical Zimm plot showing macromolecular information obtainable *via* static light scattering.

Table 2 lists similar information on a series of chitosan samples that differ in their degree of acetylation[10]. For these samples, it is observed that the percentage of the chitosan that is successfully chromatographed is directly related to the degree of acetylation. Highly deacetylated chitosans are readily chromatographed under the conditions of this assay (solvent being pH 5.4 acetate buffer, 0.1 molar; column being BioRad TSK type). Whilst those samples that possess a greater percentage of acetyl groups are only partially chromatographed. Information on sample recovery could not easily be deduced without the use of the combined SEC/light scattering system and moreover as shown in Table 2 also provides chitosan average molecular size information. Independent experiments were conducted to establish that this flow-mode static light scattering device provides consistent molecular weight and size information when compared to stand alone static light scattering experiments thus confirming our confidence in the technique.

Table 2. Chitosan data.

Sample Name	Acetylation %	Recovery %	SEC / MALLS Data ± σ			Size (nm)
			M_w	M_n	P.D.	
Q0	16.6	90	190,000 ± 7,000	120,000 ± 1,000	1.58	45.8
085-260	16.6	90	250,000 ± 22,000	145,000 ± 10,000	1.72	49.0
056-552	24.4	80	375,000 ± 20,000	195,000 ± 30,000	1.92	54.6
125-280	1.0	100	350,000 ± 20,000	196,000 ± 30,000	1.79	54.5

My laboratory has considerable experience in the analysis of starch based biopolymer systems[11-19]. Our interest in these materials is again due to their end use functionalities. Two main functionalities have been examined over the past few years; these being enzymatic hydrolysis and rheology. We have used combined SEC and light scattering techniques to study the kinetics of starch depolymerization[11-18].

As most of you are well aware, starch is naturally present as both a heterogeneous mixture of linear (poly)glucose, amylose, and a branched (poly)glucose, amylopectin. These molecules are packaged within a plant as granules. In order to study functional characteristics of starch, it was necessary for us to solve several other analytical problems. One of these problems which relates to this volume is the problem of branching analysis. Our approach is shown in Fig. 3.

Here we built upon the pioneering work of Zimm and Stockmeyer[20] who defined a branching index g_M as the ratio of polymer sizes at equivalent molecular weight (branched:linear) polymers. A series of publications from my laboratory[6-8] describes how branching information can be extracted from chromatographic data at each elution volume. Central to these advancements was the capability of obtaining, simultaneous concentration and light scattering data. Basically, by simply obtaining molecular weight directly from light scattering experiments as the biopolymer elutes, it is possible to deconvolute a differential refractometer's chromatographic trace into two fractions. This is shown in Fig. 4 for a starch sample. Here one chromatographic trace represents the linear material amylose (the non-shaded region) and the other trace represents the branched material, amylopectin (the shaded region). The only additional information that is needed in these experiments is the pure component's branching index (g_M); also obtainable *via* independent light scattering experimentation. In theory a three component mixture can be similarly deconvoluted if additional experiments such as viscometry are also available. This possibility has not been experimentally verified, or for that matter, pragmatically motivated.

The utility of these light scattering experiments were demonstrated in the example of starch hydrolysis kinetics, where theoretical predictions were directly compared to experimental observations. Theory predicted that a linear relationship should exist between certain combinations of biochemical parameters (Michaelis constants and maximum reaction velocities). These predictions were verified by light scattering experiments, thus lending credibility to both the light scattering information and the kinetic models which we have developed. Key to these developments was the ability to deconvolute chromatographic traces in terms of the relative

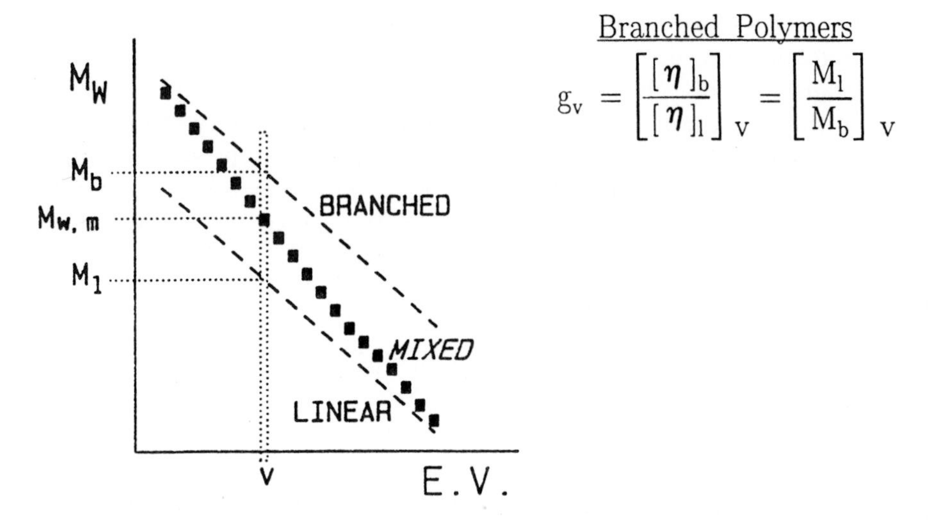

Figure 3. Definitions of branching parameters (g_m or g_v) obtained *via* the method of Zimm (g_m) or *via* SEC/LS (g_v).

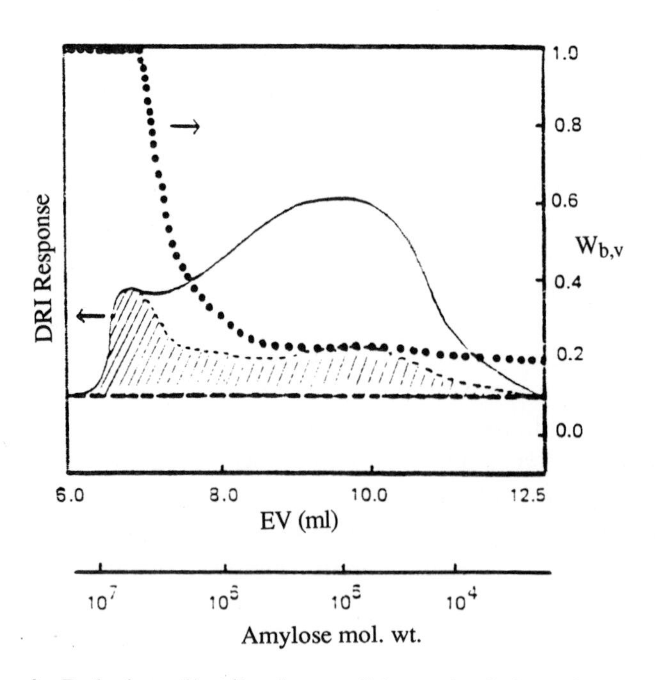

Figure 4. Relative distributions of branched (amylopectin) and linear (amylose) polysaccharides in a starch sample.

proportions of branched materials present in the reaction mixtures[14-18].

The limitations of light scattering as a means of analysis were also determined from these experiments in the case of starch-like polymers. Branching information on these biopolymer systems was readily available for materials whose molecular weights exceeded 10,000 daltons; and thus proved useful. For smaller molecular weight materials, other means of discrimination, not readily combined with chromatography were required; nuclear magnetic resonance (NMR) for example. These predictions, from theoretical grounds have led us into new pursuits beyond light scattering; but this subject is beyond the scope of this volume.

These advancements have, however, allowed us to begin to examine other physical chemical phenomena of starch based biopolymer systems. We have been able to show that the swelling and solubility of (poly)glucose materials of starches are strongly influenced by the material's granule size. Smaller granules will liberate their soluble materials more readily that their larger counter parts and the molecular weight dependency of the soluble material similarly yields predictable behaviour. This is depicted in Fig. 5. As a chemical engineer, these results are gratifying as they establish trends which seem to obey what my discipline would consider the laws of nature. Moreover, the relative proportion of branched materials is both of higher molecular weight than that of the linear amylose (as expected) and increases with increasing time/temperature contact; again following expected polymer trends. We were able to extend these analyses into the examination of the bulk rheology of starch systems, where again, obvious trends in the viscosity at a specific shear rate were apparent as we varied both the pasting time and the relative proportion of branched material; in these studies represented as the weight fraction of dent corn (a 25% amylopectin variety) mixed with waxy maize (a 100% amylopectin variety)[19]. In summary, our interests in investigating starch functionality stimulated us to develop on-line branching analysis, and this in turn provided us with the appropriate tools required by us to determine composition for mixed starch polymer systems of extreme industrial importance.

The methods of analysis chosen by my group, primarily laser light scattering also provided us with the capability to determine both biopolymer size and conformation. Again there is no need to review in detail the fundamental underlying theories as others before me in this volume have discussed these topics. Let me simply state that biopolymer molecular size can be determined from light scattering data by both batch experimentation and on-line flow through experimentation in combination with chromatography by examining the initial angular dependency at zero (or near zero) concentration of the biopolymer. When combined with a polymer separation device (such as a size exclusion chromatograph) dependencies of biopolymer molecular size *versus* column elution behaviour or biopolymer molecular weight can be established.

Biopolymer conformational information can be determined in two ways. If the biopolymer is of an appropriate molecular size (*i.e.* large enough), then conformation information can be directly determined by comparing the laser light scattering angular dependency to that expected by theory, or alternately one can infer conformational information by examining the trends in biopolymer molecular size with trends in its molecular weight. Clearly this latter method can only be performed in conjunction with a polymer separation scheme; such as that provided by a size exclusion chromatograph. More will be said of this below.

Theories of laser light scattering state that the form of the scattered light intensity function, $P(\theta)$, should vary with observational angle, polymer molecular size, and the wavelength of laser light employed[21]. How $P(\theta)$ varies will establish biopolymer conformation. We have established that the same device used on-line with our chromatographic separator is capable of such discrimination and experiment is in good agreement with theory, as shown in Fig. 6. Although these results are encouraging, there is a basic problem here. Fig. 6 shows data collected using 'model' polymer/particle systems, which are actually very large in comparison to those 'normally' encounter in chromatographic experiments. The value of the product grouping $q^2\langle R_g^2\rangle$ in these experiments is generally less than 1.5. In this range, it is nearly impossible to discriminate between different polymer conformations. This fact does not prevent us from examining polymer conformations *via* combined size exclusion chromatography and laser light scattering. Since size exclusion chromatography is capable of separating soluble polymers by molecular size (and for some pure biopolymer systems, this equates to a molecular weight separation device), it is possible to construct molecular size *versus* molecular

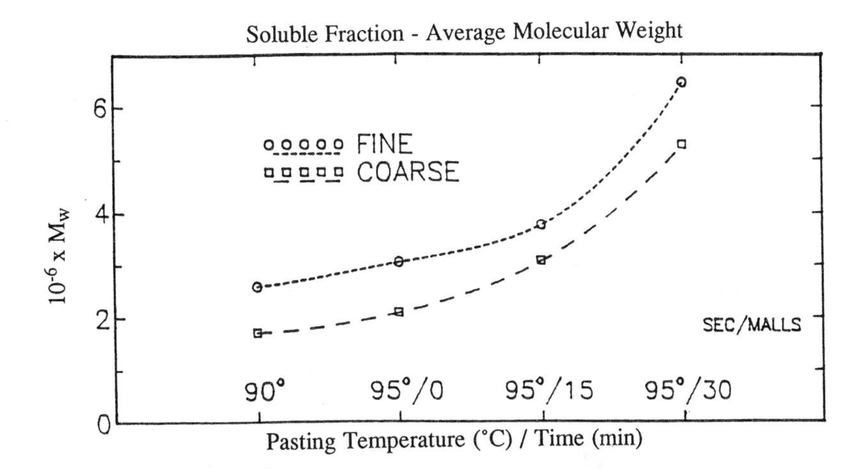

Figure 5. Distributions of branched (amylopectin) and linear (amylose) from the solubilised fraction extracted from large (coarse) and small (fine) starch granules.

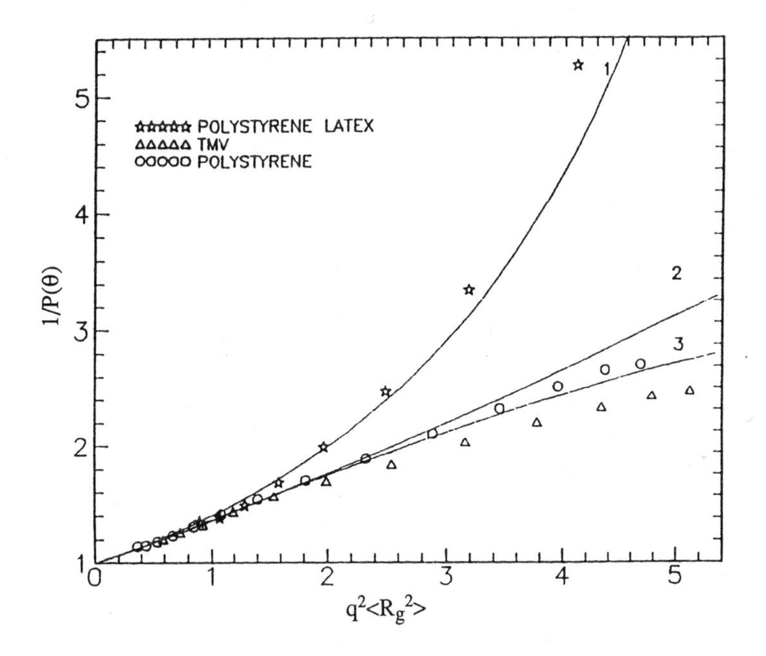

Figure 6. Comparison of reciprocal particle scattering factors for (1) spherical latex particles; (2) randomly coiled polystyrene; (3) rod-like TMV particles with their theoretical scattering curves.

weight functionalities and thus infer conformational information. Such experiments were performed with several biopolymer systems which we've examined. Shown in Fig. 7 is the relationship of amylose molecular size verses its molecular weight. Below approximately 200,000 daltons the dependency of molecular size to molecular weight is much stronger than above this value. This indicates that below 200,000 daltons, amylose is present in an extended conformation approximating that of a rod and above 200,000 daltons, amylose is more spherical in nature. These results are consistent with the findings of Burchard[22] and others[23]. Here, however, it must be stressed that the data shown in Fig. 7 was collected in approximately one hour using laser light scattering on-line with size exclusion chromatography. In summary, on-line laser light scattering photometry with SEC can readily determine biopolymer size and conformation.

As a final example, let me briefly discuss some preliminary results which we've obtained with a β 1-3 (poly)glucan slightly branched with short 1-6 glucose residues. This material is a yeast cell wall extract produced by a certain post fermentation processing strategy developed by a small biotechnology company in Worcester, Massachusetts, U.S.A.; Alpha Beta Technology. This material is purported to be able to exist in either a single helical or triple helical configuration. Our preliminary results seem to confirm this hypothesis. As shown in Fig. 8 there is clearly a discontinuity in the molecular weight elution volume data for this material which corresponds to Alpha Beta's claim of a reorientation of the β 1-3 (poly)glucan from single to triple helix. We are examining this further. The excitement here is that this reorientation or biopolymer conformation change seems to be recognized by leucocyte receptor cites. In the so-called triple helix conformation state, the white blood cells of humans exhibit phagocytic activity. If this claim is verified, then the β glucan may serve as an adjuvant and drug delivery system for targeting medicinal chemicals such as those thought active in fighting such diseases as acquired immune deficiency syndrome (AIDS). Alpha Beta has been able to establish that this beta glucan does possess bioactivity in mouse assays and human trials are now planned.

SUMMARY

What I have attempted to do here is to review several applications of

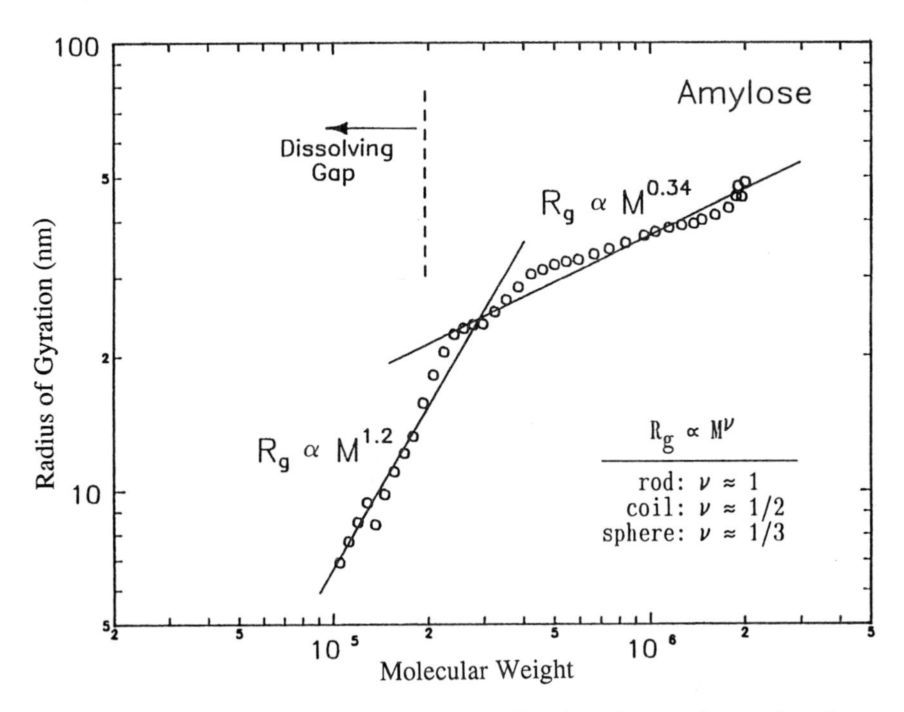

Figure 7. Relationship of amylose molecular size to its molecular weight as determined by combined SEC with multiple angle laser light scattering.

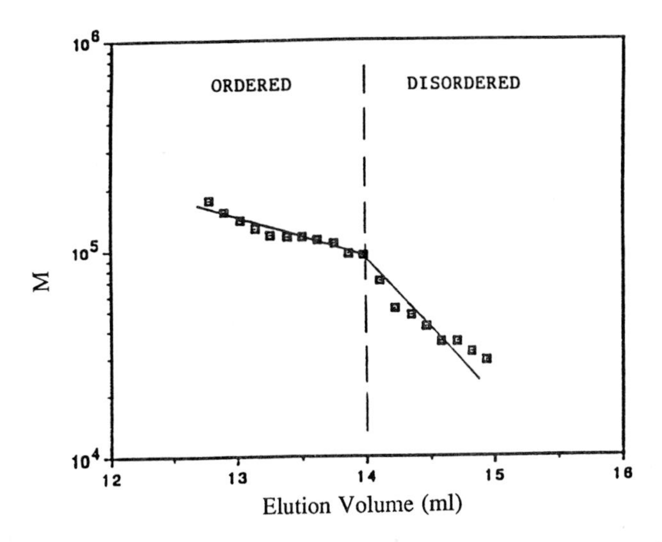

Figure 8. Molecular weight *vs.* SEC elution volume for a β 1-3 (poly)glucan showing discontinuity suggesting an order/disorder transition.

laser light scattering analysis for studies of biopolymer systems. Clearly many advancements have been made but much also needs to be done. These light scattering techniques need to be extended to other separation devices such as field flow separations. In addition, other coupled detection devices need to be developed. Laser light scattering is a powerful tool, but there are other tools that also need to be placed within the armamentarium of the biochemical scientist; particularly if he/she is to realize any of the larger more altruistic goals of this emerging science.

REFERENCES

1. J.E. Rollings, *Carbohydrate Polym.*, 1985, *5*, 37.
2. J.E. Rollings, A. Bose, M.R. Okos, and G.T. Tsao, *Adv. Chem. Series,* 1983, *203*, 345.
3. A. Bose, J.E. Rollings, J.M. Caruthers, M.R. Okos, and G.T. Tsao, *J. App. Polym. Sci.*, 1982, *27*, 795.
4. J.E. Rollings, A. Bose, M.R. Okos, and G.T. Tsao, *J. App. Polym. Sci.*, 1982, *27*, 2281.
5. G. Callec, A.W. Anderson, G.T. Tsao, and J.E. Rollings, *J. Polym., Polym. Chem.,* 1984, *22*, 287.
6. L.P. Yu and J.E. Rollings, *J. Appl. Polym. Sci.*,1987, *33*, 1909.
7. L.P. Yu and J.E. Rollings, *J. Appl. Polym. Sci.*, 1988, *35*, 1085.
8. L. Weaver, L.P. Yu, and J.E. Rollings, *J. Appl. Polym. Sci.*, 1988, *35*, 1631.
9. A. Corona, and J.E. Rollings, *Sep. Sci. Tech.*, 1988, *23*, 855.
10. R. Beri, M.S. Thesis, Worcester Polytechnic Institute, 1990.
11. J.E. Rollings, M.R. Okos, and G.T. Tsao, *ACS SYMP. SER.*,1983, *207*, 443.
12. J.E. Rollings and R.W. Thompson, *Ann. N.Y. Acad. Sci.*,1984, *434*, 140.
13. J.E. Rollings and R.W. Thompson, *Biotech. Bioeng.,* 1984, *26*, 1475.
14. J.T. Park, and J.E. Rollings, *Enzyme Microb. Tech.*, 1989, *11*, 334.
15. J.T. Park, L.P. Yu, and J.E. Rollings, *Ann. N.Y. Acad. Sci.*,1988, *542*, 53.
16. J.T. Park, M.S. Thesis, Worcester Polytechnic Institute, 1987.
17. S.W. Dean, M.S. Thesis, Worcester Polytechnic Institute, 1988.
18. J.T. Park, PhD Thesis, Worcester Polytechnic Institute, 1990.
19. A. Corona. M.S. Thesis, Worcester Polytechnic Institute, 1990.
20. B.H. Zimm and W.H. Stockmeyer, *J. Chem. Phys.*, 1949,

17,1301.
21. P.J. Flory, 'Principles of Polymer Chemistry', Cornell University Press, Ithica, N.Y. 1953.
22. W. Burchard, *Macromolecular Chem.* 1963, *64*, 110.
23. M. Kodema, H. Noda, T. Kamata, *Biopolymers*, 1978, *17*, 985.

20

Chain Rigidity of Polyuronates: Static Light Scattering of Aqueous Solutions of Hyaluronate and Alginate

By A. Gamini, S. Paoletti and F. Zanetti[1]

DIPARTIMENTO DI BIOCHIMICA, BIOFISICA E CHIMICA DELLE MACROMOLECOLE, UNIVERSITÀ DI TRIESTE, I-34127 TRIESTE, ITALY.

[1]CENTRO RICERCHE POLY-BIOS - LABORATORIO DI BIOPOLIMERI TECNOLOGICI, AREA DI RICERCA, PADRICIANO 99 - I-34012 TRIESTE, ITALY.

1. INTRODUCTION

Polysaccharides constitute major components of that part of the biological scenery which is often cumulatively called the "extracellular matrix". The matrix concept encompasses the capsule and exocellular slime of bacteria, the intercellular substances of algae, plants and animals.

The matrix biopolymers play many roles, such as - to mention only a few - to augment the mechanical stability through the formation of a three dimensional network, to ensure appropriate dynamic response to stresses and to create highly swollen environments with controlled permeability. Moreover, such matrix biopolymers participate in the immunological "intelligence" network involved in cell/cell and guest/host specific interactions. The scope of the present paper is to focus on a very limited selection of matrix polysaccharides, namely hyaluronic acid (HA) and alginates (AAs), which share several common features and biochemical challenges. They are rather "old" molecules on the phylogenetic scale and are found in such different organisms as bacteria, brown algae (AAs) and mammals (HA). It is interesting to note that in relation to man, both HA and AAs produced by bacteria can be connected to rather severe pathological states. *Pseudomonas* alginate is an active pathogenic agent in cystic fibrosis by interacting with mucin to yield "rheological clogs" in lungs[1]. In various diseases, *Streptococcus* hyaluronate acts as a more subtle masking coat, by all means similar to the HA of the invaded host, contributing to hide and protect the exogenous pathogens from immunological recognition[2].

Both HA and AAs are binary copolymers in which one of the co-monomers is a β (1 → 4) linked hexuronic acid: β-D-Glucuronic acid (β-D-GlcA) in HA and its C(2) isomer β-D-Mannuronic acid (β-D-ManA) in the alginates. Regular alternation of GlcA with a β(1 → 3) linked (2-deoxy, 2-acetamido)-D-Glucose co-monomer (*i.e.* N-acetyl-glucosamine, β-D-GlcNAc) renders HA a true regularly alternating co-polymer. Such a regular $(AB)_n$ arrangement is much rarer in the AAs, where a strictly alternating sequence of ManA with its α(1 → 4) linked co-monomer - the C(5) epimer α-L-Guluronic acid (α-L-GulA) - has been reported only in a limited number of bacterial alginates from *Pseudomonas aeruginosa* spp[3]. In bacteria, but especially so in algae, ManA, (M) and GulA, (G), residues are arranged in a blockwise pattern of homopolymeric $(M)_n$ and $(G)_n$ sequences of variable length, interspersed with (apparently) irregular alternating G/M sequences.

The interaction with bivalent ions, and in particular with the biologically relevant Ca(II) ions, discriminates between the $(G)_n$-rich alginates, on one side, and a group of polymers including HA, $(M)_n$-rich and alternating M/G alginates, on the other. In the former case, more or less rigid gels ensue, which *in vivo* play a determinant role in the build-up of the highly resistant mechanical architecture of the stipes of brown seaweeds. In the case of the latter group of polysaccharides, the tolerance to large concentrations of Me^{2+} ions is rather high, without generating a dramatic phase separation but, rather, highly viscoelastic solutions. The viscoelastic properties of aqueous solutions of HA and of some AAs are of vital importance in the *in vivo* performances of biological fluids, as well as in artificial operations like "viscosurgery" and "viscoimplementation" [4] for HA, or textile printing for AA.

There is a general consensus upon the idea that both AAs[5] and HA[6] are rather rigid and elongated polysaccharides.(Clearly, elongation and stiffness are not synonyms, although they are certainly correlated: there is often some ambiguity in the use of these concepts, but a detailed discussion cannot be given here). These extended conformations are assumed to be responsible for the peculiar rheological properties of the polymers and for their tendency to interpolymer interactions: Ca^{2+}-mediated side-by-side alignment of $(G)_n$ segments in gelling alginates and brush-like lateral attachment of proteoglycan subunits onto the long HA chain.

A second, albeit minor, aspect of interest arises from the fact that the biological importance and the increasing success of the practical applications of both HA and AAs have raised the demand for better molecular characterization of these polymers, including absolute determination of molecular weight (M) and the molecular weight distribution (MWD). The reported methods usually resort to an uncritical use of "relative" GPC calibrations: this has generated the wish to obtain reliable absolute MWD curves for these polymers to provide a critical data basis for assessing the potential pitfalls deriving from the use of conformationally very flexible polymers (dextran, pullulan) as standards in GPC experiments on HA or AAs.

In order to contribute to the verification of such a line of reasoning, we resorted to obtaining both laser light scattering (LLS) and viscosity data and to derive therefrom estimates of such parameters of stiffness as the persistence length, q, or the Kuhn length, λ^{-1}, on a series of samples including HA, G-rich- and M-rich alginates.

As an additional task, we wanted to consider the effect that an increasing acetylation of the vicinal C(2) and/or C(3) OH groups may have on the conformational properties of alginate chains. The reason for that interest stems from the observation that the presence of acetyl groups has been proved to be related to fundamental steps in the alginate biosynthesis and its *in vivo* modification to acquire the desired gelling ability[7]. In fact, the ultimate form of alginate in, say, seaweed stipes is accomplished by an *in situ* post-synthetic action of a mannuronate C(5) epimerase. A proper amount and stereochemical location of acetyl groups on (ManA)-rich alginate prevents an early epimerase activity which would dramatically convert alginate into its (GulA)-rich highly gelling form in those places (*e.g.* the fruiting bodies) where a very mobile, albeit viscous, polymer solution is required. Moreover, even a small percentage of acetyl groups was demonstrated to be able to modify the conformational behaviour of a bacterial alginate[8]. Preliminary LLS and viscosity data have been collected on a few chemically acetylated alginates in an attempt to address also that problem.

2. EXPERIMENTAL

Materials

Two alginate samples were used in this work: sample AAl was isolated from *Laminaria hyperborea* and obtained from PROTAN A/S (Drammen, Norway); the guluronate content was 72%, the fraction of GG diads 58% and that of the MG, GM and MM diads was altogether accounting for the remaining 42%. AA2 alginate was from *Macrocystis pyrifera* and obtained from the Kelco Division of MERCK (San Diego, CA, USA); the guluronate content was 42%, the fraction of GG diads 22% and that of the MG, GM and MM diads was altogether accounting for the remaining 78%. The monomer composition was determined by n.m.r. spectroscopy according to an established procedure[9].

For the preparation of the acetylated alginates a different batch of the *L.hyperborea* alginate from Protan was used. Its guluronate content slightly differed from the previous one, amounting to 68%. Details on the preparation procedure and on the methods for the determination of acetyl substitution are reported elsewhere[10].

The hyaluronic acid samples extracted from rooster comb (HA1, HA2, HA3, HA4, HA5, HA6, HA7) were kindly provided by FIDIA S.p.A. (Abano Terme, Italy).

All chemicals used were of analytical grade and MILLI-Q water (Millipore, MA, USA) was used throughout.

Methods

Light Scattering Measurements Low angle laser light scattering (LALLS) measurements were performed on an LDC-Chromatix CMX-100 apparatus operating at 632.8 nm used in the FIA (Flow Injection Analysis) mode. The same apparatus was used as a LALLS detector in the GPC experiments, coupled with a refractive index detector (Waters Model 410); the whole system was thermostatted at 25°C. A column set (TSK G6000PW + TSK G5000PW, LKB pharmacia) was used. Details of operation have been given elsewhere[11]. In the *multi-angular* mode (MALLS), the angular dependence of the intensity of the scattered light from polymer solutions was measured using a Wyatt Dawn-F photometer mounting a flow cell and a He-Ne laser source operating at 632.8 nm. The intensity of the scattered light was measured simultaneously by 15 photodiodes covering a 15°-145° range of the scattering angle, θ. Only the intensities measured in the interval 22°-120° were used to calculate the scattering function $Kc/R_{\theta,c}$ for each solution. The normalization of the photodiodes was performed before every set of measurements using a solution of low molecular weight sodium poly(styrenesulfonate), NaPSS, ($M \leq 10{,}000$) at an identical salt concentration as that of the subsequent polysaccharide solutions. During the measurements, the solutions were continuously pumped into the light scattering cell (flow rate 0.01 ml min^{-1}; B.Braun Melsungen reciprocating pump) after passing through either a 0.45 μm or a 0.22 μm Millipore filter. All polymer solutions used for light scattering measurements were dialyzed exhaustively against the proper NaCl concentration and the external solution was then used for dilutions.

Viscosity Measurements Capillary reduced specific viscosity of aqueous solutions of alginates in 0.10 M NaCl and of aqueous solutions of hyaluronates in 0.15 M NaCl were determined at 25°C in a Micro-Ubbelohde viscometer, Schott-Geräte (Type No. 53610). Intrinsic viscosity values were determined therefrom by analyzing the concentration dependence of the reduced specific viscosity and the logarithm of the relative reduced viscosity by use of the Huggins and Kraemer equations, respectively.

3. RESULTS

LLS Experiments

TILS conventional MALLS, LALLS in the FIA mode and GPC-LALLS experiments have been performed on aqueous solutions of the Na^+ form of hyaluronic acids and alginates. As an example, the Zimm plot of hyaluronate sample 7 in 0.15 NaCl and the absolute MWD curve of the AA1 alginate sample in 0.10 M NaCl are reported in Figs. 1 and 2, respectively. The macromolecular parameters as derived from LLS data are reported in Table 1a.

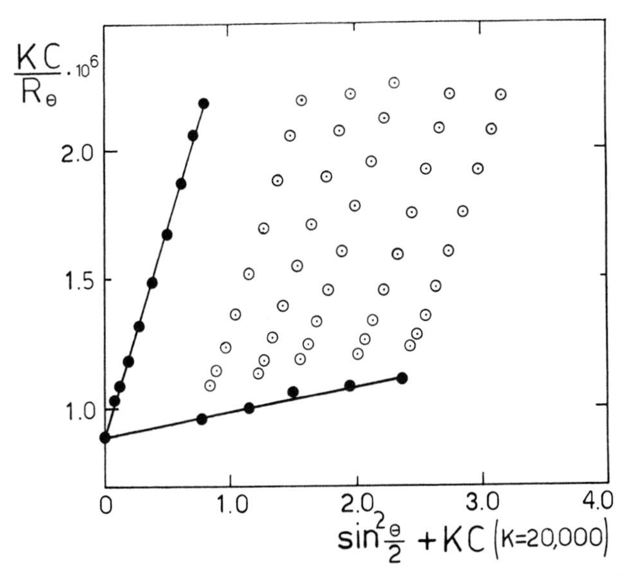

Figure 1. Zimm plot of HA7 hyaluronate sample in aqueous 0.15 M NaCl.

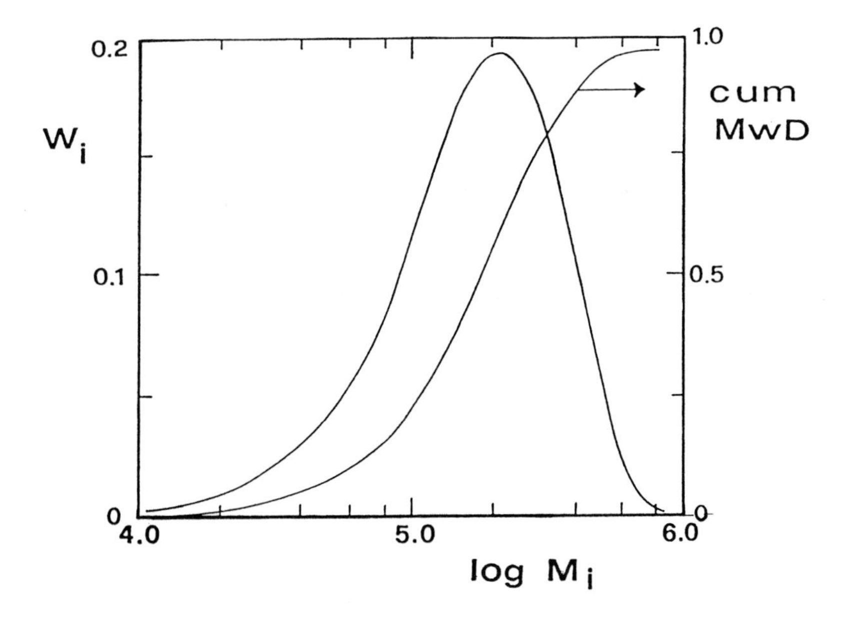

Figure 2. Absolute MWD curve for AA1 alginate sample in aqueous 0.1 M NaCl.

Table 1. Physico-chemical parameters of hyaluronate and alginate samples
(a) Parameters from TILS and GPC-LALLS experiments

Sample	$10^{-5} \times M_w$ [a]	$\langle S^2 \rangle_z^{1/2}$ (nm) [a]	$10^{-5} \times M_w$ [b]	$10^3 \times A_2$ (mol ml g^{-2}) [b]	$10^{-5} \times M_w$ [c]	$10^{-5} \times M_n$ [c]	M_w/M_n
HA 1	0.88	18	1.07	-2.6	0.80	0.54	1.48
HA 5	2.1	40	2.03	0.8	1.85	1.09	1.70
HA 6	8.6	95	8.3	1.9	7.9	4.29	1.84
HA 7	11.2	98	10.0	2.3	13.0	7.21	1.80
AA 1	2.17	59	2.14	4.7	2.06	1.27	1.62
AA 2	2.10	50	2.24	1.4	2.02	1.03	1.96

[a] From MALLS; [b] From LALLS; [c] From GPC-LALLS

(b) Viscosity parameters

Sample	$[\eta]$ (dl g^{-1})	k'	k' + k"	$10^{-5} \times M_v$
HA 2	3.17	0.63	0.52	1.12
HA 3	3.70	1.89	1.08	1.32
HA 4	3.96	0.72	0.57	1.48
HA 5	4.95	0.37	0.51	1.89
HA 6	14.95	0.42	0.53	7.41
HA 7	20.15	0.43	0.54	10.7
AA 1	7.11	0.386	0.518	2.10
AA 2	5.48	0.421	0.544	2.00

Capillary Viscosity

Capillary viscosity experiments provided the data for the evaluation of the $[\eta]$ and the Huggins and Kraemer constants for the same systems, which are reported in Table lb. From intrinsic viscosity and molecular weight data, the Mark-Houwink-Kuhn-Sakurada (MHKS) coefficients were obtained for the three ionic polysaccharides here investigated: they are reported in Table 2. For alginates, the MHKS coefficients were obtained by a fitting procedure of the absolute MWD from GPC measurements as described elsewhere[11]. K and a of hyaluronic acid were obtained, as usual, by a linear least-squares fit of the (log $[\eta]$) data plotted against (log M).

Table 2. Mark-Houwink-Kuhn-Sakurada parameters of hyaluronate and alginate samples.

Sample	NaCl (M)	a	K [a]
HA (2 to 7)	0.15	0.81	2.63×10^{-4}
AA 1	0.10	1.13	6.90×10^{-6}
AA 2	0.10	0.92	7.30×10^{-5}

[a] $[\eta]$ in dl g^{-1}

Calculation of the Stiffness Parameters

From the dependence of the radius of gyration, $\langle S^2 \rangle^{1/2}$, on the molecular weight, M, the rigidity of semiflexible polymers can be estimated by using the Doty-Benoit equation[12] obtained for worm-like chains in unperturbed conditions:

$$\frac{\langle S^2 \rangle}{q^2} = \frac{x}{3} - 1 + \frac{2}{x} - 2\,\frac{1-e^{-x}}{x^2} \qquad (1)$$

where q is the Porod persistence length and x = L/q, L being the chain contour length.

$\langle S^2 \rangle_z^{1/2}$ and M data obtained on hyaluronate solutions in 0.15 M NaCl from light scattering experiments are reported in Fig. 3 together with the theoretical curves calculated from Equation 1 for different values of the Kuhn statistical segment length λ^{-1} (= 2q). The mass per unit length, M_L, was fixed at 410 nm^{-1} corresponding

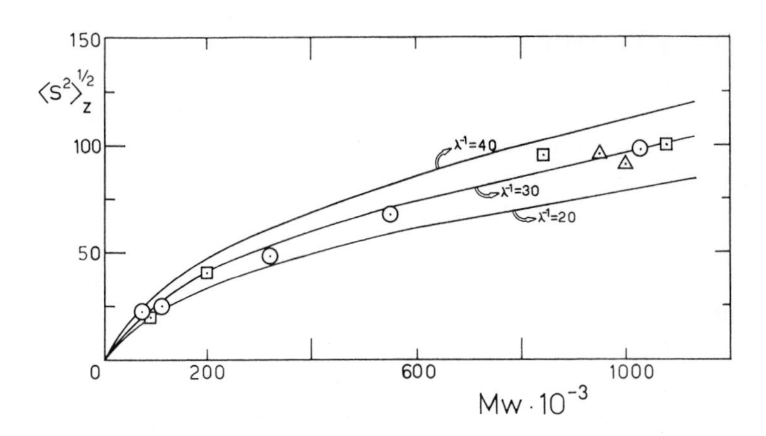

Figure 3. $<S^2>_z^{1/2}$ dependence on M_w of HA samples:
square data of this work (NaCl 0.15 M);
O data from ref. 14 in 0.5 M NaCl;
Δ data from ref. 15 in 0.2 M NaCl.

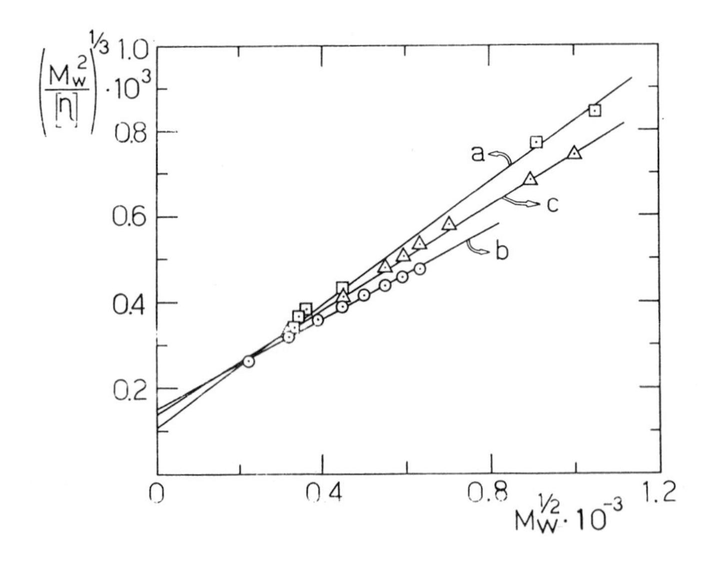

Figure 4. $(M_w^2/[\eta])^{1/3}$ dependence on $M_w^{1/2}$:
a hyaluronate samples in 0.15 M NaCl;
b AA1 sample in 0.1 M NaCl;
c AA2 sample in 0.1 M NaCl.

to the theoretical value of the single helix hyaluronate structure[13]. Data from refs. 14 and 15 are also reported for comparison.

Fig. 3 shows that all experimental points are in reasonable agreement with the theoretical values calculated with a Kuhn length of 30 nm.

If $[\eta]$ data as a function of the molar mass M are available the Kuhn statistical segment length can be obtained using the plot of $(M^2/[\eta])^{1/3}$ *vs.* $M^{1/2}$ proposed by Bohdanecky[16]. Such a plot is shown in Fig. 4 for the three ionic polysaccharides here investigated. To construct the plot of Fig. 4, the data of Table 1 were used for hyaluronate samples, while the Mark-Houwink-Kuhn-Sakurada coefficients of Table 2 were used for the two alginates.

The experimental points of Fig. 4 are reasonably fitted by lines whose intercepts, A_n, and slopes, B_n, are related to M_L and λ^{-1}, respectively, by:

$$A_n = \Phi_{0,\infty}^{-1/3} A_0 M_L \tag{2}$$

$$B_n = \Phi_{0,\infty}^{-1/3} B_0 (\lambda^{-1}/M_L)^{-1/2} \tag{3}$$

according to Bohdanecky analysis, with $\Phi_{0,\infty} = 2.86 \times 10^{-23}$ (ref. 17).

Both A_0 and B_0 tabulated in ref. 16 are functions of the reduced diameter $d_r = d/\lambda^{-1}$, although B_0 is nearly constant.

In Table 3 the fitted parameters M_L and λ^{-1} are reported together with the theoretical M_L values.

The unperturbed dimensions of monodisperse polymers can be calculated from data obtained in the presence of excluded volume effects by means of the equation proposed by Berry[18]:

$$1.42 \times 10^{-24} A_2 M^{1/2} = -A^3 + \frac{6A\langle S^2\rangle}{M} \tag{4}$$

where:

$$\lambda^{-1} = A^2 M_L \tag{5}$$

in the limit of infinite chain length. The Kuhn length values for HA, AA1 and AA2 have been calculated from the experimental second

Table 3. Calculated parameters for the evaluation of chain stiffness.

Sample	NaCl (M)	λ^{-1} (nm)	Method	M_L^{ex} (nm^{-1})	M_L^{th} (nm^{-1})	10^3 x dr
HA	0.15	22	2	415	410	5
(present work)		30	1	-	410	-
		18	3	-	410	-
HA [a]	0.20	18	(see text)	-	-	-
HA [b]	0.50	8	4	-	400	-
AA1	0.10	52	2	520	480-510	3
		40	3	-	500	-
G-rich-alginate[c]	0.5	38	2	(650)	500	(3)
		140-180	1	-	456	-
AA2	0.10	29	2	463	456	5
		29	3	-	456	-
M-rich-alginate[c]	0.20	30-28	2	(see text)	456	2-40
	0.50	26-23	2	(see text)	456	1-40
	0.50	100	1	-	456	-
M-rich-alginate[d]	0.10	34	-	-	-	-

1. Values calculated according to eq. 1; 2. Values calculated according to eqs. 2-3; 3. Values calculated according to eqs. 4-5; 4. Values calculated according to ref. 17.
[a] Data taken from ref. 15; [b] Data taken from ref. 19; [c] Data taken from ref. 20; [d] Data taken from ref. 5.

virial coefficients, A_2, and radii of gyration, $<S^2>$ (see Table 1a); they are reported in Table 3, for the comparative case of M=200,000.

For comparison purposes, Kuhn lengths estimated or taken from different sources are also tabulated in Table 3 (see further in the Discussion).

Determination of MWD Curves

All MWD curves obtained from GPC-LALLS (not all reported here for clarity) are monomodal, the polydispersity index M_w/M_n being always rather small, indicating a similarity of effects of the purification procedures which was the same for all samples (see Table 1a). Dextrans (M_w/M_n = 1.2÷2.0, Pharmacia) in 0.15 M NaCl and pullulans (M_w/M_n = 1.06÷1.14, Polymer Laboratories) in 0.10 M NaCl were used as standards to construct universal calibration plots to evaluate in the GPC mode the MWD curves of HA and AAs, respectively. The results of the calculated M_w values for HAl, and AA2 are reported in Table 4.

Table 4. Calculated M_w values of hyaluronic acid and alginate from GPC methods.

Sample	M_w		
	GPC-LALLS	Universal calibration	
		Pullulans	Dextrans
HA 1	80,000	75,000	500,000
AA2	202,000	880,000	2,400,000

Acetylated Alginate Derivatives

GPC-LALLS and capillary viscosity experiments data taken from ref 21 carried out on four alginate samples at different degree of acetylation (D.A.) have produced M_w and $[\eta]$ values reported in Table 5.

Table 5. Physico-chemical parameters of acetylated alginates in 0.1 M NaCl.

% Acetylation	0	5.3	10.3	117
$10^{-3} \times M_w$	211.7	172.0	162.2	247.1
$[\eta]$ (dl g^{-1})	5.82	6.26	7.21	7.43[a]
k'	0.49	0.42	0.41	0.46[a]
k' + k"	0.59	0.54	0.53	0.55[a]
$10^5 \times [\eta]/M_w$	2.8	3.64	4.45	3.0

[a] Obtained by extrapolation to 0.1 M NaCl.

4. DISCUSSION

Evaluation of the Stiffness of HA and AAs

Both hyaluronic acid and alginate are commonly considered to behave, in aqueous solutions, as extended polymers. In particular, the rigidity of alginates was theoretically[22] and experimentally[20,22] correlated with the chemical composition of the polymer, as the stiffness of the three types of sequence in alginate increases in the order: MG≤MM<GG.

The MHKS parameters reported in Table 2 are in very good agreement with those previously reported in the literature for hyaluronates[14,23] and for alginates of chemical composition comparable to that of the AA1[20] and AA2[5,20] samples. Among the alginates the larger a exponent value of the AA1 sample reported in Table 2 can be well accounted for by its higher content of guluronate residues. Overall inspection of the data of Table 2 indicates that the flexibility of the polymer chain should increase in the order AA1<AA2<HA.

This finding is confirmed by the Kuhn statistical segment length values, λ^{-1}, reported in Table 3.

The M_L values calculated from the viscosity data by use of Eqs. 2 and 3 agree with the theoretical ones to within 10%.

For hyaluronate samples, the λ^{-1} value ($\approx$ 22 nm) derived using the same equations and viscosity data is not too different from that of 30 nm obtained from light scattering experiments (Fig. 3 and Table 3). The slight difference between the two λ^{-1} values can be reasonably accounted for by the small, albeit not negligible, polydispersity of the samples.

Recently, a value of 18 nm was reported for the Kuhn length of hyaluronate in 0.2 M NaCl [15]. In this case only a single pair of values of $\langle S^2\rangle_z^{1/2}$ and M_w was used and a gaussian coiled structure was assumed to calculate λ^{-1}.

It is worth noting that λ^{-1} values ranging from 50 nm to 5 nm can be found in the literature (see Table 3 and ref. 15). A Kuhn length of 25 nm - 30 nm should be safely accepted not only in view of the agreement between the λ^{-1} value calculated from light scattering and that from viscosity, but also because of the agreement between our light scattering data and those from the literature obtained at higher ionic strengths, reported in Fig. 3.

The present data of Table 3 clearly indicate that the AA1 alginate sample is stiffer than the alginate with a lower content of guluronate residues, AA2, as expected. Table 3 also reports stiffness parameters calculated or taken from literature data on G-rich[20] and M-rich[5,20] alginates. The agreement is encouraging both with the value of ref. 5 and with those derived from the viscosity data of ref. 20, using the Bohdanecky procedure. In this case, the Kuhn length was obtained by the only use of Eq. 3 keeping M_L fixed to the theoretical value. Both the non-negligible differences in the ionic strength and in the chemical composition which likely exist between our alginates and those of the literature can well account for the small differences between the Kuhn length values. λ^{-1} estimates have also been obtained by using the radii of gyration of several fractions of different molecular weight reported in ref. 20 for the same G-rich and M-rich alginates. However, in this case an analysis of the $\langle S^2\rangle_z^{1/2}$ dependence on the weight-average molecular weight at 0.5 M NaCl yields unrealistically high Kuhn lengths (values $\geq$100 nm, reported in Table 3). This discrepancy seems to derive more from an internal inconsistency of the data of ref. 20 than from a conflict between viscosity and light-scattering derived λ^{-1} values. In fact, the comparison of the Kuhn lengths calculated for our HA, AA1 and AA2 samples using [η] and <S^2> data is by far more satisfactory.

In conclusion, although the data collected in this Chapter show that the stiffness of the three polysaccharides considered is only moderately high, the results safely enable one to look upon these polymers as semi-flexible ones (in the increasing order of stiffness HA<AA2<AA1), to which the "worm-like" chain model can be well applied. The Kuhn lengths here reported, very similar to those of cellulose derivatives ($\lambda^{-1} \approx$ 34 nm, ref. 24), fall in between such a small value as that of amylose ($\lambda^{-1} \approx$ 2.1 nm, ref 25) and those of much stiffer polysaccharides like xanthan ($\lambda^{-1} \approx$ 240 nm, ref. 26) and schizophyllan ($\lambda^{-1} \approx$ 400 nm, ref. 27). Not surprisingly, on passing from amylose to schizophyllan one can note an increasing tendency to give structures of higher molecularity. Although a higher molecularity is a source of stiffness by itself, an "intrinsic" conformational stiffness of the isolated chain is likely to be necessary to facilitate chain-chain association by reducing the loss of conformational entropy.

Influence of the Conformational Stiffness on the Determination of the MWD Curves

The tendency of hyaluronate and alginates to assume extended conformations in aqueous solution makes the molecular dimensions determined by GPC measurements through relative calibration curves unreliable. As an example, M_w values for HAl (M_w 80,000) and AA2 (202,000) obtained using a relative GPC calibration based on pullulans were 230,000 and 1,200,000, respectively. A similar use of dextrans yielded M_w values practically devoid of any physical meaning! It is surprising that even for the more flexible hyaluronate sample the analysis of the relative MWD curve did not work out, especially in consideration of the fact that, as the data in Fig. 3 show, the electrostatic excluded volume effects should be in this case negligible. Better estimates of M_w are obtained by using the "universal calibration" method and fairly well monodisperse pullulan samples: the data of Table 4 indicate that whereas the calculated M_w of HAl is very well evaluated, there is still a discrepancy for AA2 for which the plausible explanation is again traced back to the higher stiffness of that polymer with respect to HA. Absolute methods based on LS techniques once more have proved to be the only ways to provide reliable M_w and MWD values.

Acetylated Alginate Derivatives

Even at a very preliminary level, the M_w and $[\eta]$ data of Table 5 show that on passing from 0% to about 10% acetylation, a strong increase of chain dimensions develops, which is even more pronounced if one takes into account the decrease of M_w. In a purely operational way and as a zeroeth order approximation, one may use the ratio of $[\eta]$ to M_w to monitor the change in stiffness: $[\eta]/M_w$ increases by about 50% on passing from 0% to 10.3% acetylation. A further, major increase of the acetyl content to 117% brings the ratio of $[\eta]$ to M_w (3×10^{-5}) back to about the value for the deacetylated polymer (2.8×10^{-5}). At the present stage, one might somewhat speculatively assume that the initial insertion of acetyl groups ($0 \leq$ D.A. ≤ 0.10), which are almost exclusively confined to the ManA groups[10], introduces severe steric hindrances on these rather flexible sugars. In the last case, D.A.= 117%, on the other hand, the exogenous ester groups are largely distributed also on the GulA residues. In alginate, the latter are known to be mostly present in the stiff $^1C_4$ [22] conformation and therefore one might propose that the relative decrease in the hydrodynamic dimensions stems from an increased conformational flexibility. It is possible to envisage at least two explanations for the reported loss of stiffness upon extensive acetylation: either a relaxation of the strained ribbon-like local conformation of (GulA) sequences, or a shift of the conformational equilibrium of GulA from the $^1C_4$ to the $^4C_1$ form. Simple inspection of mechanical models shows that a severe steric hindrance is produced when the acetyl group is at position C(3) of $^1C_4$ GulA, which can be relaxed only by flipping the ring to the $^4C_1$ conformation. Acetylation on C(2) does not seem to require any chair conformational change. The hypothesis of a change in the ring conformation agrees with the values of the characteristic ratio, C_∞, calculated for the GulA homopolymers in the $^1C_4$ and in the $^4C_1$ conformations, which are 251 and 31, respectively[22]. On the other hand, conformational calculations give for the ideal copolymeric chain, $(\text{ManA-GulA})_n$, the values of C_∞ of 13 and 25 for GulA in $^1C_4$ and $^4C_1$ chair forms, respectively[22]. If the hypothesis of the shift of the conformational equilibrium of GulA from $^1C_4$ to $^4C_1$ is valid, then one should conclude that for high D.A. an acetyl substitution on C(3) occurs to a large extent on the homo-polymeric GulA sequences, rather than on the copolymeric ones. This might find support from the observed substantial decrease in Ca^{2+}-binding ability and gel strength of highly acetylated alginates, since the $(GulA)_n$ sequences are held responsible for those binding properties. Further extensive work is needed to prove this hypothesis, which might have some bearing on the biological significance of the presence of acetyl groups in bacterial alginates.

ACKNOWLEDGEMENTS

The authors gratefully acknowledge Prof. V. Crescenzi for helpful discussions, the financial support of the University of Trieste (Italian MURST Ministry funds) and that of FIDIA S.p.A., Abano Terme (Italy) (for F.Z.).

REFERENCES

1. F.A. Mian, T.R. Jarman and R.C. Righelato, *J. Bacteriol.*, 1978, *134*, 418.
2. Q.N. Myrvik, 'The Biology of Hyaluronan', John Wiley & Sons, Chichester (Ciba Foundation Symposium *143*), 1989, 165.
3. G. Skjåk-Bræk, H. Grasdalen and B. Larsen, *Carbohydr. Res.*, 1986, *154*, 239.
4. E.A. Balazs, 'The Biology of Hyaluronan', John Wiley & Sons, Chichester (Ciba Foundation Symposium *143*), 1989, 265.
5. O. Smidsrød and A. Haug, *Acta Chem. Scand.*, 1968, *22*, 797.
6. J.Scott, 'The Biology of Hyaluronan', John Wiley & Sons, Chichester (Ciba Foundation Symposium *143*), 1989, 6.
7. G. Skjåk-Bræk, B. Larsen and H. Grasdalen, *Carbohydr. Res.*, 1985, *145*, 169.
8. F. Delben, A. Cesaro, S. Paoletti and V. Crescenzi, *Carbohydr. Res.*, 1982, *100*, C46.
9. H.Grasdalen, B.Larsen and O. Smidsrød, *Carbohydr.Res.*, 1979, 23.
10. G. Skjåk-Bræk, S. Paoletti and T. Gianferrara, *Carbohydr.Res.* 1989, *165*, 119.
11. A. Martinsen, G. Skjåk-Bræk, O. Smidsrød, F. Zanetti and S.Paoletti, *Carbohydr. Polymer*, 1990, *15*, 171.
12. H. Benoit and P. Doty, *J. Phys. Chem.*, 1953, *57*, 958.
13. E.D.T. Atkins, D. Meader and J.E. Scott, *Int. J. Biol. Macromol.*, 1982, *2*, 318.
14. R.L. Cleland and J.L. Wang, *Biopolymers*, 1970, *9*, 799.
15. W.F. Reed, C.E. Reed and X. Li, *Biopolymers*, 1989, *28*, 1981.
16. M. Bohdanecky, *Macromolecules*, 1983, *16*, 1483.

17. H. Yamakawa and M. Fujii, *Macromolecules*, 1974, *7*, 128.
18. G.C.Berry, *J. Chem. Phys.*, 1966, *44*, 4550.
19. R.Cleland, *Archives of Biochemistry and Biophysics*, 1977, *180*, 57.
20. W. Mackie, R. Noy and D.B. Sellen, *Biopolymers*, 1980, *19*, 1839.
21. G. Skjåk-Bræk, F. Zanetti and S. Paoletti, *Carbohydr. Research*, 1989, *185*, 131.
22. O. Smidsrød, R.M.Glover and S.G.Whittington, *Carbohydr. Res.*, 1973, *27*, 107.
23. E. Shimada and G. Matsumura, *J. Biochem.*, 1975, *78*, 513.
24. N.S. Schneider and P.Doty, *J. Phys. Chem.*, 1954, *58*, 762.
25. W. Banks and C.T. Greenwood, *Makromol. Chem.*, 1963, *67*, 49.
26. T. Sato, T. Norisuye and H. Fujita, *Macromolecules*, 1984, *17*, 2696.
27. T. Yanaki, T. Norisuye and H. Fujita, *Macromolecules*, 1980, *13*, 1462.

21

Laser Light Scattering Studies of Polysaccharide Gels

By P.M. Burne and D.B. Sellen

ASTBURY DEPARTMENT OF BIOPHYSICS, THE UNIVERSITY, LEEDS. LS2 9JT U.K.

1. INTRODUCTION

Several authors have investigated the Rayleigh linewidth of light scattered from the gels of cross-linked flexible polymers such as polyacrylamide[1-4]. The width of the spectral broadening varies as $\sin^2(\theta/2)$, and diffusion coefficients may be calculated in the same way as for a macromolecular solution[5]. According to the theory of Tanaka *et al.*[1], this corresponds to freely diffusing fluctuations in polymer segment density and is equal to the ratio of the longitudinal elastic modulus to the force per unit volume required to maintain unit relative velocity between the polymer network and solvent. The degree of spectral broadening has, in general, been found to be small enough (*ca.* 10%) for the optical beating to be regarded as largely heterodyne, most of the scattered light arising from a stationary component in the density fluctuations.

For gels of stiff polymer chains, such as polysaccharides with a randomly branched fibrous structure, the degree of spectral broadening is much smaller (*e.g.* less than 0.01% for a 4% agarose gel), and the spectral linewidth does not vary as $\sin^2(\theta/2)$, much less variation being observed[6,7]. These gels are regarded as being almost completely stationary at a molecular level. The virtual absence of spectral broadening facilitates the measurement of well defined diffusion coefficients of compact macro-molecules such as globular proteins or dextran fractions within these gels, thus enabling the interstitial spaces to be investigated and estimates made of the mean mass per unit length of the fibrous network[6,7]. To date calcium alginate[6] and agarose[7] gels have been investigated in this way.

Similar measurements have been made with polyacrylamide gels[4] but here there is a complex interaction between the diffusion coefficients of the gel and the diffusing macromolecule - a paper detailing the fundamental differences between the two types of gel has been published[8].

The aim of the present work has been to extend the polysaccharide studies to those of microbial origin and results for deacetylated gellan will be given here.

2. EXPERIMENTAL

Photon Correlation Spectroscopy Measurements

Measurements were made with equipment previously described[9,10] modified for photon counting[4] with standard electronics from Malvern Instruments, including a 64-channel multi-bit correlator in which the penultimate 8 channels were replaced by a 64-channel delay card. The optical part of the apparatus was mounted on an air bed to minimize spurious oscillations in the autocorrelation function resulting from oscillations in the specimen[7,11]. The laser was a Spectra Physics type 125A (He-Ne 50mW).

The spectrally unbroadened component of light scattered from gels, which usually accounts for most of the scattered intensity, gives rise to a stationary component in the speckle pattern. Thus if measurements are made in the normal manner, the relative contribution of the heterodyne and homodyne terms to the temporal autocorrelation function, and also the zero delay time value relative to the baseline, depends upon which part of the stationary speckle pattern is being observed[4]. For a solution of freely diffusing macromolecules where the speckle pattern is entirely mobile, the measured autocorrelation function is given by

$$G^{(2)}(\tau) = N_S \langle n \rangle_t^2 [1 + \beta(g^{(1)}(\tau))^2] \quad (1)$$

if a full multi-bit correlator is used. N_S is the number of samples, $\langle n \rangle_t$ is the time-averaged counts per sample time, τ the correlation time, $g^{(1)}(\tau)$ is the first-order normalized correlation function of the scattered electric field, and β is the spatial coherence factor, which has a maximum value of unity and depends on the area of speckle pattern observed by the scattered light receiver system. For a single diffusing species $g^{(1)}(\tau) = \exp[-K^2 D\tau]$ where D is the diffusion coefficient and $K = (4\pi/\lambda)\sin(\theta/2)$. For a polydisperse system such as dextran, $g^{(1)}(\tau)$ is the summation of such terms weighted proportional to the scattered intensities, the initial slope of a plot of log $[g^{(1)}(\tau)]$ against τ yielding D_z. For a gel exhibiting partial spectral broadening equation (1) becomes:

$$G^{(2)}(\tau) = N_s\langle n\rangle_t^2 [1 + \beta\{2r(1-r)g^{(1)}(\tau) + r^2(g^{(1)}(\tau))^2\}] \quad (2)$$

where $r = \langle n_b\rangle_t/\langle n\rangle_t$. $\langle n_b\rangle_t$ is the time averaged counts per sample time associated with the broadened component and $\langle n\rangle_t$ is the time averaged total counts. However, because there is a stationary component in the speckle pattern, $\langle n\rangle_t$ and hence r vary spatially. In order to find the true degree of spectral broadening it is necessary either to scan angles of scatter to either side of the desired angle, or to make separate measurements using incoherent light. On the assumption of pure heterodyne beating, the second term in eq. (2) becomes $2r\beta g^{(1)}(\tau)$ so that the normalized function $g^{(1)}(\tau)$ may readily be determined. r and hence $\langle n_b\rangle_t$ may be found if β has previously been determined by making measurements on a solution (eq.(1)). The method previously described[4] in which the autocorrelation function is recorded whilst scanning angles of scatter so as to obtain all the parameters from one experiment was not used in the present work, as it induced an unacceptable level of oscillation in the specimen, although a limited number of scans of total intensity were made to ascertain degrees of spectral broadening.

Preparation of Gels

Gels were prepared by dissolving deacetylated gellan plus dextran sufficient to give an appropriate level of scatter in a hot aqueous solution of 0.025 M NaCl and 0.01 M sodium azide, and allowing to cool slowly. At gellan concentrations of less than 0.7% the gels were allowed to fill the light scattering cell which consisted of a 20 mm diameter weighing bottle. At higher concentrations, gels were cast as cylinders 6 mm in diameter and supported with their axis vertical in the light scattering cell, in which they were surrounded by a dextran solution of the same concentration as the dextran within the gel. No detectable change in volume of any of the gels was observed on storage. The decrease in gel dimensions with gel concentration is necessary to reduce secondary scatter to an acceptable level; the criterion being that a single plot of $R_\theta\lambda^4$ against $\sin^2(\theta/2)$ $(1/\lambda^2)$ is produced when total intensity light scattering experiments are performed with incoherent light of wavelengths 436 nm and

546 nm [4,6,7]. This indicates that there is no significant redistribution of scattered intensity with angle.

Dextrans in the molecular weight range 40 kg mol^{-1} to 8300 kg mol^{-1} were used and their hydrodynamic diameters were calculated from diffusion coefficient measurements in solution. Whereas dextrans are not ideal as marker macromolecules, they represent the only readily available system which produces a suitable range of hydrodynamic diameters.

3. RESULTS AND DISCUSSION

The degree of spectral broadening of light scattered from deacetylated gellan in the absence of diffusing dextran was found to be *ca.* 0.3%, with a very broad range of relaxation times. In this respect therefore gellan gels are like the gels of the other polysaccharides investigated and do not have a highly mobile structure.

Fig. 1 shows a typical plot of $g^{(1)}(\tau)$ for dextran diffusing in solution and in gellan gel. As has recently been shown to be the case with calcium alginate gels[12], the distribution of diffusion coefficients is much larger in the gel than in solution, and plots on a scale of log τ are in general necessary to represent all the data. The implication is that there is a wide distribution of localized diffusion coefficients within the gels due to their heterogeneity. However, it is difficult to postulate any distribution which would lead to the width of the $g^{(1)}(\tau)$ against τ curves observed, since on the basis of current concepts the dextran should partition within the gel in favour of the faster diffusing regions. However our theoretical work so far has not taken

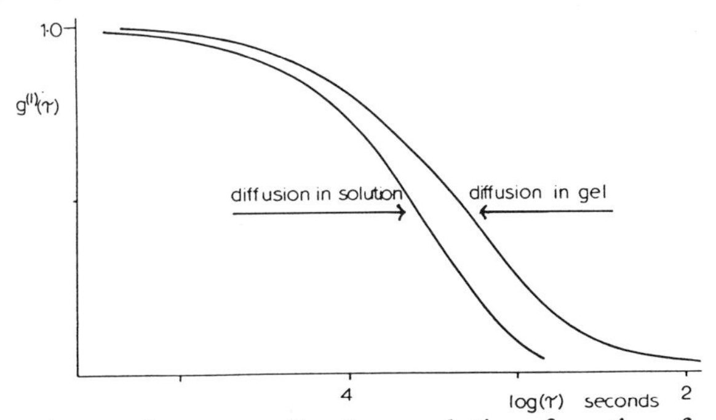

Figure 1. First order normalized correlation function for a dextran of molecular weight 1400 kg mol^{-1}, diffusing in solution and in a deacetylated gellan gel of 0.5% concentration.

account of the polydispersity of the dextran fractions and work along these lines is continuing. It must be emphasized that the broadening of the curves increases with the overall reduction in diffusion coefficient, and that in the case of the lower molecular weight fractions, which are not impeded by the gel, identical $g^{(1)}(\tau)$ plots are obtained in gel and solution.

In order to obtain a complete $g^{(1)}(\tau)$ curve it is necessary with the present equipment to make five determinations of $G^{(2)}(\tau)$ with differing time scales. These are then matched up automatically with a microcomputer (BBC model B). It has been found for all polysaccharide gels so far that there are some very long term (of the order of minutes) fluctuations in scattered intensity of uncertain origin, and the range of $G^{(2)}(\tau)$ has to be extended far enough for the baseline to be determined from the shape of the curve. Most of the data presented in the present report have however been obtained by running the autocorrelator alternately with different sample times, one sixteen times the other, with the appropriate scaling factor being introduced automatically. The integration time for each run was of the order of several seconds. Channel contents were then accumulated for successive runs on the microcomputer, taking the shorter time for the first 48 channels, and the longer sample time for the last eight. In this way the effect of the delay card was increased sixteen fold. A second order polynomial fit was made to the logarithm of the data after subtracting the baseline determined from the delayed channels, and D_z determined. This also facilitated extrapolation back to zero correlation time and the determination of the exp(-1) points in $g^{(1)}(\tau)$. Corrections for variation in room temperature were made by assuming the validity of the Stokes-Einstein equation.

Data were analyzed using the method of gel concentration/dextran hydrodynamic diameter superposition previously described[7]. Figs. 2a and 2b show plots respectively of D_z/D_{0z} (the ratio of z-average diffusion coefficient in gel to that in solution), and ratio of exp(-1) points in $g^{(1)}(\tau)$, against logarithm of hydrodynamic diameter (found from D_{0z}). The data at differing concentrations have been displaced horizontally to yield a unified plot at a concentration C_0=0.7%. It must be emphasized that each experimental point necessarily represents a different gel preparation, and it is the variability in these, rather than experimental errors, which gives rise to the scatter in the points. The fact that a unified

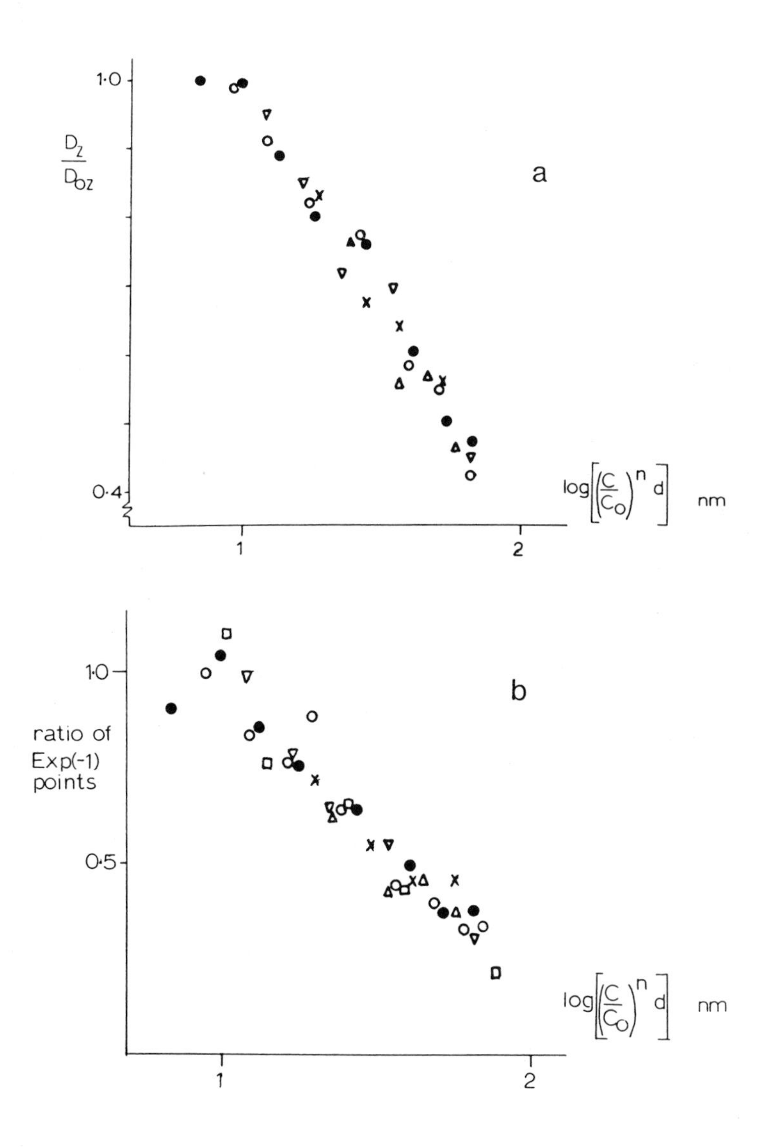

Figure 2. (a) z-average diffusion coefficient D_z relative to that in solution D_{0z}, and (b) corresponding ratio of exp(-1) points in $g^{(1)}(\tau)$ curves, for dextran fractions within deacetylated gellan gels of various concentration C, as a function of dextran hydrodynamic diameter d. The plots at various concentrations have been superimposed by horizontal displacement (C_0=0.7% see text) x 0.2%, Δ 0.3%, O 0.4%, ● 0.5%, ∇ 0.7%, square 1.0%.

plot may be obtained means that the functions describing both curves are of the form

$$f\{g(C)\ d\} \tag{3}$$

and g(C) may be found from the horizontal shifts required. Fig. 3 shows that g(C) is proportional to $C^{0.43}$. By dimensional analysis both functions must be a function of $\rho_l^{1/2}d$, or $(C/m)^{1/2}d$, where ρ_l is the length of the fibrous network per unit volume of gel, and m is the mass per unit length of the fibre. Thus m must increase slightly with gel concentration proportional to $C^{0.15}$.

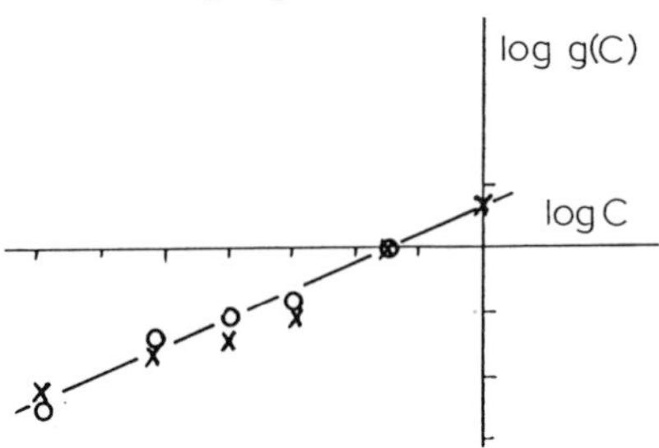

Figure 3. Horizontal shifts necessary to construct the plots shown in Figs. 2a O and 2b x. The slope of the line is 0.43.

Because of the variation in width of the $g^{(1)}(\tau)$ curves, D_z/D_{0z} is greater than the ratio of the exp(-1) points for a given hydrodynamic diameter and gel concentration. According to Ogston[13] the ratio of diffusion coefficients in gel to solution is:

$$\exp\left[-(\pi\rho_l)^{1/2}\ (d/2)\right] \tag{4}$$

The form of the variation of both functions is in reasonable agreement with this expression, and if the ratio of the exp(-1) points is taken as being the more representative of the change in diffusion coefficients, then a value for the mass per unit length of the fibrous structure of 15 kg mol^{-1} nm^{-1} at 0.7% gel concentration is obtained, with the concentration variation already indicated.

Comparison with the earlier work on agarose[7] is not straightforward since this was carried out with analogue equipment in which $g^{(1)}(\tau)$ was not accurately determined, since it necessarily ignored shorter values of τ. In this regard the exp(-1) points probably represent the better comparison. Also expression 4 was not previously found to apply - a much faster transition from the freely diffusing to complete immobilization being found. This might also be attributable to the forced fits. If the exp(-1) point in the transition is taken and expression 4 applied, then a value of 110 kg mol^{-1} nm^{-1} for the agarose fibre is obtained, which also happens to be in good agreement with the value obtained for beaded agarose by electron microscopy[14]. Similar considerations applied to calcium alginate gels[6] yields a value of 5 kg mol^{-1} nm^{-1}, which also happens to be in good agreement with later work[12]. However, further work with agarose, using the present technique, with commercially available preparations classified as being optically clear would be highly desirable.

4. CONCLUSION

The positive conclusions so far are that deacetylated gellan gels have a fibrous structure which is not highly mobile, and that the mass per unit length of the fibre increases proportional to concentration to the power 0.15, the value at 0.7% concentration being 15 kg mol^{-1} nm^{-1}.

ACKNOWLEDGEMENTS

This work was supported by Agricultural and Food Research Council Grant FG24/244. Gellan was kindly supplied by Dr R Moorhouse of Kelco-AIL, U.S.A.

REFERENCES

1. T. Tanaka, O. Hocker and G.B. Benedeck, *J. Chem. Phys.*, 1973, *59*, 5151.
2. A.M. Hecht and E.Geissler, *J.Phys. (Paris)*, 1978, *39*, 631.
3. E. Geissler and A.M. Hecht, *Macromolecules*, 1981, *14*, 185.
4. D.B. Sellen, *J. Polym. Sci. Polym. Phy. Ed.*, 1987, *25*, 699.
5. R. Pecora, *Discuss Faraday Soc.*, 1970, *49*, 222 and references cited therein.
6. W. Mackie, D.B. Sellen and J. Sutcliffe, *Polymer*, 1978, *19*, 9.
7. P.Y. Key and D.B. Sellen, *J. Polym. Sci. Polym. Phys. Ed.*, 1982, *20*, 659.
8. D.B. Sellen, *British Polymer Journal*, 1986, *18*, 28.
9. D.B. Sellen, *Polymer*, 1970, *11*, 374.
10. D.B. Sellen, *Polymer*, 1973, *14*, 359.
11. S.L. Brenner, R.A. Gelman and R. Nossal, *Macromolecules*, 1978, *11*, 202.
12. D.B. Sellen and P.M. Burne in 'Dynamic Behaviour of Macromolecules, Colloids, Liquid Crystals and Biological Systems by Optical and Electro-optical methods' 1989, p.193, ed. H.Watanabe, Hirokawa Publishing Co., Tokyo.
13. A.G. Ogston, B.N. Preston and J.D. Wells, *Proc. Roy. Soc. London*, 1973, *333*, 297.
14. T.K. Attwood, B.J. Nelmes and D.B. Sellen, *Biopolymers*, 1988, *27*, 201.

22

Light Scattering Studies of DNA Condensation

By Victor A. Bloomfield, Siqian He, An-Zhi Li and Patricia G. Arscott

DEPARTMENT OF BIOCHEMISTRY, UNIVERSITY OF MINNESOTA, 1479 GORTNER AVENUE, ST. PAUL MN 55108 USA

1. INTRODUCTION

DNA is commonly found *in vivo* in highly concentrated or tightly packaged states, for example in bacterial virus capsids, in chromatin, and in bacterial nucleoids. The energetics, structures and formation mechanisms of these compact states of DNA have been the topics of much research. It was therefore of considerable interest when model systems for studying DNA condensation were developed. One of these is the Ψ-DNA (**p**olymer and **s**alt-**i**nduced condensation) system developed by Lerman[1]. Another is the condensation of DNA in low salt by the addition of multivalent cations. It is this latter system that is our topic in this paper.

DNA condenses into compact particles when mixed with dilute solutions of trivalent cations such as spermidine or hexaammine cobalt(III) (CoHex)[2-5]. These particles are approximately the size of bacteriophage capsids, and generally have a toroidal or rodlike morphology. Remarkably, the size of the condensed particles is quite insensitive to the size of the DNA[4,6]. Particles formed with T7 or lambda phage DNA (M_r = 25-30 million) are the same size as those formed with plasmid DNA (M_r = 1-2 million). It appears that, at least in the early stages of condensation, the larger DNAs form monomolecular condensates, while the smaller DNAs form condensates incorporating an average of 10-30 molecules[6]. At later stages, more extensive aggregates may form[7]. The critical cation concentration for DNA condensation, and the morphology of the condensates (*i.e.* whether they are predominantly toroids or rods), depends somewhat on cation structure[8].

This paper reviews some of our previous work on electron microscopic and light scattering studies of DNA condensation, and presents a progress report on our current efforts. These efforts include the formulation of a statistical thermodynamic theory of condensate size distribution, light scattering measurements of condensation kinetics over a wide range of DNA and CoHex concentrations, and interpretation of the kinetics to obtain a model for the mechanism of condensation.

2. MATERIALS AND METHODS

DNA

In the light scattering studies of the kinetics of condensation reported here, we used plasmid pBluescript KS(-). This is a closed circular DNA molecule containing 2964 base pairs (bp). Previous work[6,8] used a closely related plasmid, a derivative of pUC12. Our earliest work[3] used the considerably larger (approximately 41,000 bp), linear T7 DNA.

Electron Microscopy

Electron microscope methods are described in Arscott *et al.*[6]

Light Scattering

Our light scattering apparatus consists of a Lexel Model 95-2 Ar^+ laser, a Malvern thermostatted sample holder and goniometer, ITT FW-130 photomultiplier, and Langley-Ford 1096 correlator interfaced to a Macintosh computer. The laser was operated at 488 nm at a power of 150 mW. Measurements were generally done at 90° scattering angle. Total scattering intensity was measured with an Ortec rate meter, and calibrated relative to a benzene standard. Quasielastic light scattering autocorrelation functions were analyzed by cumulant analysis to obtain the z-average diffusion coefficient[9]. This was converted into an effective hydrodynamic radius using the Stokes-Einstein equation.

Condensation Conditions

In our light scattering kinetics experiments, DNA concentration ranged from 1-15 μg/ml, and CoHex cation from 5 to 125 μM. In all cases, the buffer contained 1 mM NaCl and 10 mM TrisHCl, pH 7.5. Condensation was achieved by rapid hand mixing of buffered DNA and CoHex solutions. The first light scattering measurements could generally be made about 10 s after mixing. Condensation conditions in earlier experiments are described in the original papers.

3. RESULTS

Electron Microscopic Results

There are two particularly striking results that emerge from electron microscope visualization of the products of DNA condensation by multivalent cations. First, the condensates form highly ordered toroidal or rodlike particles, rather than the amorphous aggregates that might have been expected on the basis of polymer theory[7]. Second, regardless of the size of the DNA (over a 100-fold variation in length, from 400 bp to 40,000 bp), the toroidal particles are all generally about the same size[4,6]. A typical distribution of inner and outer radii or toroids is shown in Fig. 1. Analysis of this distribution assuming hexagonal close packing of DNA molecules gives 13 ± 4 molecules of 2700 bp DNA, and 26 ± 11 molecules of 1350 bp DNA. Thus in both cases there is an average of 35,000 bp in a toroid, very close to the number of base pairs in a single molecule of T7 or λ phage DNAs, which appear to condense in a monomolecular fashion. It is one of our goals, described in more detail in the Discussion, to understand the energetic and statistical factors leading to this highly reproducible size distribution.

Despite the general similarity of condensate size and morphology in most condensation reactions, there are some reproducible differences. One is that CoHex condenses DNA at approximately 5-fold lower ligand concentration than does spermidine[4], even though the binding constants of the two compounds are virtually identical[10,11]. Another is that CoHex tends to produce somewhat smaller or more compact toroids than spermidine[12], which is due at least in part to the smaller interhelical spacing produced by CoHex[13].

A third interesting difference is that, in a comparison of CoHex, spermidine, and permethylated spermidine, only the last compound produced a large fraction of rods relative to toroids (Fig. 2).[8] (Rodlike particles have previously been observed in ethanolic

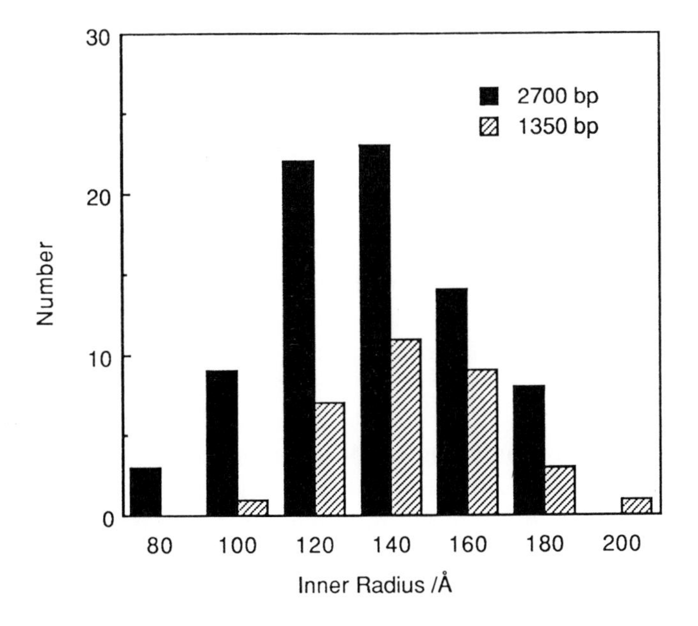

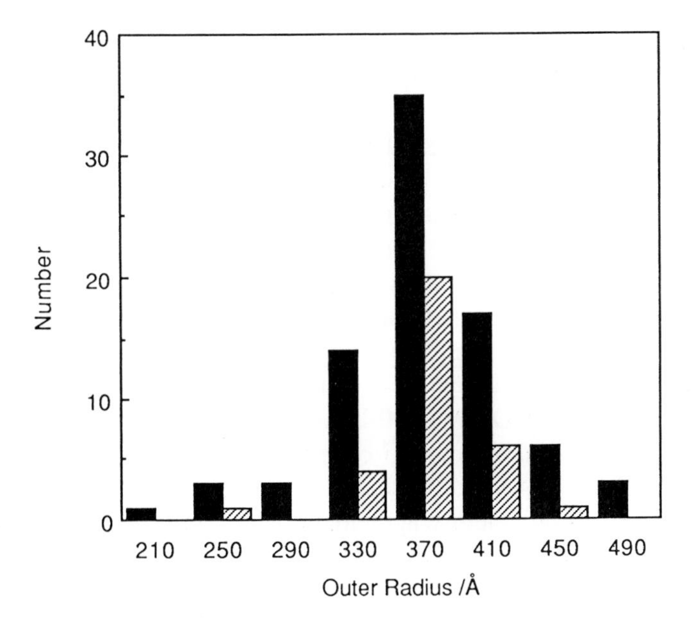

Figure 1. Distribution of inner and outer radii of toroids from 2700 bp and 1350 bp DNA molecules. Data from ref. 6.

condensates of DNA[14].) As the concentration of me$_8$-spermidine increases, so does the percentage of rods. This compound is a considerably poorer condensation agent than CoHex or spermidine. It requires higher concentrations (400 μM - 600 μM compared to 75 μM - 100 μM CoHex and 200 μM spermidine in 10 mM NaCl) and condensation is slower. The reason that me$_8$-spermidine favors rod formation is not known. It may stabilize the kinks or sharp bends formed during the folding back of DNA helices. Also, since such kinks or bends are high energy distortions, their formation must involve high activation energy and consequent slow kinetics relative to toroidal bending. The slow reaction of me$_8$-spermidine may therefore give sufficient time for kinking to occur.

Despite their morphological differences, toroids and rods have similar distributions of dimensions. That is, the distribution of toroid circumferences is similar to the distribution of rod lengths; and the distribution of toroid thicknesses (outer radius - inner radius) is similar to the distribution of rod diameters[6]. This suggests that the energetics of the two forms are similar (*e.g.* that the energy of constant gentle bending in toroids is similar to the energy of infrequent sharp bending in rods), and that the form adopted by any given particle is largely determined by kinetic factors.

Light Scattering Studies of Condensation

Comparison of Electron Microscopic and Solution Dimensions It is important to demonstrate that toroid formation is not an artifact of electron microscope conditions, and that similar particles are also formed in solution. We have used QLS to measure diffusion coefficients D and hydrodynamic radii R_h of condensed DNA solutions[3,6]. Values of R_1 and R_2 were measured by EM. Calculated values of D for 2700 bp and 1350 bp condensates were 6.47×10^{-8} cm^2/s and 6.35×10^{-8} cm^2/s, respectively, compared with measured values of 4.37×10^{-8} cm^2/s and 4.41×10^{-8} cm^2/s [6]. These are in reasonably satisfactory agreement, when one considers that the solution value of D is a z-average[9] which will strongly weight larger particles such as small aggregates of condensates, and also that there are generally some rods as well as toroids in solution, since D_{rod}/D_{toroid} is about $^{11}/_{12}$ [16,17].

Dependence of Toroid Dimensions on DNA and CoHex Concentrations The experiments described above were

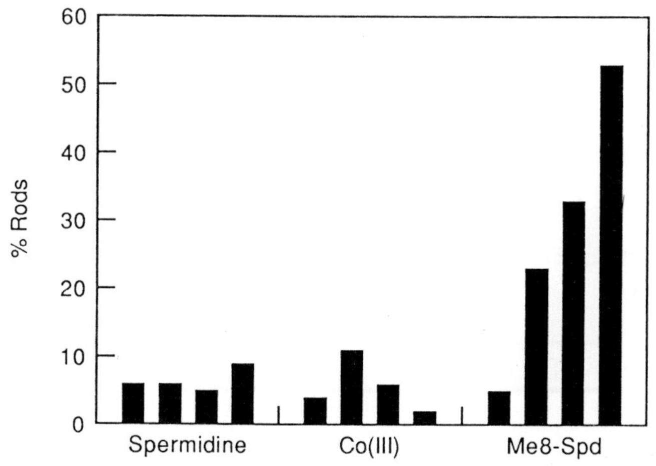

Figure 2. Percentage of rods produced by three trivalent condensing agents. Concentration of condensing agent increases from left to right in each series. Data from ref. 8.

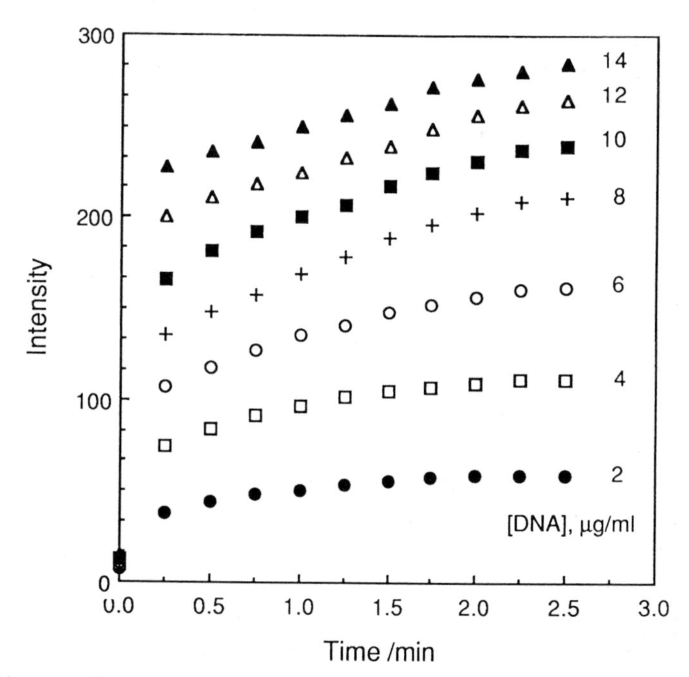

Figure 3. Total scattering intensity as a function of time after addition of 60 μM CoHex to a solution of the indicated DNA concentration.

carried out on solutions close to the onset, or near the midpoint, of the rather sharp, highly cooperative condensation transition. In total scattering intensity measurements of the transition, we find that condensation occurs only above a critical CoHex concentration for a given DNA concentration. These values are given in Table 1.

Table 1. Critical DNA and CoHex concentrations for the onset of condensation.

[DNA] μg/ml	[CoHex] μM
3	5.0
6	7.5
9	12.5
12	17.5
15	22.5

For a fuller understanding of the phenomenon, we need to know the dimensions of particles formed at higher DNA and/or CoHex concentrations, when the driving force for condensation is stronger. Fig. 3 shows some typical curves of total scattering intensity I as a function of time after addition of 60 μM CoHex to a DNA solution of the indicated concentration. There are two observable phases. There is an initial very rapid rise of I, which is greater in magnitude and rate as [DNA] increases, followed by a slower increase over some minutes. We concentrate first on the slow phase, which typically reaches a plateau at about 30 minutes.

Fig. 4 shows the ratio of the scattering intensity at the plateau condition, relative to the scattering from the original uncondensed DNA solution, as a function of DNA and CoHex concentrations. The ratio increases with increasing [DNA]; it initially increases and then levels off with increasing [CoHex]. At any point on this diagram, we may calculate the weight-average molecular weight if we know the particle size and scattering form factor, from the standard equation $I = K\langle M_w \rangle cP(q)$ where K is a constant, c the mass concentration of DNA, and P(q) the scattering form factor which depends on scattering vector q. For simplicity in this calculation, we have assumed that the condensates are spheres whose radii equal the

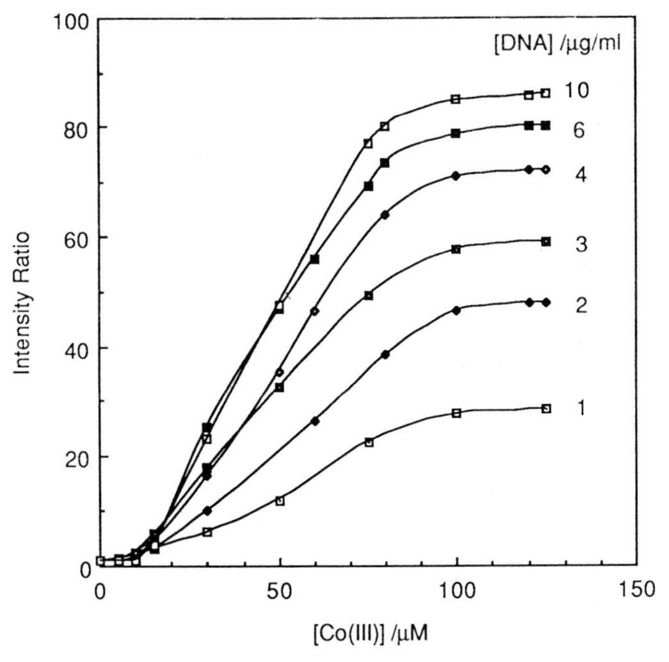

Figure 4. The ratio of the scattering intensity at the plateau condition, relative to the scattering from the original uncondensed DNA solution, as a function of DNA and CoHex concentrations.

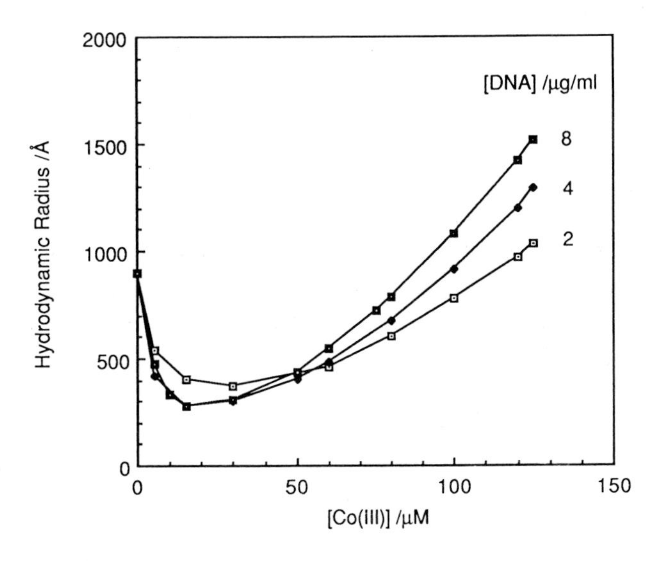

Figure 5. Dependence of R_h, measured under plateau conditions, on DNA and CoHex concentrations.

hydrodynamic radii R_h obtained from QLS diffusion measurements. The dependence of R_h, measured under plateau conditions, on DNA and CoHex concentrations is plotted in Fig. 5. The resulting condensate molecular weights, expressed as number of DNA molecules per toroid, is plotted in Fig. 6. It is evident that at very high DNA or CoHex concentrations, the particles become very large; EM often shows large aggregates under these conditions. However, under conditions typical of the minimum in Fig. 5, close to the transition midpoint, the average number of DNA molecules per toroid is in the range 10-20, in good agreement with EM results.

Dependence of Initial Condensation Rates on DNA and CoHex Concentrations We turn now to consideration of the initial rates of light scattering increase, *i.e.* the first points in Fig. 3. These points were measured at about 10 s, as soon as possible after hand mixing of DNA with CoHex. (There may well be significantly earlier processes occurring which we are unable to measure, as studied by Porschke[18].) Fig. 7 shows the dependence of the initial intensity jump on DNA concentration for several CoHex concentrations. These plots are accurately linear and pass through the origin. The slopes of these lines are linear functions of [CoHex], intersecting the [CoHex] axis at the threshold for each DNA concentration (Table 1). Thus above the critical concentrations, the initial rate of DNA condensation is first order in both DNA and condensing ligand. This result will be important in kinetic modelling of the condensation mechanism.

4. DISCUSSION

Forces leading to condensation

Our major interest in undertaking this work is to understand the molecular mechanism of DNA condensation by multivalent ligands. We view the experimental determinations of condensate size and shape, and kinetics of condensation, as consistency checks on any proposed mechanism.

In a previous paper[19], we listed DNA bending or kinking, polymer chain entropy, and coulombic repulsion between DNA phosphates as the major sources of free energy resisting condensation. To this one may add repulsive hydration structure forces[20] and fluctuation-enhancement of coulombic and hydration

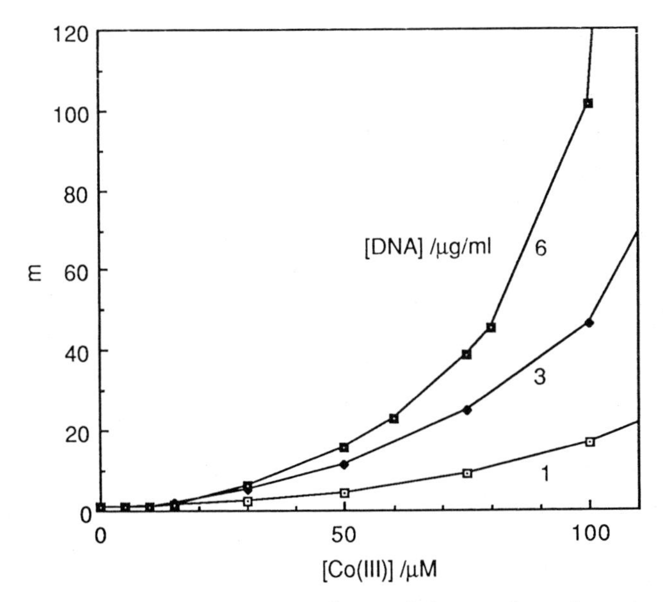

Figure 6. Condensate molecular weights under plateau conditions, expressed as number of DNA molecules per toroid, as functions of DNA and CoHex concentrations. Calculated as explained in text from the data in Figs. 4 and 5.

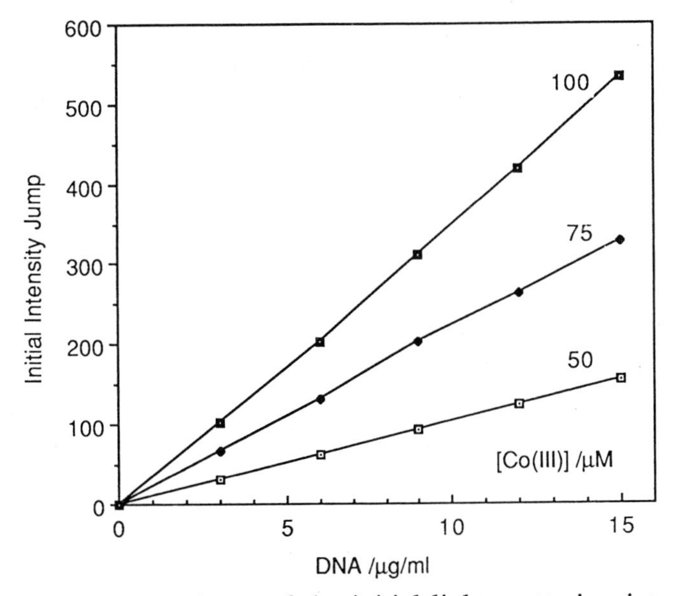

Figure 7. Dependence of the initial light scattering intensity jump in Fig. 3 on DNA concentration for several CoHex concentrations.

forces[21] as probably significant items. To compensate these, attractive forces giving rise to condensation may include cross-linking by bound ligands[4,13], intrinsic curvature of charge-neutralized DNA[22], ion atmosphere dispersion forces[23], attractive hydration structure[24], and intermolecular base pairing consequent on distortion of base pairing and stacking by condensing ligands[25].

In an initial attempt to assess the significance of these various terms, we have considered electrostatics, bending, and hydration structure as the dominant forces. We used the Rau-Parsegian[24] equation for hydration structure free energy (which has both attractive and repulsive components)

$$\frac{\Delta G_{hyd}(D)}{kT} = \left\{\frac{2\pi l\lambda^2}{K_0(a/\lambda)}\right\}$$

$$\left\{\frac{P_r^2 K_0(D/2\lambda)}{K_0(a/\lambda)I_0(D/2\lambda) - K_0(D/2\lambda)I_0(a/\lambda)} - \frac{P_a^2 K_1(D/2\lambda)}{K_1(D/2\lambda)I_0(a/\lambda) + K_0(a/\lambda)I_1(D/2\lambda)}\right\} \quad (1)$$

the Brenner and Parsegian[26] equation for electrostatic interaction between two parallel rods

$$\frac{\Delta G_{elec}(D)}{kT} = \frac{4f^2e^2K_0(\kappa D)}{\varepsilon l[a\kappa K_1(\kappa a)]^2} \quad (2)$$

and the standard expression[19] for the free energy of bending

$$\frac{\Delta G_{bend}}{kT} = \frac{l_p l}{2r_c^2}. \quad (3)$$

We have also included a rather arbitrary nonlinear term

$$\frac{\Delta G_{nonlinear}}{kT} = \psi L^p \quad (4)$$

since all the other terms are linear in the amount of DNA, and this obviously cannot lead to a maximum in the size distribution.[27] The sum of these free energies per base pair, multiplied by the number of base pairs per DNA molecule and by the number of DNA molecules per condensate, is put into the equation derived by Tanford[27] for micelle formation

$$\ln X_m = -m\Delta G°(m)/kT + m \ln X_1 + \ln m \quad (5)$$

to obtain the distribution of condensate sizes m.

In these equations, the free energy is per base pair. The variables, their meanings and the values used are :

D	interhelix distance	28 Å
a	helix radius	10 Å
l	length per base pair	3.4 Å
bp	base pairs/DNA molecule	1350 or 2700
m	DNA molecules/condensate	2-50
L	total length of DNA	m x bp x l
l_p	persistence length	500 Å
r_c	radius of curvature of toroid	400 Å
κ	Debye-Hückel inverse length	0.01 $Å^{-1}$
ε	dielectric constant	80
f	fraction unneutralized DNA charge	0.104
λ	decay length of hydration force	4.5 Å
P_r^2	repulsive hydration coefficient	4.4×10^7 ergs/cm^3
P_a^2	attractive hydration coefficient	6.2×10^7 ergs/cm^3
p	power of nonlinear repulsion	2.2 - 3.0
ψ	coefficient of nonlinear repulsion	depends on p: 7.1 x 10^{-9} at p=2.5
c	concentration of DNA	10 μg/ml
X	mole fraction of DNA	1-2 x 10^{-11}

With this choice of parameters, all of which are tightly constrained by other measurements or by the convergence of the calculation, we find that the most probable number of DNA molecules is about 10 for 2700 bp DNA and 20 for 1350 bp DNA, in good agreement with experiment. The size is thus predicted to be independent of DNA length (at least over this narrow range). However, the calculated distribution is essentially monodisperse, in contrast to the broader dispersion observed. Perhaps the calculated distribution will be broadened by inclusion of other free energy terms (*e.g.* entropy of packing), but they will have to be nonlinear in the amount of DNA. The fraction of uncondensed DNA X_1 needed to obtain convergence of these calculations is generally 0.1 to 0.5. This has not been measured, but the result seems reasonable. Also, in order to obtain convergence, f must be slightly greater than 0.10, in excellent agreement with the observed dependence on condensing ligand and salt concentrations if the DNA neutralization is interpreted by counterion condensation theory[3].

The nonlinear term ψL^p has no physical basis. A simple physical argument based on the electrostatic self-energy of a charged spherical condensate might have predicted a term proportional to L^2, but a significantly higher power p was required to obtain a most probable size.

These calculations show that condensation is an exquisitely balanced, highly cooperative process. Since there are 2700 x 13 bp in the average toroid, and each base pair experiences an average of about 5 interactions with its neighbors, there are about 2×10^5 interactions. However, for the transition to take place under measurable conditions, the total free energy change (in units of kT) should be about 20. Thus each interaction has a net attraction of about 10^{-4} kT!

Model of Condensation Mechanism

We are currently attempting to fit the data in Figs. 3-7 to a kinetic model which will define the molecularity of condensation with respect to CoHex and DNA and which will also be consistent with the equilibrium results. At the time of this conference, we have not yet succeeded. It seems likely, on the basis of scattering and electron microscopic evidence, that three stages must be considered: (1) association of a few DNA molecules without condensation; (2) condensation of this paucimolecular complex and recruitment of additional molecules to the condensate; (3) long-time aggregation of condensates into larger structures.

ACKNOWLEDGEMENT

This research was supported in part by NIH Grant GM 28093.

REFERENCES

1. L.S. Lerman, *Proc. Natl. Acad. Sci. USA*, 1971, *68,* 1886.
2. D.K. Chattoraj, L.C. Gosule and J.A. Schellman, *J. Mol. Biol.*, 1978, *121*, 327.
3. R.W. Wilson and V.A. Bloomfield, *Biochemistry*, 1979, *18*, 2192.
4. J. Widom and R.L. Baldwin, *J. Mol. Biol.*, 1980, *144*, 431.
5. J. Widom and R.L. Baldwin, *Biopolymers*, 1983, *22*, 1595.

6. P.G. Arscott, A.-Z. Li and V.A. Bloomfield, *Biopolymers*, 1990, in press.
7. C.B. Post and B.H. Zimm, *Biopolymers*, 1982, *21*, 2123.
8. G.E. Plum, P.G. Arscott and V.A. Bloomfield, *Biopolymers*, 1990, in press.
9. D.E. Koppel, *J. Chem. Phys.*, 1972, *57*, 4814.
10. W.H. Braunlin, T.J. Strick and M.T. Record Jr., *Biopolymers*, 1982, *21*, 1301.
11. G.E. Plum and V.A. Bloomfield, *Biopolymers*, 1988, *27,* 1045.
12. T.J. Thomas and V.A. Bloomfield, *Biopolymers*, 1983, *22*, 1097.
13. J.A. Schellman and N. Parthasarathy, *J. Mol. Biol.*, 1984, *175,* 313.
14. T.H. Eickbush and E.N. Moudrianakis, *Cell*, 1976, *13,* 295.
15. S.A. Allison, J.C. Herr and J.M. Schurr, *Biopolymers*, 1981, *20*, 469.
16. C.-M. Tchen, *J. Appl. Phys.*, 1954, *25*, 463.
17. R. Zwanzig, *J. Chem. Phys.*, 1966, *45*, 1858.
18. D. Porschke, *Biochemistry*, 1984, *23*, 4821.
19. S.C. Riemer and V.A. Bloomfield, *Biopolymers* , 1978, *17,* 785.
20. D.C. Rau, B. Lee and V.A. Parsegian, *Proc. Natl. Acad. Sci. USA*, 1984, *81*, 2621.
21. R. Podgornik, D.C. Rau and V.A. Parsegian, *Macromolecules,* 1989, *22*, 1780.
22. G.S. Manning, *Biopolymers*, 1981, *20*, 1261.
23. F. Oosawa 'Polyelectrolytes', Marcel Dekker, New York, 1971.
24. D.C. Rau and V.A. Parsegian, *J. Mol. Biol.*, 1991, in press.
25. D.A. Knoll, M.G. Fried and V.A. Bloomfield, in 'Structure & Expression. DNA and Its Drug Complexes' R.H. Sarma and M.H. Sarma, eds., Adenine Press, New York, 1988, p. 123.
26. S.L. Brenner and V.A. Parsegian, *Biophys. J.,* 1974, *14*, 327.
27. C. Tanford, *Proc. Natl. Acad. Sci. USA*, 1974, *71*, 1811.

23

Dynamics and Structures of DNA

By J. Michael Schurr, Ug-Sung Kim, Lu Song, Bryant S. Fujimoto, Clement E. Furlong[1] and Joseph A. Sundstrom[1]

DEPARTMENTS OF CHEMISTRY (BG-10) AND [1]GENETICS (SK-50), UNIVERSITY OF WASHINGTON, SEATTLE, WA 98195 U.S.A.

1. INTRODUCTION

Long-Range Effects of Sequence on Secondary Structure

The primary question addressed here is whether the base-sequence at one point in a DNA molecule can exert a long-range influence on the secondary structure of its flanking sequences. Previous investigations have identified two kinds of long-range correlation between sequence and secondary structure. (1) There may exist non-nearest-neighbour correlations that cannot be described by the nearest-neighbour Ising model. Evidence that non-nearest-neighbour effects can extend over 35 base-pairs or more has come from melting studies on synthetic DNAs containing block sequences[1]. (2) There may also exist long-range correlations between sequence and secondary structure that can be ascribed to co-operativity within the nearest-neighbour Ising model. Such co-operativity is purely statistical and arises from the high free energy, hence infrequent occurrence, of junctions between coexisting domains of different secondary structure. The co-operativity lengths manifested in alcohol-induced transitions between A, B, and Z structures were estimated to be in the range 10-37 base-pairs by using nearest-neighbour Ising theory[2]. The existing evidence regarding either kind of long-range correlation between sequence and secondary structure is somewhat indirect, as it is based entirely on the analysis of transition profiles using the nearest-neighbour Ising model. The approach here is to show directly that a particular change in sequence over a small region of a large restriction fragment, specifically incorporating 16 base-pairs (bp) of alternating GC near the centre of an ~1100 bp fragment, induces disproportionately large relative changes in properties that reflect the secondary structure of the entire molecule.

Long-Range Effects of Ligand Binding on Secondary Structure

We also address the related question of whether a bound ligand can exert a long-range effect on its flanking sequences. Specifically, we show that, *under high salt conditions*, one added ethidium dye per 300 base-pairs induces an extensive change in secondary structure of the DNA molecule bearing the $(GC)_8$ insert, but has no effect on control fragments without the insert.

The Possibility of Transient Fluctuations Between Distinct Alternate Secondary Structures

DNA is known to exhibit several distinct families of regular secondary structure, such as A, B, C and Z[3], as well as a high-salt form described further below, and the transitions between some of these are known to be highly co-operative[2]. Still other regular secondary structures have been predicted[2], but not yet observed. Even sequences that exhibit irregular structures, such as permanent bends[4], may also exhibit regular structures under some conditions. The disappearance of gel migration anomalies with increasing temperature may reflect an equilibrium between irregular bent and regular straight states. In view of the evident polymorphism and co-operativity of DNA, the correct description of its secondary structure in solution should lie between the following two extremes. (1) A given sequence adopts a more or less constant configuration determined entirely by its constituent and closely neighbouring base-pairs. (2) A given sequence transiently fluctuates among two or more distinct secondary structures, which may belong to much larger domains of variable position and size, and whose relative stabilities are governed by distant as well as close neighbouring base-pairs. Most workers today are presently inclined toward the first (constant, local) description of the secondary structure. However, that is quite incompatible with the previously discussed non-nearest neighbour effects extending over 35 base-pairs[2] and with the presently observed long-range effects of the $(GC)_8$ insert. If, instead, the second (fluctuating long-range) description of secondary structure prevails, then it might be manifested in several ways. (1) In cases where the equilibrium prevalence of two or more secondary structures is delicately balanced, a particular change in sequence over a small critical region might shift that equilibrium over a much larger domain, resulting in a long range effect of sequence on the *average* secondary structure. (2) If a bound ligand significantly perturbs the structure, especially of that critical sequence, then it too might exert a long range effect on the *average* secondary structure. (3) If permanent bends are associated with one or more of the fluctuating secondary structures, or with junctions between those, which is most likely, then such fluctuations would contribute to the equilibrium root mean square curvature and flexibility of the DNA. Because fluctuations among co-operative secondary structures are expected to be rather slow, they are not expected to contribute to the dynamic flexibility at very short times (*e.g.* $< 1\ \mu s$). In other words, the dynamic bending rigidity (κ_d) that governs flexing at very short times might significantly exceed the static bending rigidity (κ_s) inferred from the equilibrium persistence length (P_s). We present evidence below that is consistent with each of these possible manifestations of the second (fluctuating, long-range) description of DNA secondary structure. We have previously reported evidence for long-range (*i.e.* more or less global) switching of the secondary structure in supercoiled DNAs by changing buffer- ions[4], by CAP binding[5], by Mg^{2+} binding[6], and by changing superhelix density[7], but this is the first time that we have encountered long-range changes in the secondary structure of a linear DNA.

The Dynamic Bending Rigidity of DNA

Recent advances in theory[8,9] allow us to determine the *dynamic* bending rigidities, or equivalently the dynamic persistence lengths, $P_d = \kappa_d/k_BT$, from the reported initial relaxations in the off-field decays of the transient electric dichroism[10,11] or birefringence[12] of restriction fragments containing 95 to 250 bp. As indicated in Table 1, the dynamic value, P_d = 2100 Å, exceeds the consensus static persistence length, P = 700 Å, under the prevailing millimolar salt conditions by about a factor of three. In 0.1 M NaCl, the estimates of P_d range from 1250 Å to 1700 Å. The former value results from Brownian dynamics simulations of transient photodichroism data for a 209 bp fragment[13], and will almost certainly increase when a more realistic (lower) torsion constant is assumed in the simulation. The latter value comes from analysis of epr spectra of spin-labelled bases in DNAs containing 12 to 96 base-pairs (B.H. Robinson and E.

Hustedt, personal communication). The hyperfine tensor is preaveraged by the rapid bending motions, so only the equilibrium theory[14] is required. Evidently, in 0.1 M NaCl, P_d is also about 3 times the consensus static value, P = 475 Å - 500 Å. These results imply that DNA exhibits long-lived (compared to 10^{-5} s) bent states, either transient or permanent, that contribute to the apparent static flexibility. This finding is consistent with the second (fluctuating, long-range) description of DNA secondary structure, but does not preclude the first (constant, local) description. However, if the first (constant, local) description applies, then it remains to explain why the bending potential between base-pairs is either non-simple (exhibits multiple minima with barriers comparable to k_BT) or biased (permanently bent).

Table 1. Comparison of dynamic and static persistence lengths of DNA.

N (bp)	[NaCl] (mM)	P_d (A)[a]	P (A)[b]
92-250	1	2100	700
12-96	100	1700	500
209	100	1250	500

(a) $P_d = \kappa_d/k_BT$ governs bending dynamics during a time $t \lesssim 10^{-6}$ s.
(b) P characterizes equilibrium rms curvature.

Probes of Secondary Structure

We apply a variety of techniques that are sensitive to secondary structure of the DNA. Although space limitations preclude a detailed discussion of each, a brief explanation of their relevant aspects is appropriate.

The apparent torsion constant α between base-pairs is investigated by measuring the time-resolved fluorescence polarization anisotropy (FPA) of ethidium intercalated in DNA[4,6,7,15]. In this work, ethidium is added to a level of 1 dye per 200 bp, where depolarization by excitation transfer is insignificant[16]. In 0.1 M NaCl, ethidium is found to introduce no rigidity weaknesses and to have no effect on the torsional rigidity of the DNA up to 1 dye per 5 bp [16]. Moreover, there appears to be no dependence of the torsion constant on overall base-composition[17], although linearized pBR322 and pUC8 DNAs do exhibit anomalously high values[17]. The α-values reported here are proportional lower bounds that faithfully reflect any relative differences or changes in torsion constant that may arise from differences or changes in secondary structure. An estimate of the dynamic bending rigidity is required to determine the absolute value of α, which typically is about 1.35 times its lower bound[6,17]. Here we use the lower bound α simply as a qualitative indicator of differences or changes in the underlying secondary structure. α reflects relatively long wavelength torsional motions from about 50 bp at t = 1 ns to 550 bp at t = 120 ns, and in that way samples extended zones surrounding the binding site. The FPA measurements also provide an indication of the relative amounts of intercalated and non-intercalated dye, and thus probe approximately the binding affinity for intercalation.

Changes in the circular dichroism (CD) spectrum reflect changes in the chiral electronic environment around each base. The CD spectra represent a uniformly weighted average over all base-pairs.

Dynamic light scattering (DLS) measurements at small scattering vector ($K \rightarrow 0$) yield the translational diffusion coefficient (D_0) of the centre of mass[18-21], which reflects the tertiary structure of the molecule. In the absence of permanent bends, that depends only upon the static persistence length[20,22]. At larger scattering vectors, DLS yields an apparent diffusion coefficient ($D_{app}(K)$) that increases with K to a quasi-plateau value (D_{plat}) near $K^2 = 20 \times 10^{10}$ cm^{-2} [19-21,23]. For weakly bending rods, D_{plat} is independent of the torsion constant and increases with decreasing bending rigidity[24]. For very long strongly bending filaments, crankshaft motions dominate the short wavelength displacements and D_{plat} becomes very sensitive to the torsion constant[20,21]. Until now, D_{plat} and α for such long DNAs were always observed to change in the *same* direction. The restriction fragments studied here are much too long to qualify as weakly bending rods, but still might not be sufficiently long that D_{plat} is very sensitive to changes in α. In any case, we use D_{plat} as a qualitative indicator of differences or changes in secondary structure, even though we cannot say with certainty whether it probes primarily the bending or twisting rigidity.

The optical melting profile yields the midpoint transition temperature (T_m), which reflects the enthalpic stability of the average structure. At high resolution it yields information about melting domains within the molecule.

S1 nuclease generally attacks duplex DNAs at a far slower rate than single-strand DNAs, and is often used empirically to detect structural anomalies in duplex DNAs. Here we use the difference in cleavage rates by S1 nuclease (under the same conditions) to distinguish secondary structures of DNAs with slightly different sequences.

2. MATERIALS AND METHODS

Preparation of the DNA Samples

The DNAs studied here are restriction fragments obtained from a genetically engineered derivative of pBR322, or from pBR322 itself, as indicated schematically in Fig. 1. The restriction fragment containing the 16 bp $(GC)_8$ insert is prepared as follows. The circular plasmid pBR322 is restricted (cut) at positions -2 and +29 by EcoRI and HindIII, respectively, and the large piece is electrophoretically separated from the small piece. This destroys the gene for resistance to tetracycline and removes a TaqI site. After dephosphorylation at the 5'-ends by alkaline phosphatase, the large piece is combined with a synthetic oligonucleotide containing the 16-bp insert sequence $(GC)_8$, two spacer base-pairs, and single-strand EcoRI and HindIII linkers at opposite ends. Circular DNAs containing the synthetic insert spontaneously self-assemble, and are ligated to the covalently closed form by T4 ligase. This recombinant plasmid is then used to transform *E. coli* strain MM294, and the clone (colony) with the insert is selected by resistance to ampicillin, but susceptibility to tetracycline. The insert clone plasmid is further assayed for the presence of a unique TaqI restriction fragment, which results from removal of the TaqI site. This recombinant plasmid is then transferred into *E. coli* strain HB101 for amplification. Plasmid pBR322 is also transferred into a separate line of *E. coli* HB101 for amplification. Details of cell maintenance, growth, amplification, harvesting, and isolation of these plasmid DNAs are presented elsewhere.

Restriction Fragments

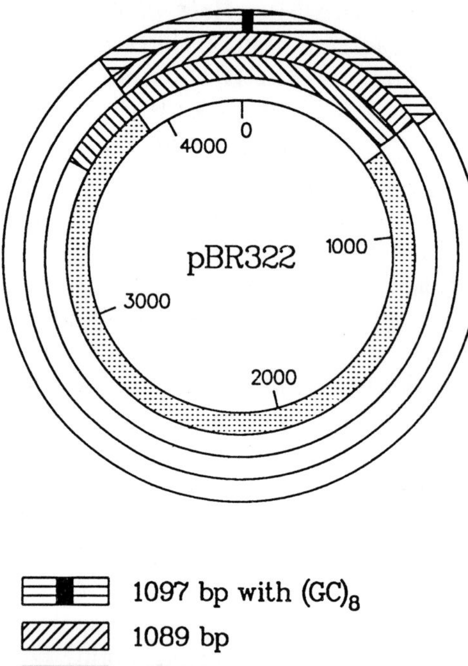

Figure 1. Schematic diagram of the restriction fragment locations in circular pBR322. The thin black rectangle is centered at the position of the 16 bp $(GC)_8$ insert, but is much wider than the insert sequence would be on this scale.

The *insert* fragment containing 1097 bp with 16 bp of $(GC)_8$ near its centre is prepared by restricting the recombinant plasmid with HincII (two sites), and separating the 1097 bp fragment from its 3256 bp complementary piece by HPLC using a Nucleogen column. The separation was laborious, due to inefficient separation of such large DNAs. Nevertheless, upon completion of our experiments on the first sample, a second insert sample was prepared in basically the same way from the HB101 clone, and the critical experiments were repeated. The control fragment containing 1089 base-pairs without the insert is prepared by restricting pBR322 with HincII, HinfI, and PstI, and separating the products by HPLC. The latter two enzymes are employed to cut the large complementary fragment into smaller pieces more amenable to HPLC separation. HinfI also removes 21 bp inward from the 3'- end of the ~1100 bp HincII fragment. The sequence of the 1097 bp insert fragment differs from that of its 1089 bp control by less than 30 bp between the 5'-end and its HinfI site. A second control fragment containing 1383 bp without the insert is prepared by restricting pBR322 with HinfI and PstI, and separating the fragments by HPLC. Here, too, overlap with the sequence of the insert fragment is substantial, as indicated in Fig. 1. Besides these control fragments, whose sequences heavily overlap that of the insert fragment, some studies were also performed on the 3256 bp complementary fragment, and on whole linearized pBR322.

Characterization of the DNA Samples

All samples migrated as single sharp bands in gel electrophoresis. The insert fragment was simultaneously compared with the 1089 bp control and with a 123 bp ladder by low resolution gel electrophoresis in 1.5% agarose. The mobilities of the insert and 1089 bp fragments were practically indistinguishable, so the insert causes no significant change in mobility. Both fragments show a very slight retardation with respect to the 123 bp ladder, but run within 18 bp of their expected positions. Resolution of these gels is about ±7 bp. In our method of construction, the synthetic insert can be incorporated only as odd multimers consisting of 1, 3, 5 ... monomer units. The possible occurrence of a multimers higher than one is precluded by the preceding results, which are compatible with the 1097 bp expected for the monomeric insert, but not with the 1142 bp expected for a trimeric insert. The size of the monomeric unit of the insert is determined to a resolution of 1 bp in 12% polyacrylamide, by comparing a double digest of the insert fragment by EcoRI and Hind III with an MspI digest of PBR322. The insert monomer runs at the expected position for an 18 bp duplex with four base linkers at opposite ends.

These restriction fragments are subjected to fluorescamine tests for contaminating polyamines and proteins. The resulting signal is indistinguishable from the background level. The A_{260}/A_{280} ratios of all samples lie between 1.90 and 2.00.

A sedimentation velocity measurement on the 1097 bp insert fragment by Prof. David Teller yielded a sedimentation coefficient $s_{20,w}$ = 9.27 S. When combined with the diffusion coefficient D_0 = 6.5×10^{-8} cm^2 s^{-1} measured by DLS, and the density increment, $(\partial\rho/\partial c)_\mu$ = 0.468, a molecular weight $M_r = (7.42 \pm 0.3) \times 10^5$ is obtained, in excellent agreement with the expected value $M_r = 7.26 \times 10^5$. The observed sedimentation band was extremely sharp, which argues strongly against aggregation of any kind. The measured sedimentation and diffusion coefficients lie within the range of values expected for ordinary DNAs containing 1097 bp. Both this fact and the normal gel mobility imply that the tertiary structure of the insert fragment is normal.

Experimental Procedures

Time-resolved FPA measurements are performed essentially as reported in recent work[7]. The instrumentation and data analysis procedures were recently reviewed[6]. Here the first 3 ns are typically omitted from the fit of the difference data ($D(t) = I_{\parallel}(t) - I_{\perp}(t)$) for two reasons. A short- lived ($\lesssim$ 1 ns) fluorescent impurity from the Nucleogen column was present in some samples and was not completely removed by dialysis, filtration, and centrifugation through sephadex G-50. At NaCl concentrations above 1 M, the ratio of intercalated to non-intercalated ethidium falls below 12, so the amplitude of rapid decay of the non-intercalated dye ($\lesssim$ 2 ns) is no longer negligible. Omission of the first 3-4 ns from the fit of the difference data is known to yield valid torsion constants[6]. Fits of the difference data to the intermediate zone formula[25,26] were satisfactory with reduced χ^2 values ≤ 1.2 in almost all cases. These torsion constants are measured at 20°C.

CD measurements are performed at 20°C and converted to molar ellipticities [θ] in deg-cm^2/dmol, as described earlier[7].

DLS photon correlation functions are measured and fitted over 8 relaxation times to a single-exponential plus base-line, and $D_{app}(K)$ is calculated from the best-fit relaxation time in the usual way[7,21]. Unusually low DNA concentrations (25 μg/ml), the modest molecular weight and loss of scattering power with increasing salt concentration (and refractive index) above 1.0 M NaCl combine to make these experiments rather difficult. It was necessary to use a specially designed water bath to avoid heterodyning from flare at the lower scattering angles. The $D_{app}(K)$ values presented here are measured at 21°C, and corrected to 20,w conditions. K is calculated using the appropriate refractive index for each salt solution.

Optical melting experiments are performed essentially as described in earlier work[27]. After smoothing, the data for fraction melted are differentiated to obtain the derivative melting profile.

S1 nuclease digestions are carried out for various times at 37°C, and the products are analyzed by electrophoresis in 1.5% agarose gels.

3. RESULTS AND DISCUSSION

Comparison of the Insert Fragment with the Control Fragment(s) in 0.1 M NaCl

Apparent torsion constants for the various DNAs in 0.1 M NaCl are compared in Fig. 2. The α-value of the 1097 bp insert fragment is smaller than that of the control DNAs by a factor of nearly 1.35. The lifetimes of the intercalated ethidiums are identical in all cases (~23 ns), and the ratios of intercalated to non-intercalated dye are similarly comparable. Hence, there is no indication of any difference in dye binding sites between the insert fragment and the controls. Within experimental error, the second sample of the 1097 bp insert fragment yields the same torsion constant as the first. The essential point is that a change in sequence over 3% of the total length reduces the torsion constant by a factor of 1.35.

Molar ellipticities near the maximum at 273 nm ($[\theta]_{273}$) are compared in Fig. 2. Again the 1097 bp insert fragment differs

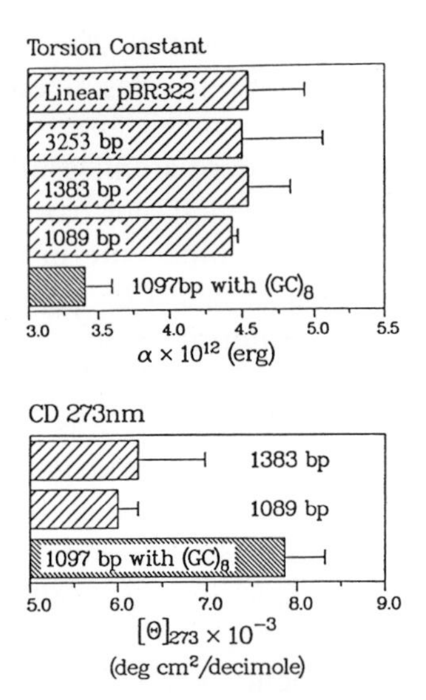

Figure 2. Comparison of the 1097 bp fragment containing the $(GC)_8$ insert with the control fragments in 'physiological' buffer. Top: Torsion constant α. Bottom: Molar ellipticity $[\theta]_{273}$ near the maximum of the positive circular dichroism band at 273 nm. The 'physiological' buffer is 0.1 M NaCl, 10 mM Tris, 1 mM Na_2EDTA, pH 7.8, T = 293 K.

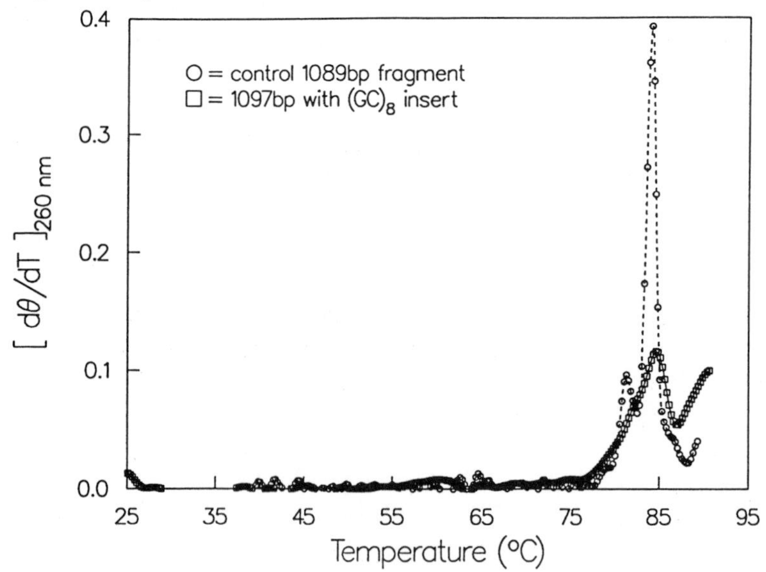

Figure 3. Differential optical melting profiles in 'physiological' buffer. The derivative of the fraction (θ) of transition completed is plotted *vs*. temperature. Circles are for the 1089 bp control and squares are for the 1097 bp insert fragment containing $(GC)_8$. The buffer is the same as in Fig. 2.

substantially from the control fragments. Its [θ] value is larger by a factor of about 1.30 at 273 nm, and smaller (in absolute value) by a factor of about 1.15 at 244 nm. The important point is that a change in sequence over 3% of the total length increases $[\theta]_{273}$ of the entire fragment by a factor of 1.30.

Derivative optical melting profiles are compared in Fig. 3. Here, too, the insert fragment differs substantially from its control. The midpoint temperatures for the overall transitions are $T_m \cong$ 83.6°C for the 1089 bp control, and $T_m \geq 86.8$°C for the 1097 bp insert fragment. The 1089 bp control exhibits a weak early transition at 81.3°C and its main transition at 84.1°C. The 1383 bp control fragment is similar. The 1097 bp insert fragment exhibits two broader transitions of comparable magnitude, the first at 84.6°C and the second at or above 90°C. Melting of the 1089 bp control is practically reversible, in the sense that the profiles obtained during the ascending and descending temperature runs very nearly coincide. The upper melting transition of the 1097 bp insert fragment is similarly reversible, but the lower transition is not. As yet we have no predicted melting profiles for these sequences from nearest-neighbour theory with which to compare these data. One may employ the empirical relation between T_m and overall % GC[28], $T_m = 16.6 \log [M] + (0.41)(\% GC) + 81.5$, where [M] is the counterion concentration, to estimate that T_m should change by only about 0.53°C as the GC content is increased by 1.3% from the 1089 bp control (51.2% GC) to the 1097 bp insert fragment (52.5% GC). The observed change in T_m ($\geq$ 3.2°C) exceeds the expected change by at least a factor of 6. Similarly, the domain size of the second transition, which is induced by changing only 3% of the total sequence, amounts to about half of the total helix-coil transition.

Under appropriate conditions, S1 nuclease attacks the 1097 bp insert fragment much more rapidly (parent molecule completely cleaved) than the 1089 bp control fragment (parent molecule only partially cleaved) (data not shown). In both cases the cleavage appears to be largely non-specific, although there is some slight preference for cutting the insert fragment near the insert. Evidently, changing 3% of the total sequence affects the susceptibility to S1 nuclease of a substantial fraction of the molecule.

The preceding examples all indicate that the 16 bp$(GC)_8$ insert induces unexpected and disproportionately large relative changes in properties of the whole molecule. Indeed, the observed changes exceed any plausible estimates by an order of magnitude or more. If α and $[\theta]_{273}$ of the secondary structure induced by the insert differ from normal by not more than a factor of two, then these results imply that the $(GC)_8$ insert influences secondary structure over 30% or more of the total sequence, or at least several hundred base-pairs flanking the insert.

$D_{app}(K)$ for the 1097 bp insert fragment is plotted *vs.* K^2 in Fig. 4. Corresponding data for the 1089 bp control essentially coincide with the curve in Fig. 5 (data not shown). However, the relatively low precision of the data at large K^2 does not preclude a small difference in D_{plat}. The question arises why $D_{app}(K)$ is largely unaffected by the insert, while all other probes of secondary structure exhibit rather substantial changes. A probable explanation is that these fragments are simply not long enough for D_{plat} to be very sensitive to the torsional rigidity.

The High Salt Form of Ordinary Duplex DNA

It has long been known that $[\theta]_{273}$ declines with increasing NaCl concentration above 1.0 M [29,30], but additional information regarding this state of the DNA is lacking. The present DNAs also show a pronounced decrease in $[\theta]_{273}$ with increasing NaCl, as indicated in Fig. 5. In contrast, the D_0 values of the 1089 bp control and the 1097 bp insert fragment (after correction to 20,w conditions) are independent of [NaCl] from 0.1 M to 4.3 M, as indicated in Figs. 6, 7 and 8. The same holds for the 1383 bp control. This suggests that the equilibrium persistence length does not *decrease* appreciably with increasing [NaCl]. D_{plat} for the 1089 bp control increases slightly and more or less linearly with [NaCl], as shown in Fig. 6, and the same holds for the 1383 bp control. Although the difference in torsion constant between the 1097 bp insert fragment and the controls persists to very high salt concentration, as shown in Fig. 9, all of these DNAs exhibit the same trend, namely an increase in α between 0.1 and 1.0 M NaCl, followed by a steady decrease between 1.0 and 4.30 NaCl. Moreover, with increasing [NaCl], the ratio of intercalated to non-intercalated ethidium decreases by a factor of about 6 between 0.1 M to 1.0 M, but then turns about and actually increases by a factor of about 2.5 between 1.0 M and 4.3 M (data not shown). The obvious reversal in slope of these properties (*i.e.* α and ratio of intercalated to non-intercalated dye) and the obvious

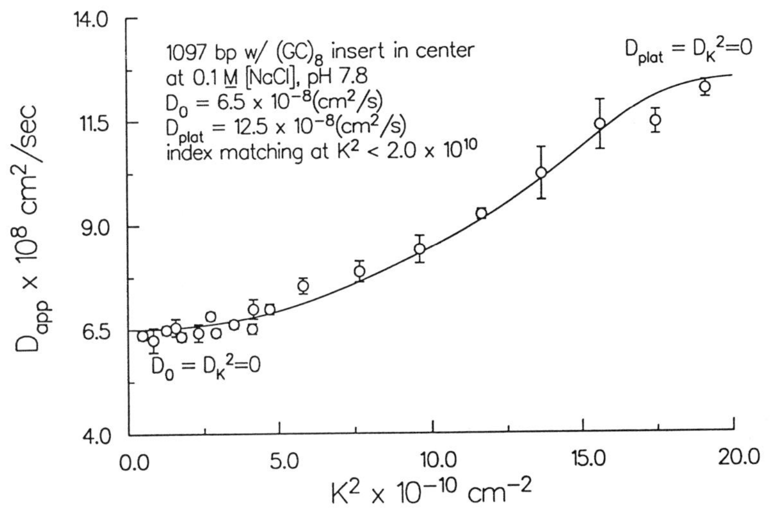

Figure 4. $D_{app}(K)$ *vs*. K^2 for the 1097 bp insert fragment containing $(GC)_8$. The buffer is the same as in Fig. 2. The curve is simply drawn to guide the eye through the data points. The corresponding data for the 1089 bp control (not shown) essentially superimpose on the curve drawn through these data. These data are measured at 21°C and corrected to 20,w conditions.

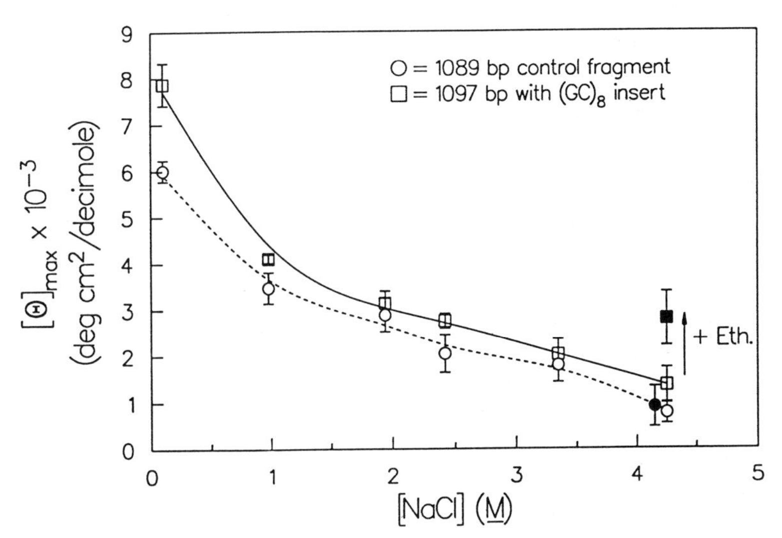

Figure 5. Molar ellipticity at the maximum $[\theta]_{max}$ near 273 nm *vs*. [NaCl]. Circles are for the 1089 bp control fragment and squares are for the 1097 bp insert fragment containing $(GC)_8$. Apart from [NaCl], the buffer conditions are the same as in Fig. 2. Note the prominent change in slope near 1.0 M NaCl, which is also observed for the 1383 bp control (not shown).

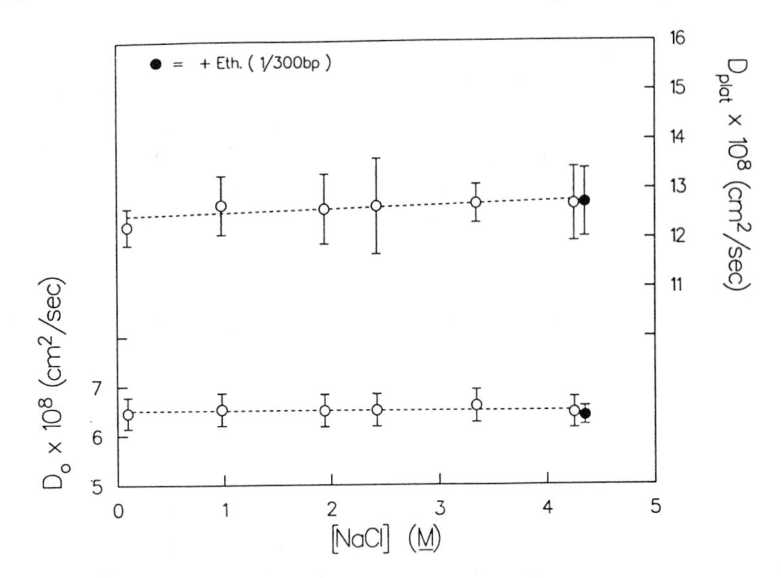

Figure 6. D_0 and D_{plat} *vs.* [NaCl] for the 1089 bp control fragment. D_{plat} is just the extrapolated $D_{app}(K)$ at $K^2 = 20 \times 10^{10}$ cm^{-2}. Apart from [NaCl], the buffer components are the same as in Fig. 2. These data are measured at 21°C and corrected to 20,w conditions.

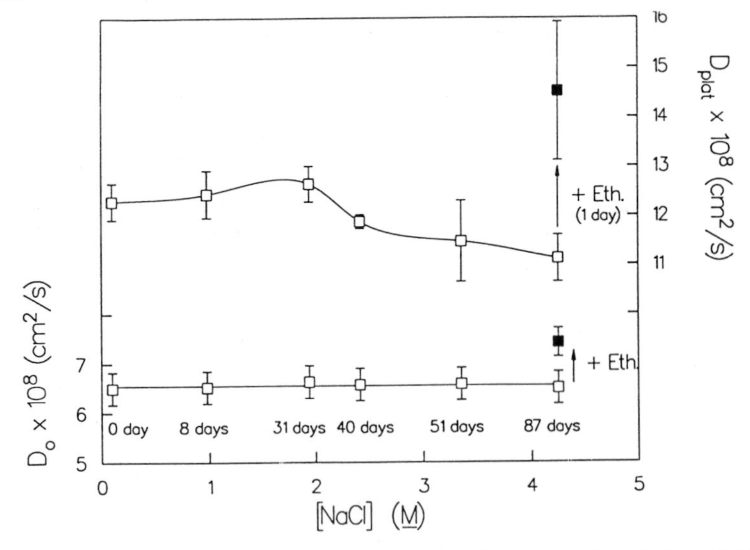

Figure 7. D_0 and D_{plat} *vs.* [NaCl] for the first sample of the 1097 bp insert fragment containing $(GC)_8$. The elapsed time in days between the measurements at each salt concentration are indicated at the bottom of the figure. Buffer and temperature conditions are the same as in Fig. 6. The solid squares denote the values measured 1 day after addition of 1 ethidium dye per 300 base-pairs to the 4.3 M NaCl solution.

change in slope of $[\theta]_{273}$ with increasing [NaCl] near 1.0 M suggests that the *control* DNA actually is beginning to adopt an alternative secondary structure, which we call simply the high salt form. This new state, which sets in with increasing [NaCl] above 1.0 M, exhibits a low $[\theta]_{273}$, a low torsion constant, an approximately normal persistence length, and an enhanced affinity for intercalated ethidium.

The data in Figs. 6-9 provide another example of the insensitivity of D_{plat} to changes in α.

Salt Induced Transition of the Insert Fragment

It is apparent from the D_{plat} data in Figs. 7 and 8 that the 1097 bp insert fragment undergoes a structural transition between 2.0 M and 2.5 M NaCl that presumably involves formation of Z-helix at the $(GC)_8$ sequence. The difference in magnitude of the transition between the two different samples of the 1097 by insert fragment is attributed to the different rate of passage through the transition region. The first sample was equilibrated for 9 days at 2.5 M NaCl before performing the measurements, whereas the second sample was equilibrated for only 2 days before taking data. The molecular property responsible for the decrease in D_{plat} is not yet certain. In principle, that could be ascribed to an *increase* in bending rigidity, or perhaps to a decrease in torsion constant. The latter possibility seems less probable, because these DNAs are very likely too short for D_{plat} to be very sensitive to changes in α, as noted above. These relatively short DNAs are sufficiently free-draining that a modest *increase* in bending rigidity only negligibly decreases D_0, but substantially decreases D_{plat}. Our combined D_0 and D_{plat} data preclude any substantial *decrease* in bending rigidity, whether global or local, in particular at the boundaries of Z-helix regions, in agreement with previous work[11]. Assuming that the bending rigidity below the transition is more or less uniform, only an extensive or global increase in that could significantly alter D_{plat}. An increase in bending rigidity over a small region of an otherwise uniform filament could not significantly affect D_{plat}. Hence, our data suggest that this salt-induced transition between 2.0 M and 2.5 M involves much more of the sequence than just the $(GC)_8$ insert *per se*.

The absence of any change in α corresponding to the observed decrease in D_{plat} is contrary to our previous experience, and

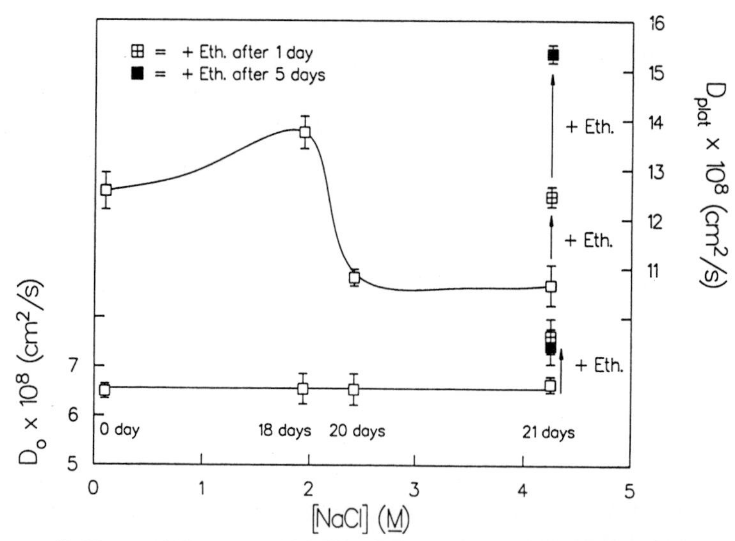

Figure 8. D_0 and D_{plat} *vs*. [NaCl] for the second sample of the 1097 bp insert fragment containing $(GC)_8$. The elapsed time in days between the measurements at each salt concentration are indicated at the bottom of the figure. Buffer and temperature conditions are the same as in Fig. 6. The squares with the plus (+) inside denote values measured 1 day after addition of 1 ethidium dye per 300 base-pairs to the 4.3 M NaCl solution, and the solid squares denote values measured 5 days after the ethidium addition.

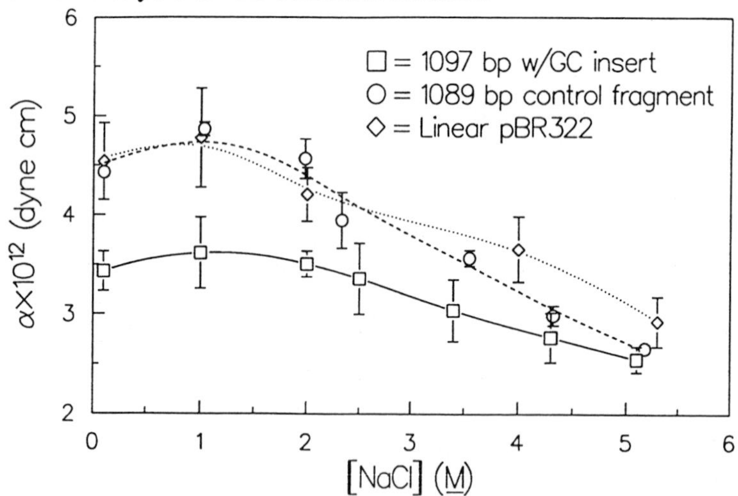

Figure 9. Torsion constant α *vs*. [NaCl]. Squares denote the 1097 bp insert fragment containing $(GC)_8$. Circles denote the 1089 bp control fragment. Diamonds denote the whole linearized pBR322. Apart from [NaCl], the buffer and temperature conditions are the same as in Fig. 2.

therefore puzzling. Any change in secondary structure leading to an increase in bending rigidity would likely also affect the torsional rigidity. This aroused our suspicion that the presence of ethidium dye in the FPA sample might be interfering with this salt-induced structural transition, and led to the experiments described in the next section.

Long-Range Effect of Ethidium on DNA Secondary Structure in 4.3 M NaCl

In 4.3 M NaCl, the addition of 1 ethidium per 300 bp has no significant effect on D_0, D_{plat}, or $[\theta]_{273}$ of the 1089 bp control, as indicated in Figs. 5 and 6. The same holds for the 1383 bp control. However, for the 1097 bp insert fragment, the same addition of 1 ethidium per 300 bp causes significant increases in D_0 and D_{plat}, and more than doubles $[\theta]_{273}$, as indicated in Figs 5, 7 and 8. This ethidium-induced transition occurs slowly on a time scale of days, as indicated in Fig. 8. When the dye and salt are removed by dialysis, the insert fragment returns to its original behaviour in 0.1 M NaCl, so the system is reversible in that sense. In 0.1 M NaCl, ethidium has no significant effect on D_0, D_{plat}, or $[\theta]_{273}$ of either the 1089 bp control or the 1097 bp insert fragment. These data show clearly that ethidium binding to the insert fragment in 4.3 M NaCl effects a long-range change in secondary structure that extends far (hundreds of bp) beyond its binding site.

4. CONCLUSIONS

The present data show that the 16 bp $(GC)_8$ insert exerts a pronounced influence on the *average* secondary structure of its flanking sequences that must extend over at least several hundred base-pairs, if not the entire fragment. With increasing [NaCl] the insert induces a transition between 2.0 M and 2.5 M NaCl that probably also involves a substantial fraction of the total sequence. In 4.3 M NaCl, less than 1 bound ethidium per 300 bp induces a long range alteration of the *average* secondary structure of the insert fragment, but not of the control fragments.

These results and the 3:1 ratio of the dynamic to static persistence length support the idea that a given local sequence fluctuates among two or more distinct secondary structures that extend over much larger domains, and whose relative stabilities are governed by distant as well as close neighbouring base-pairs. Our data further suggest that the 16 bp sequence $(GC)_8$ significantly shifts the prevailing equilibrium toward one of those structures.

ACKNOWLEDGEMENT

This work was supported in part by grants from the National Institutes of Health and the National Science Foundation.

REFERENCES

1. R.D. Wells, R.W. Blakesly, J.F. Burd, H.W. Chan, J.B. Dodgson, S.C. Hardies, G.T. Horn, K.F. Jensen, J.E. Larson, I.F. Nes, E. Selsing and R.M. Wartell, *CRC Crit. Revs. Biochem.*, 1977, *4*, 305.
2. V.I. Ivanov, L.E. Minchenkova, E.E. Minyat and A.K. Schyolkina, *Cold Spring Harbor Symp. Quant. Biol.*, 1983, *47*, 243.
3. S. Arnott and R. Chandrasekaran, 'Biomolecular Stereodynamics' (ed. R. Sarma) Adenine Press, New York, 1981, vol. 1, p. 99.
4. J.H. Shibata, J. Wilcoxon and J.M. Schurr, *Biochemistry*, 1984, *23*, 1188.
5. A.S. Benight, J. Langowski, P.-G. Wu, J. Wilcoxon, J.H. Shibata, B.S. Fujimoto, N.S. Ribeiro and J.M. Schurr, 'Laser Scattering Spectroscopy of Biological Objects,' ed. J. Stepanek, P. Anzenbacher, and B. Sedlacek, 'Studies in Physical and Theoretical Chemistry,' Elsevier, Amsterdam, 1987, vol. *45*, p. 407.
6. J.M. Schurr, B.S. Fujimoto, P.-G. Wu and L. Song, to appear in 'Fluorescence Spectroscopy,' ed. J.R. Lakowicz, Plenum Publ. Corp., 1991, in press.
7. L. Song, B.S. Fujimoto, P. Wu, J.C. Thomas, J.H. Shibata and J.M. Schurr, *J. Mol. Biol.*, 1990, *214*, 307.
8 L. Song, S.A. Allison and J.M. Schurr, *Biopolymers,* 1990, *29*, 1773.
9. L. Song and J.M. Schurr, *Biopolymers*, 1990, *29*, 1792.
10. S. Diekmann, W. Hillen, B. Morgenmeyer, R.D. Wells and D. Pörschke, *Biophys. Chem.*, 1982, *15*, 263.
11. D. Pörschke, W. Zacharias and R.D. Wells, *Biopolymers,* 1987, *26*, 1971.

12. D. Eden and C. Sunshine, in 'Dynamic Behavior of Macromolecules, Colloids, Liquid Crystals, and Biological Systems by Optical and Electro-Optical Methods', ed. H. Watanabe, Hirokawa, Tokyo, 1990.
13. S.A. Allison, R.H. Austin and M.E. Hogan, *J. Chem. Phys.*, 1989, *90*, 3845.
14. P.-G. Wu, B.S. Fujimoto and J.M. Schurr, *Biopolymers*, 1987, *26*, 1463.
15. J.C. Thomas, S.A. Allison, C.J. Appellof and J.M. Schurr, *Biophys. Chem.*, 1980, *12*, 177.
16. P. Wu, B.S. Fujimoto, L. Song and J.M. Schurr, *Biophys. Chem.*, in press.
17. B.S. Fujimoto and J.M. Schurr, *Nature*, 1990, *344*, 175.
18. B.J. Berne and R. Pecora, 'Dynamic Light Scattering: With Applications to Biology, Chemistry and Physics,' Wiley, New York, 1976.
19. J.M. Schurr, *CRC Crit. Revs. Biochem.*, 1977, *4*, 371.
20. J.M. Schurr and K.S. Schmitz, *Annu. Rev. Phys. Chem.*, 1986, *37*, 271-305.
21. J. Wilcoxon and J.M. Schurr, *Biopolymers*, 1983, *22*, 2273.
22. H. Yamakawa and M. Fujii, *Macromolecules*, 1973, *6*, 407.
23. J.M. Schurr, *Biopolymers*, 1983, *22*, 2207.
24. L. Song, U.-S. Kim, J. Wilcoxon and J.M. Schurr, *Biopolymers*, 1991, *31*, 547.
25. S.A. Allison and J.M. Schurr, *Chem. Phys.*, 1979, *41*, 35.
26. J.M. Schurr, *Chem. Phys.*, 1984, *84*, 71.
27. A.S. Benight, P.F. Flynn, B.R. Reid, D.E.Wemmer and J.M. Schurr, *J. Mol. Biol.*, 1988, *200*, 377.
28. C. Schildkraut and S. Lifson, *Biopolymers*, 1965, *3*, 195.
29. W.A. Baase and W.C. Johnson, Jr., *Nucleic Acids Res.*, 1979, *6*, 797.
30. A. Chan, R. Kilkushie and S. Hanlon, *Biochemistry*, 1979, *18*, 84.

24

DNAs as Models of Rigid and Semi-rigid Rodlike Macromolecules

By R. Pecora

DEPARTMENT OF CHEMISTRY, STANFORD UNIVERSITY, STANFORD, CALIFORNIA 94305-5080, U.S.A.

1. INTRODUCTION

The study of the dynamics of rigid and semi-rigid rodlike macromolecules in solution is beset by several difficulties. Foremost among these is the lack of sets of well-defined, monodisperse, homologous series of model macromolecules. It is generally very difficult to prepare such a series for synthetic macromolecules, and, although it it is possible to prepare very narrowly disperse molecules for many biological macromolecules and particles (*e.g.* virus particles), it is difficult in most cases to prepare homologous series. A notable exception to this is the DNAs. Very short DNAs ranging from the monomer to about 100 base pairs in length ("oligonucleotides") may be relatively easily prepared with the aid of DNA synthesizers, and larger molecules may be made by using genetic engineering techniques to prepare appropriate bacterial plasmid DNAs from which monodisperse, blunt-ended fragments may be cut using restriction enzymes ("restriction fragments"). It is also likely that recently developed polymerase chain reaction (PCR) techniques may be used to produce oligonucleotides and fragments even more cheaply than has been done heretofore.

In our laboratory we have been utilizing a range of DNAs as model systems for studying macromolecular dynamics. The smaller oligonucleotides (less than about 30 base pairs in length) may be used as model rigid rods to test hydrodynamic theories of translational and rotational diffusion for rodlike molecules of relatively small length to diameter ratios. Such information about overall translational and rotational motion may also be used to help extract information about local internal motions within the molecules from various experimental techniques that are sensitive to a combination of overall and local motion (NMR, fluorescence polarization decay, *etc.*)[1,2]. It is also clear that the scattering techniques used to study the overall rotational and translational dynamics are quite sensitive to overall molecular dimensions and may be used to follow processes that result in a change of molecular size or shape (*e.g.* for oligonucleotides, B to Z transitions, binding of proteins and anti-cancer drugs, *etc.*).

The restriction fragments (most of which are in the range from several hundred to several thousand base pairs in contour length) are used to model the dynamics of single, semi-rigid chains in relatively dilute solutions[3]. In addition to linear double helical fragments, superhelical and relaxed circles may be studied[4,5]. The "flexibility" (defined as the ratio of contour length to persistence length) varies with molecular weight for a given series of molecules (*e.g.* the linear ones) and solution conditions. The flexibility of molecules with given molecular weights may be varied by changing solution ionic strength. Dynamic properties that are followed include the scattered intensity time autocorrelation function that is measured in polarized dynamic light scattering studies and the decay function of the electric birefringence. Total intensity light scattering, and X-ray scattering are used to study structural parameters. These directly measured properties can be related to macromolecular translational diffusion coefficients and rotational and long range internal chain relaxation times. The use of DNA restriction fragments avoids many of the problems of data interpretation associated with polydispersity, such as is found in other model polyelectrolytes (*e.g.* the polystyrene sulphonates). A disadvantage of the restriction fragments is that they are relatively difficult to produce in large amounts, so that they have heretofore been used only to study dilute solution properties and only with techniques that require relatively small volumes of sample. The production techniques, while commonplace in biology laboratories, are also generally unfamiliar to polymer physicists.

2. OLIGONUCLEOTIDES

The rotational and translational diffusion coefficients of oligonucleotides are studied by, respectively, depolarized dynamic light scattering (DDLS) and polarized dynamic light scattering (DLS)[6,7]. In DDLS the frequency distribution of the scattered light is measured using a Fabry-Perot interferometer equipped with confocal plates as the predetection filter. The scattered light spectrum, after deconvolution from the instrumental line shape, is a Lorentzian whose half-width is six times the molecular rotational diffusion coefficient. The polarized DLS is obtained by standard methods. The time correlation function of the polarized scattered intensity is measured and analyzed using the programs CONTIN and DISCRETE[8-12]. The intensity time correlation function is a single exponential with decay constant equal to two times the product of the scattering vector length squared and the molecular translational diffusion coefficient.

The rotational and translational diffusion coefficients of three duplex, B-DNA oligonucleotides were measured as well as the rotational diffusion coefficient of a hairpin molecule with an almost spherical shape. The duplex oligonucleotides which constitute a homologous series have base sequences $d(CG)_n$, where n = 4 (octamer) and 6 (dodecamer) and

d[CGTACTAGTTAACTAGTACG]

(20-mer). The hairpin molecule d(CGCGTTGTTCGCG) consists of 13 bases (tridecamer) and is folded such that the four bases at each end are paired with each other. The TTGTT sequence forms a loop[13].

To illustrate the sensitivity of the technique, the rotational diffusion coefficient of all four oligonucleotides are shown in Fig. 1 as a function of concentration. Note that the molecule with the largest rotational diffusion coefficient is the hairpin molecule. It rotates faster than the B-form duplexes because of its compact almost spherical shape. Note that the measurement of rotational diffusion coefficients is sensitive enough to distinguish between different conformations of oligonucleotides and also between B-form duplexes that differ by only a few base pairs in length.

The translational and rotational diffusion coefficients may be combined to test the consistency of hydrodynamic theories for rotation and translation of short rodlike particles and to obtain a hydrodynamic diameter of DNA in water based buffer solutions[14-16]. Our studies indicate that the hydrodynamic subunit models of Garcia de la Torre *et al.* give a consistent explanation of the rotational and

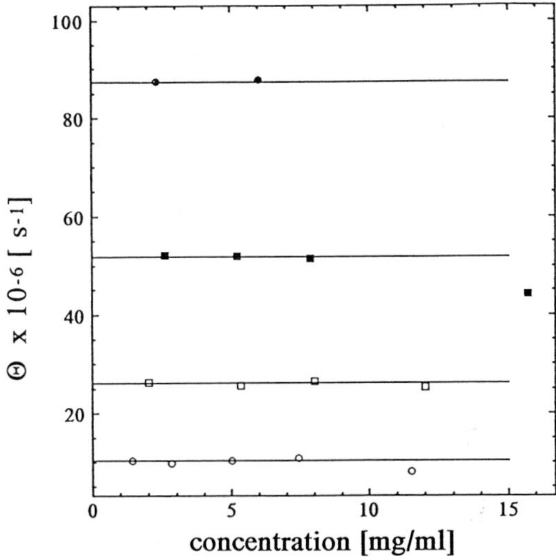

Figure 1. Rotational diffusion coefficients *versus* concentration corrected to 20°C for the three oligonucleotides. Plot symbols: filled circles = tridecamer (hairpin), filled squares = 8-mer, open squares = 12-mer, and open circles = 20-mer. The highest concentration points for the 8-mer and 20-mer are indicative of aggregation. Note that the hairpin (which is almost spherical in shape) has the largest rotational diffusion coefficient. (Data from refs. 1 and 2.)

translational diffusion coefficients of the three duplex B-form oligonucleotides. The data indicate a rise per base pair between 3.4Å and 3.5Å. The three oligonucleotides have an average hydrodynamic diameter equal to (20 ± 1.5) Å. This diameter is smaller than the 24Å - 26 Å usually given for the diameter of DNA in solution[17].

3. RESTRICTION FRAGMENTS

Our strategy in using DNA restriction fragments as model systems was to first use genetic engineering techniques to clone DNA fragments of varying length into bacterial plasmid DNAs[3]. The bacteria containing the modified plasmid DNAs are then allowed to multiply, and the plasmids extracted. The desired fragments are then cut out of the plasmids using various restriction enzymes, separated from any remaining part of the plasmid that may be present and prepared for the appropriate experiment. The fragments that are being used in the majority of our experiments are 367, 762, 1010 and 2311 base pairs in contour length. These lengths were chosen so that they could be conveniently studied by both polarized DLS[7] and transient electric birefringence (TEB)[18]. Superhelical, relaxed circle and linear duplex forms of the 2311 bp fragment are under study. The 367, 762 and 1010 bp fragments under study are all in the linear B-DNA form.

Only a very brief outline of some of the general features of the results to date are given here. Both DLS[19,20] and TEB[18, 21-23] studies have been performed on the linear forms of all four fragments. DLS experiments have been done on the relaxed circle form of the 2311 bp fragment[24] and both DLS and TEB have been performed on the superhelical form of this same fragment[4]. The TEB studies were generally done at low ionic strengths (1.5 to 3 mM in sodium) and low concentrations (about 5 - 10 $\mu g\ ml^{-1}$), while most of the DLS studies were done at higher ionic strengths (100 mM in sodium) and higher concentrations (about 100 $\mu g\ ml^{-1}$). Since the persistence length depends on ionic strength, it is likely that in the DLS experiments a fragment of a given molecular weight is more flexible than in the corresponding TEB experiment. DLS experiments at lower ionic strengths (as well as at higher concentrations) are now in progress.

Both the polarized DLS intensity autocorrelation functions and the TEB decay functions were analyzed by the programs CONTIN and DISCRETE[8-12]. The TEB experiments generally showed more than one relaxation process, an indication that intramolecular flexing motions contribute to the spectra. At low scattering angles, the DLS time correlation functions were single exponentials as expected from theory[25] while at higher angles they were generally composed of two or more significant exponentials. The time constant for the low angle decay is proportional to the macromolecular mutual diffusion coefficient. (The mutual diffusion coefficient is equal to the self diffusion coefficient when the macromolecule concentration is low)[6]. The bi- or multiexponential decay function at higher angles is an indication that rotational and/or intramolecular flexing motions are influencing the time correlation functions. By use of a combination of studies varying the scattering angle and simulations of the relaxation time distributions for models[19,20,22], it is possible to obtain the translational diffusion coefficient and the relaxation time of the first normal relaxation mode of the fragment.

An example of a CONTIN analysis of a DLS time correlation function from the 1010 bp DNA fragment is shown in the middle portion of Fig. 2. The product of the scattering vector length q and the molecular radius of gyration is 1.5. This corresponds to a

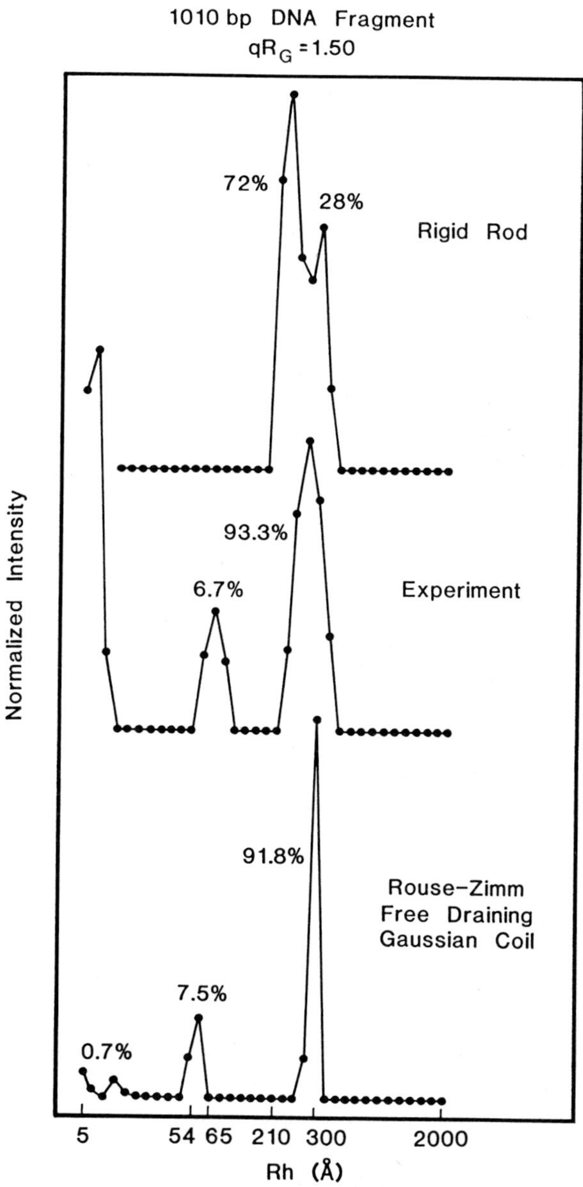

Figure 2. Distribution of hydrodynamic radii (proportional to the relaxation times) from CONTIN analysis of the DLS time correlation function for the 1010 bp linear, B-duplex restriction fragment (middle). Also shown are CONTIN analyses of calculated time correlation functions for a rod model (top) and free draining Gaussian coil model (bottom) of the same fragment. (From ref. 20.)

scattering angle of 90°. The plot shows the fraction of the scattered correlation function (actually the scattered electric field correlation function) that relaxes with a given apparent hydrodynamic radius. The hydrodynamic radius is simply defined in terms of the reciprocal of the apparent diffusion coefficient D *via* the Stokes-Einstein relation:

$$R_h = \frac{k_B T}{6\pi\eta D}$$

where k_B is Boltzmann's constant, T, the temperature in Kelvin and η, the solvent viscosity. The large peak with 93.3% of the scattered intensity represents the translational diffusion of the molecule. The peak with 6.7% of the area represents both translational diffusion of the molecule and a further, faster contribution from rotational and/or long range internal motion of the chain. It is from this peak the relaxation time for the longest mode internal relaxation time of the chain may be obtained. The third peak at very small hydrodynamic radius (not included in the percentage calculations of the other peaks) is only a few percent of the total relaxation function and is most likely an artifact of the data analysis method[19, 20]. A peak of this type is often observed in both simulated and real data for a wide variety of systems at the low hydrodynamic radius end of the CONTIN window. If the CONTIN window is moved to even lower R_h, this peak usually follows it.

Also shown in Fig. 2 are CONTIN analyses of calculated correlation functions for a rigid rod model (top) and a free draining Gaussian coil model (bottom) of this fragment[20, 25]. These models have as inputs the measured value of the translational diffusion coefficient from low angle DLS (for both the rod and coil), the theoretical radius of gyration for a wormlike coil based on the persistence length (coil model), and the length and rotational diffusion coefficient calculated for a straight rod with the given number of base pairs (rod model). Note that the free-draining Gaussian coil distribution gives good agreement with experiment .

In general the DLS experiments for all four restriction fragments give translational diffusion coefficients that are in reasonable agreement with those predicted by the Yamakawa-Fujii theory for translational diffusion of wormlike chains.[26] The internal times for the longest three fragments (the 367 bp fragment is too small to measure internal times by DLS) , however, are in best accord with those predicted by the free-draining Gaussian coil model with input parameters consisting of the experimental translational diffusion coefficient and the radius of gyration of a wormlike coil with the persistence length and contour length of the given DNA. In fact, as illustrated in Fig. 2, the entire relaxation time distribution is well approximated by this model. This apparent agreement with the free draining model is likely to be the result of an approximate cancellation of effects due to hydrodynamic interaction and chain stiffness, both of which are included in free draining Gaussian coil model calculations in a rather indirect way[22]. The TEB studies show similar agreement with the free draining model for the 2311 base pair fragment, but the shorter fragments show significant deviations from free draining Gaussian coil behaviour. This is most likely due to the fact that the TEB studies were done at lower ionic strength where the persistence of DNA is higher and the molecules are stiffer.

REFERENCES

1. W.Eimer, J.R. Williamson, S.G. Boxer and R. Pecora, *Biochemistry*, 1990, *29*, 799.
2. W. Eimer and R. Pecora, *J. Chem. Phys.*, 1991, *94*, 2324.
3. R.J. Lewis, J.H. Huang and R. Pecora, *Macromolecules,* 1985, *18*, 1530.
4. R.J. Lewis, J.H. Huang and R. Pecora, *Macromolecules*, 1985, *18*, 944.
5. G. Voordouw, Z. Kam, N. Borochov and H. Eisenberg, *Biophys. Chem.*, 1978, *8*, 71.
6. B.J. Berne and R.Pecora, 'Dynamic Light Scattering', Wiley-Interscience, New York, 1976.
7. R. Pecora, ed., ‘Dynamic Light Scattering: Applications of Photon Correlation Spectroscopy’, Plenum, New York, 1985.
8. S.W. Provencher, *Comput. Phys. Commun.,* 1982, *27*, 213, 229.
9. S.W. Provencher, 'CONTIN User's Manual', Technical Report ENBL-DA02, European Molecular Biology Laboratory, Heidelberg, 1980.
10. S. W. Provencher, *Makromol. Chem.,* 1979, *180*, 201.
11. S.W. Provencher, J. Hendrix, L. De Maeyer and N. Paulussen, *J Chem. Phys.,* 1978, *69*, 4237.
12. S.E. Bott , Ph.D. Thesis, Stanford University, Stanford, CA, 1984.
13. J.R. Williamson and S.G. Boxer, *Biochemistry*, 1989, *28*, 2819.

14. M.M. Tirado and J. Garcia de la Torre, *J. Chem. Phys.* 1979, *71*, 2581; 1980, *73*, 1986.
15. M.M. Tirado, M.C. Lopez Martinez and J. Garcia de la Torre, *J. Chem. Phys.*, 1984, *81*, 2047.
16. J. Garcia de la Torre, M.C. Lopez Martinez and M. M. Tirado, *Biopolymers* , 1984, *23*, 611.
17. M. Mandelkern, J.G. Elias, D. Eden and D.M. Crothers, *J. Mol. Biol.*, 1981, *152*, 153.
18. R.J. Lewis, R. Pecora and D. Eden, *Macromolecules* , 1986, *19*, 134.
19. S.S. Sorlie and R. Pecora, *Macromolecules*, 1988, *21*, 1437.
20. S.S. Sorlie and R. Pecora, *Macromolecules* , 1990, *23*, 487.
21. R.J. Lewis, S. A. Allison, D. Eden and R. Pecora, *J. Chem. Phys.*, 1988, *89*, 2490.
22. S.A. Allison, S.S. Sorlie and R. Pecora, *Macromolecules* , 1990, *23*, 1110.
23. R.J. Lewis and R. Pecora, *Macromolecules* , 1986, *19*, 2074.
24. J. Seils and R. Pecora, *Macromolecules*, in press.
25. R. Pecora, *J. Chem. Phys.*, 1964, *40*, 1604; 1968, *48*, 4126; *49*, 1032.
26. H. Yamakawa and M. Fujii, *J. Chem. Phys.*, 1976, *64*, 5222.

Part III: Macromolecular Assemblies

25

Total [illegible]nd Quasi-elastic Light [illegible]ications in Microbiology

By Stephen E. Harding

UNIVERSITY OF NOTTINGHAM, DEPARTMENT OF APPLIED BIOCHEMISTRY AND FOOD SCIENCE, SUTTON BONINGTON, LE12 5RD, U.K.

1. INTRODUCTION

The previous chapters in this volume have considered the advances in laser light scattering theory and methodology followed by specific applications to systems of biological macromolecules. In this first chapter on "large" macromolecular assemblies we will survey some of the many recent applications of laser light scattering to problems in microbiology and focus on some of the work that we have been specifically involved with on viral and bacterial systems. The review element of this Chapter is intended to supplement an earlier article from this laboratory[1]. The reader is also referred to a more general review by Bloomfield[2] which includes a consideration of quasi-elastic light scattering (QLS) applications in virology and to the study of microbial motility, and an earlier article by Wyatt[3] which considers total intensity light scattering (TILS) applications in bacteriology.

Laser light scattering applications in microbiology can be broken down conceptually into two parts: 1. studies on the macromolecular components of microbes: this has essentially been covered earlier in this volume: many of the macromolecular systems considered there are microbial in origin; 2. studies on whole micro-organisms: this will be considered here.

Studies on "whole" microbes offer certain advantages and certain disadvantages as far as laser light scattering techniques are concerned. The advantages include (i) the greater signal to noise ratio (that is to say also the so-called "dust problem" is not as severe[4]); (ii) the correspondingly smaller concentrations generally required to give a sufficient signal; (iii) since the wavelength of the incident laser light can be of the same order as the maximum dimension of the microbe (Fig. 1), internal structural information can, in principle, be obtained from the nature of the "resonances" in the angular scattered intensity envelopes [3].

The relatively large size - compared to macromolecular systems - of microbial scatterers also brings problems, namely that we are on the limits, and in many cases go beyond them, of the applicability of the relatively simple "Rayleigh-Gans-Debye" representations of the scattering data. These criteria, already considered by Wyatt[5] in this volume are summarised by the two equations:

$$\left| \frac{n}{n_0} - 1 \right| \ll 1 \tag{1}$$

$$(4\pi n a/\lambda_0)|(n/n_0)| \ll 1 \tag{2}$$

a is the maximum dimension of the scatterer, n, n_0 the refractive indices of scatterer and surrounding medium respectively and λ_0 the wavelength of the incident light. These criteria are usually satisfied for smaller microbes, such as most viruses, and the highly useful limiting case or "Zimm plot" can also be applied. They can also be satisfied for larger micro-organisms - for example vegetative bacterial cells suspended in an aqueous media where $|n-n_0|$ is very small - although because of the resonances, the Zimm plot and related representations become inapplicable. For the highly dehydrated and refractile bacterial spores - particularly air-borne particles - the criteria are not satisfied, and for this case, interpretation of the scattering records can be rendered opaque because of the complexity of the mathematical representations ("Lorenz-Mie" theory) involved[6]. A further problem for the larger or denser microbes is that sedimentation phenomena can also obscure this interpretation. Nonetheless, if one allows for these reservations laser light scattering methods, both total intensity, or quasi-elastic, can provide powerful - and unlike for example electron microscopy, non-destructive - probes into the structure and dynamics of microbial systems, as we will consider now. In this consideration, details of the theory and instrumentation behind both the total intensity ("static") and quasi-elastic ("dynamic") light scattering methods will not be given: this has been extensively covered in the earlier Chapters.

2. APPLICATIONS TO VIRUS SYSTEMS

Total Intensity Light Scattering (TILS)

For smaller microbes like many viruses the simpler Rayleigh-Gans-Debye (RGD) theory is applicable and provided that $qR_g \ll 1$ (where q, R_g are the Bragg wave vector and the "radius of gyration" respectively) it is possible to use the TILS or "static"[6] Zimm plot to obtain (i) the (weight average) molecular weight M_w, (ii) the radius of gyration, R_g, and (iii) the thermodynamic second virial coefficient, B (or A_2).

The static Zimm biaxial plot is based on the following relation:

$$Kc/R_\theta = (1/P(\theta))\,[(1/M) + 2Bc + O(c^2)] \quad (3)$$

where c is the particle concentration (g/ml), θ is the scattering angle, K is an experimental constant which includes the square of the refractive index increment, R_θ is the Rayleigh excess ratio and P(θ) is the "form factor" which, in the limit of the scattering angle, $\theta \rightarrow 0$ takes the form

$$P(\theta) = 1 - (16\pi^2/3\lambda_0^2)\,R_g \sin^2(\theta/2) \quad (4)$$

Examples of Zimm plots applied to virus systems can be found for vaccinia virus (molecular weight, $M \sim 3x10^9$) [7], R17 virus ($M \sim 4x10^6$)[8] and tobacco mosaic virus, TMV ($M \sim 40x10^6$) [9,10].

The radius of gyration from the static-Zimm plot, if used in conjunction with other techniques such as sedimentation analysis and the translational diffusion coefficient from QLS can be used to model the conformation of viruses and other macromolecular assemblies in terms of arrays of spherical particles or "bead models": this procedure has proved particularly useful for the modelling of bacteriophages (see, *e.g.*, refs 11, 12). Alternatively "whole body" or triaxial ellipsoid modelling can be employed, involving combinations of R_g, B (after allowance for charge effects) and the intrinsic viscosity[13]. If a specific conformation of the virus is assumed (*e.g.* a rod), alternative relations for P(θ) are available permitting for example mass per unit length estimations[6]: this has been applied, for example, to a study of the rod shape bacteriophage fd[14].

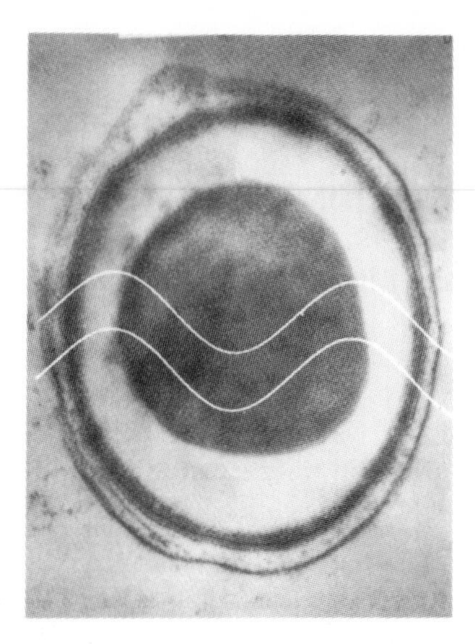

Figure 1. The maximum dimension of many microbes (in this example a spore from *B. cereus*) is of the same order as the wavelength of the incident (visible) radiation.

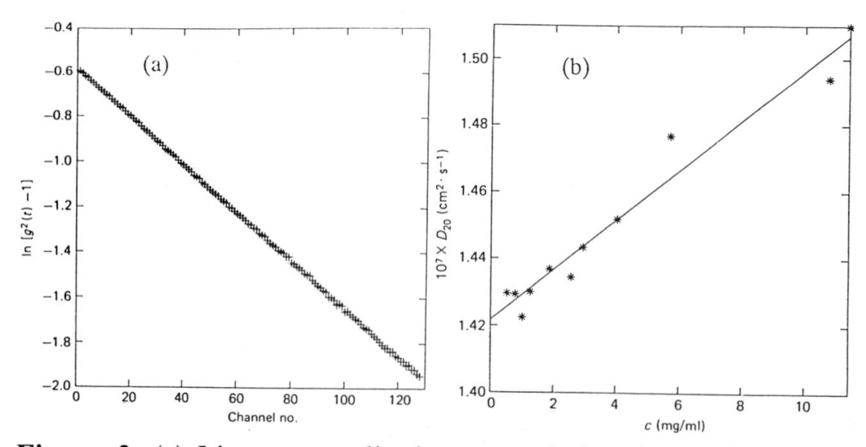

Figure 2. (a) Linear normalised autocorrelation decay plot for a quasi-spherical virus (TYMV). Channel number = delay time(τ)/sample time (τ_S). τ_S = 1.0 μs; duration time = 60 s; scattering angle θ = 90°; temperature = 24.85°C; c = 4.07 mg/ml; (b) Linear extrapolation of the apparent diffusion coefficient to zero concentration for TYMV. Reproduced, with permission, from ref. 19.

Quasi-elastic Light Scattering (QLS)

QLS contributions - in terms of the translational diffusion coefficient, D_T - to virology have been in four principal areas: (i) hydrodynamic rigid-bead modelling; (ii) flexible particle modelling; (iii) molecular weight measurement and (iv) for following the dynamics of assembly or swelling processes. As we have just mentioned, hydrodynamic modelling procedures based on QLS measurements are now well established (see, *e.g.* refs 2, 11, 12) and data capture and analysis[15] is now relatively straightforward for quasi-spherical viruses such as tomato bushy stunt virus (TBSV), turnip yellow mosaic virus (TYMV) and many other plant viruses. For these quasi-sphericals, the simple single exponential term correlation equation, in terms of the normalised intensity autocorrelation function $g^{(2)}(\tau)$ - where τ is the delay time - is usually adequate[16], provided that heterogeneity is not significant:

$$[g^{(2)}(\tau) - 1] = \exp(-D_T q^2 \tau) \tag{5}$$

The translational diffusion coefficient, D_T can thus be readily obtained from a plot of ln $[g^{(2)}(\tau) - 1]$ *versus* τ (Fig. 2a). Measurements based on a single angle - usually a high one like 90° to minimise dust problems - are usually sufficient. If the system is polydisperse, D_T obtained from eq. 5 will be a z-average, and a measure of the polydispersity is usually given in the form of a "Polydispersity Factor", PF, *i.e.* the normalized z-average variance of the distribution of diffusion coefficients[16]. The D_T obtained from eq. 5 is usually corrected to standard solution conditions (water as solvent at a temperature of 20.0°C) and then extrapolated to zero concentration to give a parameter, $D^0_{T(20,w)}$ free of non-ideality effects[17]. As with macromolecules, for viruses a linear extrapolation is usually adequate (Fig 2b).

$$D_{T(20,w)} = D^0_{T(20,w)} (1 + k_d) \tag{6}$$

where k_d is the concentration dependence diffusion coefficient parameter defined by[18]

$$k_d = 2BM - k_s - \bar{v} \tag{7}$$

and k_s is the sedimentation concentration dependence parameter and $\bar{v}$ the partial specific volume. This equation explains why the concentration dependence of the diffusion coefficient is usually not as severe compared with other hydrodynamic parameters (opposing effects on the RHS of eq. 7) and if k_s is known from a separate sedimentation velocity experiment, eqs. 6 and 7 provide an alternative route to eq. 3 for obtaining B, the thermodynamic second virial coefficient.

We have used QLS in this way to compare the gross morphology of TBSV variants[19] and TYMV in various solvents[20]. QLS translational diffusion coefficients have also been used *via* the Svedberg equation to evaluate the molecular weights of the bacteriophages type 5 adenovirus[21], MS2[22] and VS11[23], and also infectious pancreatic necrosis virus[24].

For rod-shape and other non-spheroidal viruses it is necessary to correct for rotational diffusion effects by measurement at several angles followed by a subsequent extrapolation to zero angle: a procedure which has yielded both D_T and the rotational diffusion coefficient D_R for tobacco mosaic virus (TMV)[10,25,26] and bacteriophages T4B and T7[27]. D_T and D_R values have also been recently obtained at a single angle (90°) from bimodal resolution using the "SIPP" algorithm[28].

Both extrapolations of $D^0_{T(20,w)}$ to zero concentration and zero angle can now be performed in a biaxial "dynamic" Zimm plot, and this procedure has been applied to microbial polysaccharides[29]. For rod-like viruses anisotropy of D_T itself can be a problem[30]. A further complication is one of particle flexibility, and Fujime and Maeda[31] have considered this for filamentous fd virus. The effects of point (amino acid) mutations on the relative length and flexibility of fd has recently been considered[32].

Comparison of data from TILS and QLS with other hydrodynamic data

The dangers - largely because of the serious errors resulting from even trace amounts of dust and other contaminating material - of using light scattering techniques in isolation for the characterisation of macromolecular solutions are widely appreciated. For solutions of viruses the situation is not as severe compared to the situation encountered when studying smaller macromolecules and assemblies[4,33], but independent confirmatory or complementary measurements - from for example analytical ultracentrifugation or electron microscopy - are always useful. In general, agreement is good. Johnson and Brown[10] in their article discussed the good agreement of molecular weight measurements from the QLS/Svedberg equation ($M \sim 40.8 \times 10^6$) with that from chemical analysis (39.4×10^6) and sedimentation equilibrium (41.6×10^6) for TMV. Similar good agreement has been obtained between TILS ($M \sim 14.6 \times 10^6$)[14] and sedimentation equilibrium (14.2×10^6)[31] for (wild-type) filamentous fd bacteriophage.

Another good example of agreement of results from laser light scattering with independent techniques is for the case of the quasi-spherical TYMV[20]: the molecular weights evaluated from the translational diffusion coefficient and sedimentation coefficient *via* the Svedberg equation for four different solvents are in excellent agreement with those obtained independently from sedimentation equilibrium (Table 1a). Furthermore values for the second virial coefficient obtained from combination of the coefficient k_d obtained from the concentration dependence of the diffusion coefficient and k_s, the corresponding parameter from sedimentation velocity are in excellent agreement with the virial coefficient obtained from sedimentation equilibrium (Table 1b) and the hydrodynamic radii calculated from the translational diffusion coefficient are again in excellent agreement with the "thermodynamic" radii from the second thermodynamic virial coefficient from sedimentation equilibrium (Table 1c).

Despite their large size, it is possible to obtain preparations of virus solutions with very high monodispersities. This has made them highly suitable systems for testing experimentally the validity of theories representing the concentration dependence of the translational diffusion coefficient and related hydrodynamic parameters[18,20].

Analysis of dynamics of virus systems

Arguably the greatest value of laser light scattering methods - particularly QLS - in virology is for the analysis of the dynamics of self-assembly and related processes. We quote just two examples here. One (see, *e.g.*, Fig. 3a) is the analysis of the kinetics of head/tail assembly reactions for bacteriophages[34-38]; the second is the analysis of the effects of the removal of calcium ions on the dynamics of swelling of southern bean mosaic virus (SBMV), monitored by hydrodynamic diameters (Fig. 3b) and polydispersity

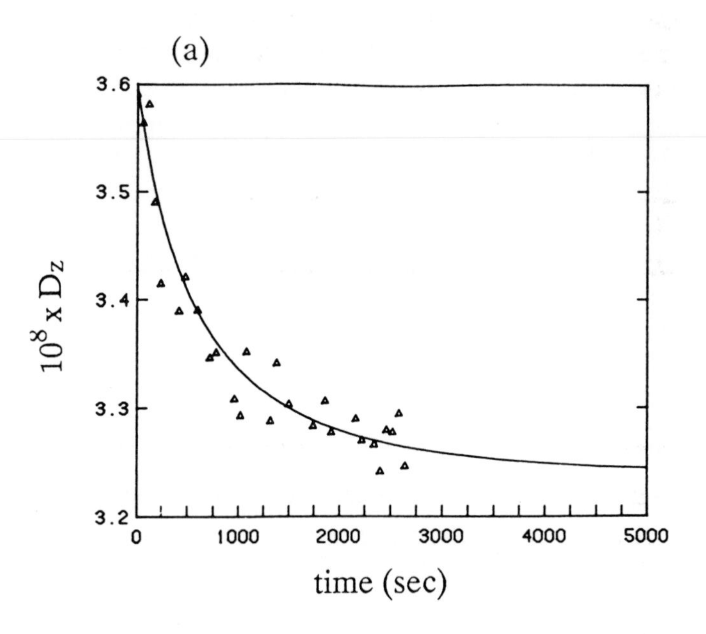

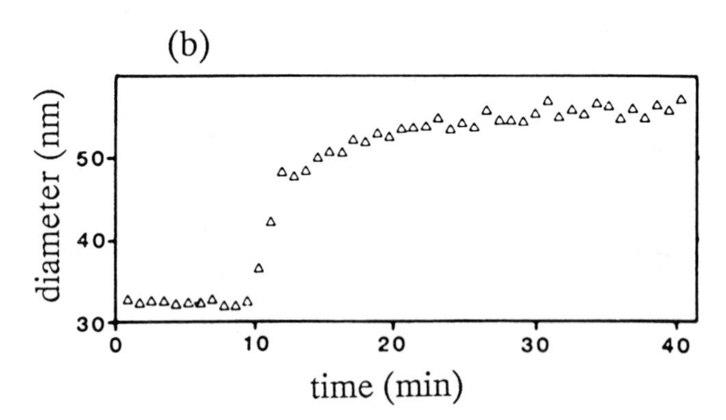

Figure 3. Following the dynamics of viral processes using QLS:
(a) Head/tail association kinetics of T4D bacteriophage. Line fitted corresponds to an association constant, k=7.9 x 10^{-6} M^{-1} s^{-1}. Reproduced with permission, from ref. 37.
(b) Swelling of southern bean mosaic virus on removal of Ca^{++} ions. Reproduced with permission, from ref. 38.

Table 1. Physical data for TYMV: Agreement between QLS/sedimentation velocity and sedimentation equilibrium for various solvent pH and ionic strength conditions

(a) Molecular Weight

pH	I [a]	10^{-6} x M_w [b] (±0.05) g/mol	10^{-6} x M_w [c] (±0.2) g/mol	10^{-6} x M_w [d] (±0.2) g/mol
7.8	0.1	5.80	5.5	5.3
6.8	0.1	5.73	5.8	5.5
6.0	0.1	5.75	5.6	5.6
4.75	0.1	5.77	5.6	5.6
6.8	0.2	5.64	5.7	5.5

(b) Thermodynamic ('osmotic pressure') second virial coefficient, B

pH	I [a]	10^6 x B [e] ±0.10 ml.mol.g^{-2}	10^6 x B [f] ±0.4 ml.mol.g^{-2}
7.8	0.1	1.30	1.3
6.8	0.1	1.21	1.5
6.0	0.1	1.25	1.3
4.75	0.1	1.10	1.3
6.8	0.2	1.32	1.7

(c) Radii of equivalent spherical particle

pH	I [a]	r_H [g] (± 0.2) nm	r_B [h] (± 0.5) nm
7.8	0.1	15.1	16.3
6.8	0.1	15.1	15.8
6.0	0.1	15.2	15.9
4.75	0.1	15.5	15.4
6.8	0.2	14.9	16.1

(a) ionic strength; (b) from the sedimentation coefficient and the (QLS) translational diffusion coefficient via the Svedberg equation; (c) whole cell weight average from sedimentation equilibrium; (d) point weight average extrapolated to zero concentration (sedimentation equilibrium); (e) from k_D and k_s; (f) from sedimentation equilibrium; (g) from the QLS translational diffusion coefficient; (h) from the (sedimentation equilibrium) value for B.

factors from QLS, in conjunction with total intensity measurements[39].

3. APPLICATIONS TO BACTERIAL AND OTHER CELLULAR MICROBIAL SYSTEMS

Both TILS and QLS have provided major inroads into our understanding of bacterial systems; perhaps most widely known is the application of QLS as a probe into the motility of bacteria and protozoa. However there has also been progress in other areas, such as the use of the total intensity method for modelling the structure - in terms of refractive index profiling - of isolated quasi-spherical bacterial spores and isolated marine micro-organisms, and also in the application of QLS for following the dynamics of suspensions of bacterial spore ensembles as a probe into their high resistance to thermal destruction; we will now survey some aspects of the progress in both these areas.

Total Intensity Light Scattering

With the larger microbes resonances in the angular scattered intensity profiles become significant - even at lower angles - and the Zimm plot and related methods become inapplicable. One ingenious solution was that of Morris *et al.*[40] who increased the wavelength of the incident radiation to the infra red, allowing application of the Zimm plot method to a particle whose hydrodynamic diameter was nearly a micron. This procedure has been applied[41] to *Serratia marcescens* ($M \sim 0.7 \times 10^{11}$), *Escherichia coli* ($M \sim 1.0 \times 10^{11}$) and *Thiobacillus ferrooxidans* ($M \sim 1.6 \times 10^{11}$) and an example is given in Fig. 4. However in general the resonances in the angular scattered intensity envelopes have to be taken into account, and indeed provide potential information about internal structure - in terms of refractive index profiles - of micro-organisms.

RGD and Lorenz-Mie TILS representations

Refractive index profiling of microbes (in terms of *e.g.* average dimensions and refractive indices of concentric regions within the microbe - Fig. 5) is achieved by model fitting *i.e.* comparing the measured angular scattered intensity envelope (Fig. 6) with a theoretical profile based on a given model, and then iterating or refining this model until adequate agreement is obtained[3]. Thus this

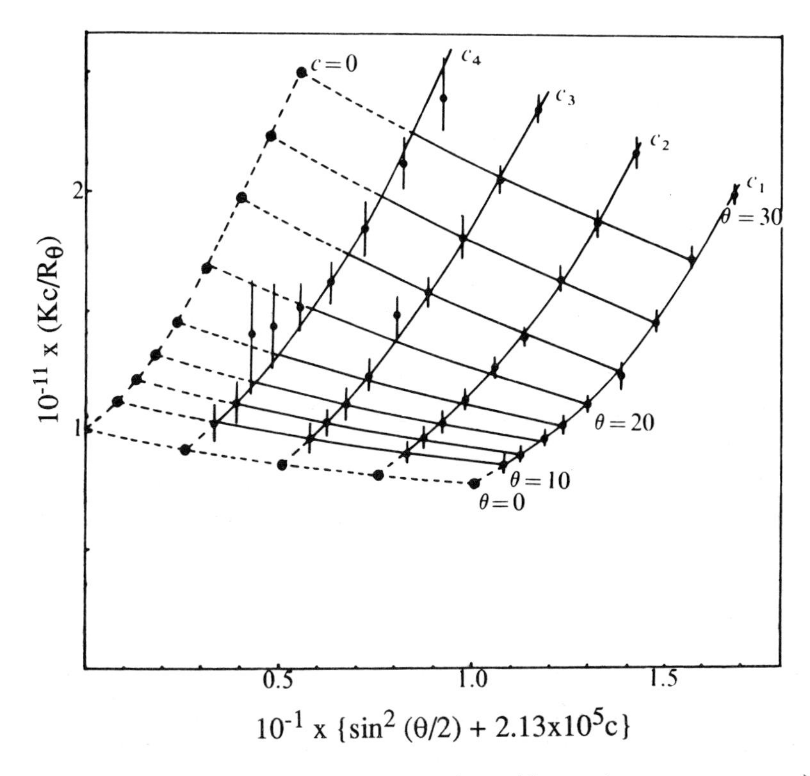

Figure 4. Zimm plot for a bacterium (*Serratia maracescens*) at λ=1.06 µm. Stock concentration (c_1) = 0.47 g/ml. Reproduced, with permission, from ref. 40.

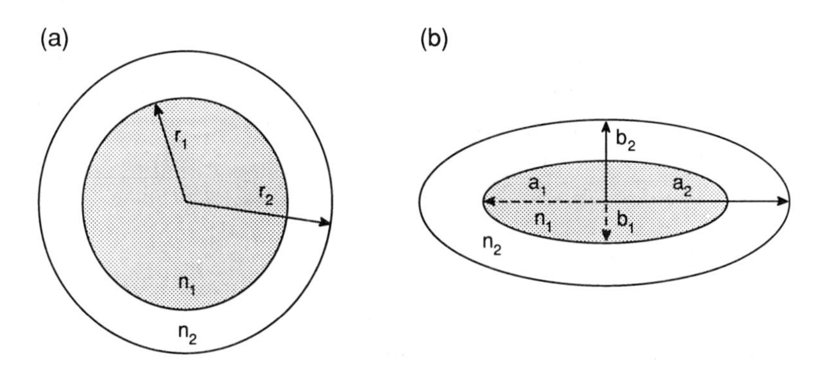

Figure 5. Refractive index profiling. (a) Coated sphere model: characterised by inner (or core) and outer (shell) radii, r_1, r_2 and inner and outer refractive indices, n_1, n_2 respectively (*cf.* Fig. 1); (b) Coated ellipsoid model: inner semi-axes a_1, b_1; outer semi-axes a_2, b_2.

procedure differs from the Zimm / radius of gyration method which does not require any *a priori* assumed conformation[6].

The RGD representations assume no change of phase or other distortions of the electric vector caused by the scattering particle: the phase differences between scattered waves from different points in a given particle are simply a function of position within that particle and not of the material of the particle[6]. [Representations are normally given for vertically polarised incident light since for horizontal or unpolarised light an extra term involving $\cos^2\theta$ can obscure these records[3,42]].

Even for larger microbes the RGD theory can still give a reasonable representation of the scattered intensity profiles, provided that the refractive index difference between scatterer and medium is small (an example of this situation could be for a vegetative bacterial cell in aqueous suspension). A good illustration of the range of validities of RGD is given by Kerker (pp. 428-429 of ref. 6). For general RGD scatterers, the form factor $P(\theta)$ can be represented by the equation[6]:

$$P(\theta) = (1/V^2)\left|\int_V e^{i\delta}\, dV^2\right| \qquad (8)$$

where the integral is over the total volume, V of a particle whose volume elements produce a scattered phase δ at a common reference plane.

Explicit expressions for $P(\theta)$ have been worked out for a wide range of particle shapes: of particular use are the concentric (2 or 3 layered) sphere and coated ellipsoid models (Fig. 5 and refs 42,43) for modelling the refractive indices and approximate sizes of cell nucleii, protoplasm and membranes.

For many microbes however - particularly for bacterial spores which have a relatively high internal refractive index - the RGD approximation is not satisfied and account has to be taken of phase changes and other distortions of the electric field caused by the scattering microbe. The theory describing such scattering behaviour (Lorenz-Mie) is considerably more complicated and solutions for the intensity *versus* angle are only available for a very limited number of particle forms. Table 2[44-53] lists those that are particularly relevant to situations involving micro-organisms. For particles with spherical symmetry, the solution is summarised by the form, for vertically polarised incident light[3,42]:

$$I(\theta) = \{I_0/(qR)^2\} \left| \sum_{l=1}^{\infty} [{}^{e}B_l\, \tau_l(\cos\theta) + {}^{m}B_l\, \pi_l(\cos\theta)] \right|^2 \quad (9)$$

where $I(\theta)$ and I_0 are the intensities of the scattered and incident light respectively, R is the distance between the scattering particle and the reference plane or detector, ${}^{e}B_l$ and ${}^{m}B_l$ are the "electric" and "magnetic" scattering coefficients (involving Bessel functions) and τ_l and π_l angular functions involving Legendre type of polynomials.

Table 2. Particle types for which Lorenz-Mie solutions are available

Type	Ref.
Homogeneous sphere	44
Coated sphere	45
Multi-layered sphere	6
Spheres with continuously varying n	46,47
Homogeneous cylinders and ellipsoids	48
Coated ellipsoids	49,50,51
Homogeneous spheres with holes	52
Homogeneous spheres with projections	53

Asymmetry and Polydispersity

Despite the elegance of these theoretical advances, both RGD and Lorenz-Mie representations of the angular intensity envelopes have been difficult to apply for two fundamental reasons: (i) Random orientations of non-spherical scatterers: this tends to smooth out the resonances in the intensity envelope[51]. Fig. 6b illustrates this problem for a series of coated ellipsoids which could conceivably model bacterial spores; (ii) A further complication is of heterogeneity within a spore ensemble[3] and the combined effects of both asymmetry and heterogeneity can lead to rather featureless angular intensity profiles as shown in Fig. 6c[54] for an ensemble of *B. subtilis* spores. It is still possible to infer some useful information from these type of curves: for example for antibiotic susceptibility testing[55] and in studies on the marine microbe *Chlorella*[56].

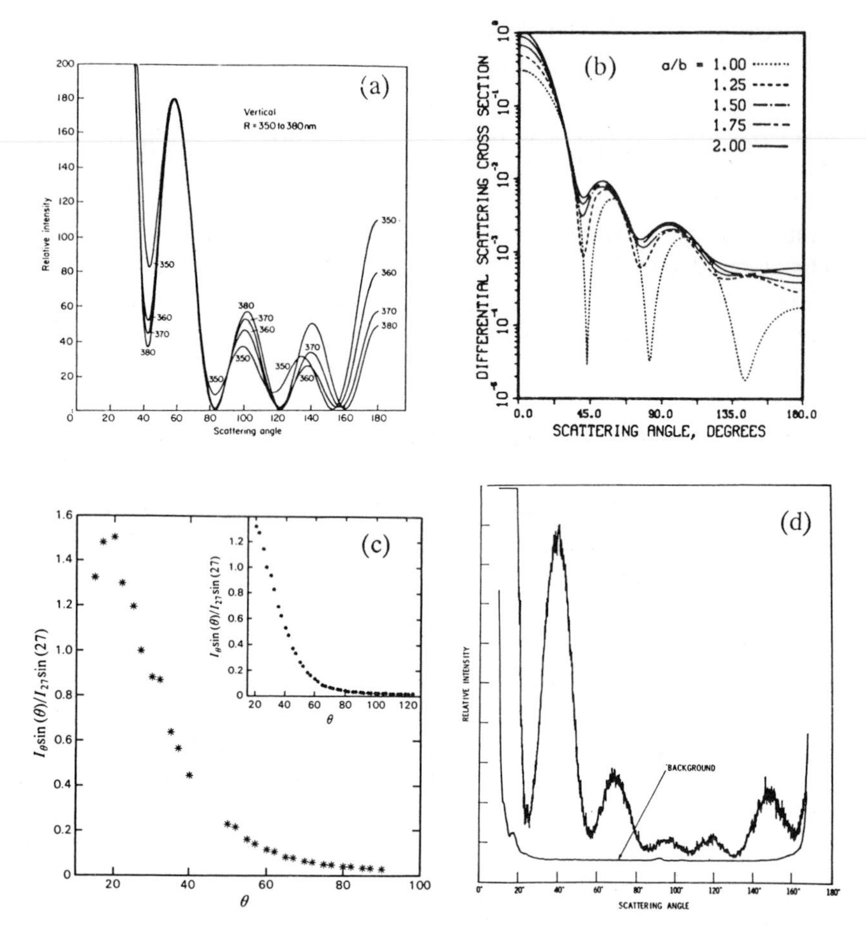

Figure 6. Angular scattered intensity envelopes from bacterial systems (vertically polarised incident radiation):

(a) Theoretical TILS patterns for coated sphere models with varying radii. Reprinted, with permission, from ref. 3;

(b) Theoretical TILS for randomly oriented ellipsoidal bacteria with varying aspect ratios. The dotted line corresponds to the normalized scattering intensity function against angle for a spherical particle showing strong structural features. As the ellipsoid aspect ratio is increased the structural features in the curve are smoothed out, making modelling approaching the impossible. Reprinted, with permission, from ref. 51;

(c) Experimental TILS for a suspension of spores of *B. subtilis*. Reprinted, with permission, from ref. 54;

(d) Experimental TILS, for a single isolated spore, suspended in air, of *B. sphaericus*. Reprinted, with permission, from ref. 57.

Single particle measurements

The "simple" solution - in principle - to asymmetry and heterogeneity is two fold. The first is to work with quasi-spherical bacterial spores. Although this greatly restricts the numbers of interesting systems that can be analysed there are some spores - such as from *B. sphaericus* - which have reasonable spherical symmetry. The second part of the solution is to record scattering intensity envelopes from isolated single spores. Wyatt and Phillips[57] successfully achieved this by using an aerosol technique to isolate single spores, giving successfully structural features which could be used for refractive index profiling (Fig. 6d). Suspending spores in air poses additional problems: for example, because of the relatively large refractive index difference between microbe and air, RGD theory is almost certainly not applicable, and more seriously the delicate osmotic balance between the interior and exterior of the microbe can be greatly disturbed - even for spores.

Instruments have been designed for multi-angle measurement of single particles in aqueous solution[58,59]. Wyatt and Jackson[59] have indicated how such a device can be used to detect different phytoplankta from the characteristic scattered intensity envelopes. Ulanowski *et al.*[60,61] have applied coated-sphere Lorenz-Mie theory to model the refractive index profiles of spores of *B. sphaericus* and basidospores of *L. pyriforme* in water, and proposed this as a way of monitoring the water contents of spores. The technqiue is however currently limited to particles with spherical symmetry - because of the random orientation problem: application to the more interesting cases of ellipsoidal spores will require ways of providing constant orientation in the laser beam for a time long enough to get a sufficient signal/noise ratio across the angular scattered intensity envelope.

Flow cytometers

Similar in principle - although designed for a different purpose - are flow cytometers, which also facilitate light scattering measurements on single microbes[62,63] and several commercial instruments are available (see, *e.g.* ref. 64). Light scattering detection from flow cells of suitable design are usually made at two scattering angles[63]: although two is too few for the modelling of internal structures, it has been inferred that data of this type can permit the identification of microbial types[64].

Quasi-elastic light scattering of bacterial spores

QLS can also give useful information on bacterial spores, and moreover ensembles thereof. For quasi spherical spores, the single term exponential form of the autocorrelation function (eq. 5) gives a good representation of the data and as an example QLS has been used in parallel with turbidity measurements to monitor the possible changes in the gross morphology of *B. subtilis* spores[54] and *B. megaterium* spores[65] during germination - particularly during the early stages.

Such information has been used to help delineate between the various theories to explain the mechanisms responsible for the strong dehydration and heat resistance of bacterial spores. A similar application has been in the use of QLS in conjunction with electron microscopy[66] to show how culture growth temperature has an important effect on the resistance to thermal destruction (at 121°C). The effects of culture temperature on resistance to destruction by disinfectants has also been studied.

Chen *et al.*[43] have shown that "scaling" behaviour - *i.e.* superposition of plots of the autocorrelation function $g^{(2)}(\tau)$ plotted *versus* $q^2\tau$ for a range of angles - is a measure of the symmetry and homogeneity of a freely diffusing scattering particle. Non-scaling behaviour has been demonstrated for spores of *B. subtilis* (Fig 7a)[54] and *B. megaterium*[65] although germinated spores of *B. megaterium* do appear to scale better, possibly a measure of the decreased asymmetry and greater homogeneity of the germinated spores.

Microbial motility and chemotaxis

The most prolific application of QLS to microbiology has arguably been in this area. Besides being Brownian diffusers many microbes (flagellates, cilliates and bacteria) possess their own motility or "free translational motion"[43,68] deriving from their own metabolism. QLS has provided a powerful non-destructive and non-invasive probe into this motility to study the velocity distributions of motile bacteria such as *E. coli* [68-73], *Salmonella* [61,74] and parasitic micoorganisms such as *Trypanosoma bruceii*[75,76]. After neglecting contributions from "non-translational motion"[77,74] the general equation describing the intensity autocorrelation function $g^{(2)}(\tau)$ of motile microbes has been given as :

$$[g^{(2)}(\tau) - 1]^{1/2} = \int_0^\infty [\exp(i\mathbf{q}.\mathbf{V})]\, P(\mathbf{V})\, d^3\mathbf{V} \tag{10}$$

where **V** is the velocity and P(**V**) is the velocity distribution. For the relatively simple case of an isotropic distribution of velocities solution of the integral in eq. 10 gives an expression linking $g^{(2)}(\tau)$ with the root mean square velocity, the fraction of motile microorganisms and D_T. Scaling procedures similar to those for free Brownian diffusion (Fig. 7a) have been developed[72,43] involving superposition of plots of $g^{(2)}(\tau)$ *versus* $q\tau$ (as opposed to $q^2\tau$ for free diffusion) for different angles as an assay for spherical symmetry and isotropicity of the scatterers. In this way for example Stock[74] has demonstrated scaling at low angles for *Salmonella typhimurium* (Fig. 7b). For typical non-scalers Chen and Hallett[78] have attempted a model to fit progressive rotational and helical movements. In recent work Wang and Chen[79] have studied the formation and propogation of bands of *E. coli* in the presence of oxygen and serine substrates, comparing two proposed formulisms for band formation.

4. DISCUSSION

Although this survey is by no means exhaustive it is hoped that it has given the Biochemist / Microbiologist an appreciation of the breadth of application to laser light scattering to microbial systems. Although it is probably fair to say that much of the attention over the last two decades has been in the application of dynamic light scattering or QLS, no less remarkable is the progress in static or total intensity light scattering for the analysis of single microbial particles in aqueous media.

Our own interest in this field is still principally in the application of laser light scattering as a probe into the heat resistance of bacterial spores: at the present time this is concerning itself with the detection of possible deleterious effects of the introduction of lux genes as a bioluminescent probe. The difficulties in applying QLS have been in terms of the inherent polydispersity of bacterial systems and also, if a dynamic process is being followed, variability in the rates at which individual spores germinate or are destroyed by chemical agents. To this end the value of combining QLS

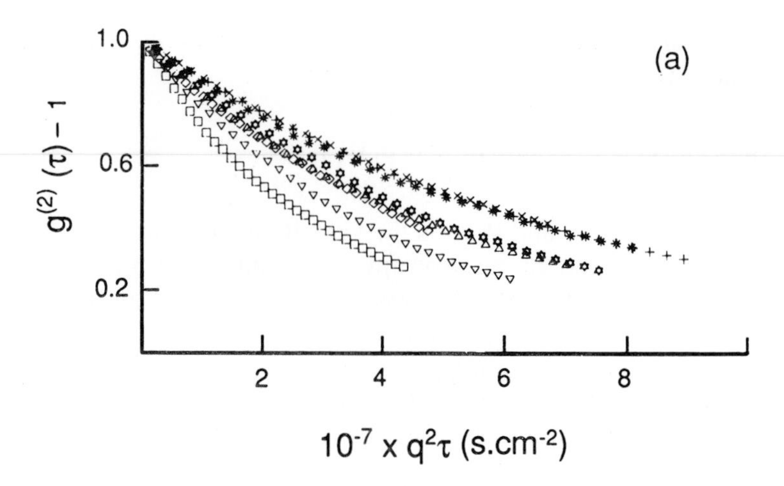

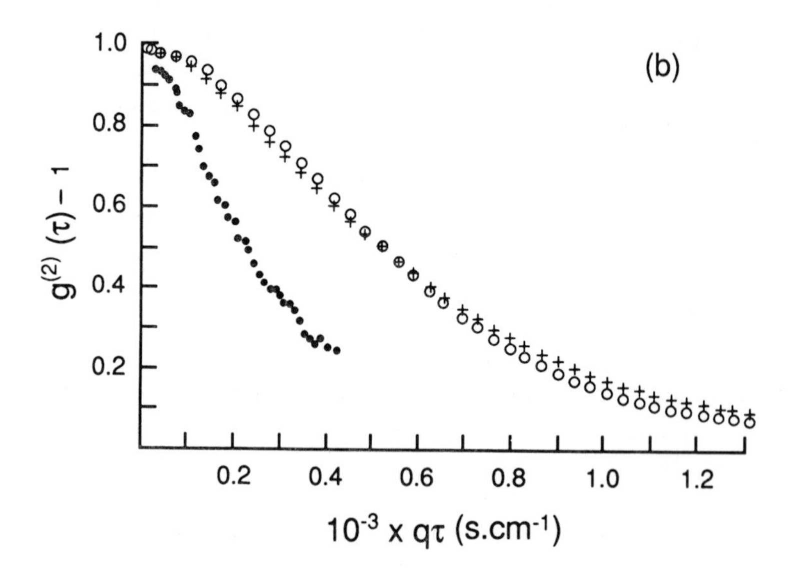

Figure 7. QLS scaling plots of bacterial systems:

(a) $q^2\tau$ scaling for spores of *B. megaterium* (from ref. 65);

(b) $q\tau$ scaling, for *E. coli* (Reprinted, with permission, from ref. 74).

The different symbols in each case correspond to different angles. In (a) no scaling is evident. In (b) scaling is evident only at low angles.

measurements on spore ensembles with measurements by TILS on single spores is clearly indicated An important development in this area will occur when a method is developed to fix the orientation of individual non-spherically symmetric microbes in a laser beam to permit refractive index profiling of many more species than can currently be realistically studied.

ACKNOWLEDGEMENTS

My interest in light scattering of microbial systems arose during my collaboration with P. Johnson and I thank him for many valuable discussions. I am grateful also to A. Molina-Garcia for his valuable contributions to our work on tomato bushy stunt virus, fd bacteriophage and *Bacillus cereus* spores.

REFERENCES

1. S.E. Harding, *Biotech. & Appl. Biochem.*, 1986, *8*, 489.
2. V.A. Bloomfield, *Ann. Rev. Biophys. Bioeng.*, 1981, *10* , 421.
3. P.J. Wyatt, *Meth. Microbiol.*, 1973, *8*, 183.
4. See, for example A.H. Sanders and D.S. Cannell in 'Light Scattering in Liquids and Macromolecular Solutions' (V. Degiorgio, M. Corti and M. Giglio eds.), pp. 173-182, Plenum, New York, 1980.
5. P.J. Wyatt, Chap. 3, this volume.
6. M. Kerker, 'The Scattering of Light and other Electromagnetic Radiation', Academic, New York, 1969.
7. R.J. Fiel, E.H. Mark and B.R. Munson, *Arch. Biochem.Biophys.*, 1970, *141*, 547.
8. R.D. Camerini-Otero, PhD Thesis, New York University, 1973.
9. H. Boedtker and N.S. Simmons, *J. Amer. Chem. Soc.*, 1958, *80*, 2550.
10. P. Johnson and W. Brown, Chap. 11, this volume.
11. J. Garcia de la Torre and V.A. Bloomfield, *Quart. Rev. Biophys.*, 1981, *14*, 83.
12. J. Garcia de la Torre, in 'Dynamic Properties of Biomolecular Assemblies' (S.E. Harding and A.J. Rowe eds.), Chap. 1, Royal Society of Chemistry, Cambridge, 1989.
13. S.E. Harding, *Biophys. J.*, 1987, *51*, 673.
14. S.A. Berkowitz and L.A. Day, *J. Mol. Biol.*, 1976,*102*, 531.

15. V.A. Bloomfield and T.K. Lim in 'Methods in Enzymology' (C.H.W. Hirs and S.N. Timasheff eds.) Vol. 48 Pt. F, pp. 414-494, Academic, Fl., 1978.
16. P. Pusey in 'Photon Correlation and Light Beating Spectroscopy' (H.Z. Cummins and E.R. Pike eds.) pp 387-428, Plenum, New York, 1974.
17. C. Tanford, 'Physical Chemistry of Macromolecules', Chap. 6, Wiley and Sons, New York, 1961.
18. S.E. Harding and P. Johnson, *Biochem. J.*, 1985, *231*, 543.
19. A.D. Molina-Garcia, S.E. Harding and R.S.S. Fraser, *Biopolymers*, 1990, *29*, 1443.
20. S.E. Harding and P. Johnson, *Biochem. J.*, 1985, *231*, 549.
21. C.J. Oliver, K.F. Shortridge and G. Belyavin, *Biochim. Biophys. Acta.*, 1976, *437*, 589.
22. P. Nieuwenhuysen and J. Clauwaert, *Biopolymers*, 1978, *17*, 2039.
23. Y. Sakaki, T. Maeda and Y. Oshima, *J. Biochem.* (Tokyo), 1979, *85*, 1205.
24. P. Dobos, R. Hallett, D.T.C. Kells, O. Sorensen and D. Rowe, *J. Virol.*, 1977, *22*, 150.
25. J.C. Thomas and G.C. Fletcher, *Biopolymers*, 1978, *17*, 2755.
26. Y. Sano, *J. Gen. Virol.*, 1987, *68*, 2439.
27. P.C. Hopman, G. Koopmans and J. Greve, *Biopolymers*,1980, *19*, 1241.
28. A.A. Timachenko, N.B. Griko and I.N. Serdyuk, *Biopolymers*, 1990, *29*, 303.
29. W. Burchard, Chap. 1, this volume.
30. J. Wilcoxon and J.M. Schurr, *Biopolymers*, 1983, *22*, 849.
31. S. Fujime and T. Maeda, *Macromolecules*, 1985, *18*, 191.
32. A.D. Molina-Garcia, S.E. Harding, F.G. Diaz, J.Garcia de la Torre, D. Rowitch and R.N. Perham (1991) mss. submitted.
33. R.E. Godfrey, P. Johnson and C.J. Stanley in 'Biomedical Applications of Laser Light Scattering' D.B. Sattelle, W.I. Lee and B.R. Ware eds.), pp 373-389, Elsevier, Amsterdam, 1982.
34. G.J. Baran and V.A. Bloomfield, *Biopolymers*, 1978, *17*, 2015.
35. J.B. Welch and V.A. Bloomfield, *Biopolymers*, 1978, *17*, 2001.
36. J.A. Benbasat and V.A. Bloomfield, *J. Mol. Biol.*, 1975, *95*, 335.
37. J.A. Benbasat and V.A. Bloomfield, *Biochemistry,* 1981, *20*, 5018.
38. B.V. Chernyak, *FEMS Microbiol. Lett.*, 1982, *13*, 113.

39. M. Brisco, C. Haniff, R. Hull, T.M.A. Wilson and D.B. Sattelle, *Virology*, 1986, *148*, 218.
40. V.J. Morris, H.J. Coles and B.R. Jennings, *Nature,* 1974, *249*, 240.
41. H.J. Coles, B.R. Jennings and V.R. Morris, *Phys. Med. Biol.*, 1975, *20*, 225.
42. P.J. Wyatt, *Appl. Optics*, 1968, *7,* 1879.
43. S.-H. Chen, M. Holz and P. Tartaglia, *Appl. Opt.*, 1977, *16*, 187.
44. G. Mie, Ann. Physik., 1908, 25, 377; see also ref. 6 for a discussion on the appropriateness of whom should be credited with this theory.
45. A.L. Aden and M. Kerker, *J. Appl. Phys.*, 1951, *22*, 1242.
46. P.J. Wyatt, *Phys. Rev.*, 1962, *127*, 1837.
47. P.J. Wyatt, *Phys. Rev.*, 1964. *134*, AB1.
48. S. Asano and M. Sato, *Appl. Opt.*, 1980, *19*, 962.
49. P.W. Barber, PhD Thesis, Univ. California, Los Angeles, 1973.
50. P.W. Barber and H. Massoudi, *Aerosol. Sci. Med.Technol.*, 1982, *1*, 303.
51. D.S. Wang, H.C.H. Chen, P.W. Barber and P.J. Wyatt, *Appl. Opt.*, 1979, *18*, 2672.
52. P. Latimer, *Appl. Optics*, 1984, *23*, 1844.
53. P. Latimer, *Appl. Optics*, 1984, *23*, 442.
54. S.E. Harding and P. Johnson, *Biochem. J.*, 1984, *220*, 117.
55. D.W.L. Hukins, J. Murray and P. Evans, *UV Spectrom. Group. Bull.*, 1980, *8*, 18. For a critical review of methods, see also J. Kiehlbauch, J.M. Kendle, L.G. Carlson, F.D. Schoenknecht and J.J. Plorde, Clinics in *Lab. Med.*, 1989, *9*, 319.
56. M.S. Quinby-Hunt, A.J. Hunt, K. Lofftus and D. Shapiro, *Limnol. Oceanog.*, 1989, *34*, 1587.
57. P.J. Wyatt and D.T. Phillips, *J. Colloid Interface Sci.*, 1972, *39*, 125.
58. I.K. Ludlow and P.H. Kaye, *J. Colloid Interface Sci.*, 1979, *69,* 571.
59. P.J. Wyatt and C. Jackson, *Limnol. Oceanog.*, 1989, *34*, 96.
60. Z.J. Ulanowski, I.K. Ludlow and W.M. Waites, *FEMS Microbiol Lett.*, 1987, *40,* 229.
61. Z.J. Ulanowski, I.K. Ludlow and W.M. Waites, *Microbiol. Research*, 1989, *93*, 28.
62. H.B. Steen and T. Lindmo, *Science,* 1979, *204*, 403.
63. G. C. Salzman, in 'Cell Analysis' (N. Catsimpoolas ed.), Chap. 5, Plenum, New York, 1980.

64. R.W. Spinrad and J.F. Brown, *Appl. Optics*, 1986, *25*, 1930.
65. S.E. Harding and P. Johnson, *J. Appl. Bacteriol*, 1986, *60*, 227.
66. A.D. Molina-Garcia, S.E. Harding, L. de Pieri, N. Jan and W.M. Waites, *Biochem. J.*, 1989, *263*, 883.
67. G.B. Stock, *Biophys. J.*, 1976, *16*, 535.
68. R. Nossal and S.-H. Chen, *Opt. Commun.*, 1972, *51*, 117.
69. R. Nossal and S.-H. Chen, *Nature*, 1973, *244*, 253.
70. D.W. Schaefer and B.J. Berne, *Biophys. J.*, 1975, 15, 785.
71. M. Holz and S.-H. Chen, *Appl. Opt.*, 1978, *17*, 1930.
72. P.C. Wang and S.-H. Chen, *Biophys. J.*, 1981, *36*, 203.
73. S.H. Chen and P.C. Wang in 'Biomedical Applications of Laser Light Scattering' (D.B. Sattelle, W.I. Lee and B.R. Ware eds.), pp 173-190, Elsevier, Amsterdam,1982.
74. G.B. Stock, *Biophys. J.*, 1978, *22*, 79.
75. R.J. Gilbert, R.A. Klein and P. Johnson, *Biochem. Pharmacol.*, 1983, *32*, 3447.
76. R.A. Klein, *Biochem. Soc. Trans.*, 1984, *12*, 629.
77. R. Nossal, S.H. Chen and C.C. Lai, *Opt. Commun.*, 1971, *4*, 35.
78. S.H. Chen and F.R. Hallett, *Q. Rev. Biophys*, 1982, *15*, 131.
79. P.C. Wang and S.H. Chen, *Biophys. J.*, 1986, *49*,1205.

26

Quasi-elastic Light Scattering Studies of Membrane Structure and Dynamics

By J.C. Earnshaw

DEPARTMENT OF PURE AND APPLIED PHYSICS, THE QUEEN'S UNIVERSITY OF BELFAST, BELFAST. BT7 1NN NORTHERN IRELAND

1. INTRODUCTION

Biophysicists and biochemists use a range of model systems in investigations of the physicochemical behaviour of membranes to avoid the intrinsic complexity of the real biomembrane. Here we restrict our considerations to two particularly simple models, the bimolecular black lipid film and the insoluble monolayer of amphiphilic molecules at the air-water interface. Other membrane models have been studied by quasi-elastic light scattering (QLS)[1] but the scattering processes involved are completely different from those concerned here. Here we consider scattering by thermally excited modes of membrane motion[2], and, to a lesser extent, intra-membrane molecular modes[3].

Light scattering has long been used as probe of structure and dynamics in supramolecular systems. The structural information usually derives from the scattered intensity, whilst the spectral content of the scattered light carries information regarding the dynamics[4]. This broad statement remains true when the objects of study are model membranes, but further useful structural information may be carried by the changes of the spectrum in cases when the statistical fluctuations of the system which scatter light are non-stationary.

We will discuss a range of examples illustrating these themes. These examples cover two simple systems: bimolecular lipid membranes (BLM) formed from glycerol monoöleate (GMO) and spread monolayers of GMO and of n-pentadecanoic acid (PDA).

These materials offer considerable advantages for exploratory studies, some practical and some of principle. GMO was originally selected because of its ability to form large stable BLM, suitable for light scattering. It has the added advantage that it has a thermotropic transition at an experimentally convenient temperature. For monolayers at the air-water interface the simple fully saturated fatty acids, particularly PDA, are widely regarded as basic test materials. This interest has recently been reinforced by the demonstration that the transitions of monolayers of PDA (and certain other simple amphiphiles) are simple first order in character[5].

2. THEORETICAL AND EXPERIMENTAL BACKGROUND

Membranes, both real and model, form interfaces between two fluids which are continually perturbed by random molecular bombardment. Membranes may support many modes, from hydrodynamic motions of the system as a whole to modes of coherent molecular fluctuation. Several of these may scatter light. We will primarily be concerned with the hydrodynamic modes, but will have occasion to consider molecular modes.

The theoretical description of light scattering by model membranes is well established[2,6]. Only two hydrodynamic modes concern us: transverse or capillary waves on the interface, governed by the transverse shear modulus of the surface (γ), and dilational or compressional modes, governed by the dilational modulus (ε) of the interface. The former dominates the scattering of light, largely because the cross-section for scattering by dilational modes includes the wavenumber (~100 cm^{-1}) multiplied by the membrane thickness (~2 nm) [7]. However, for an air-water interface the two modes are coupled, so that the capillary waves are affected by the dilational modulus[2]. The indirect effect of ε upon the light scattering causes ε to be harder to determine than γ. For a BLM the two modes decouple and only γ is accessible from light scattering[2].

The two moduli γ and ε are familiar in the low frequency ($\omega \to 0$) limit as the interfacial or surface tension and the inverse of the film compressibility: $\varepsilon = d\gamma / d\ln A$, where A is the molecular area. Both of these moduli may be expanded as response functions to allow for dissipative processes[8]:

$$\gamma = \gamma_0 + i\,\omega\,\gamma' \qquad (1)$$

$$\varepsilon = \varepsilon_0 + i\,\omega\,\varepsilon' \tag{2}$$

The two interfacial viscosities γ' and ε' are *not* the conventional surface viscosity, which refers to shear in the membrane plane: γ' is the transverse shear viscosity and ε' is the in-plane dilational viscosity. Both viscosities are perhaps best regarded as interfacial excess quantities[8].

The dynamic membrane may be analysed in terms of fluctuations which evolve in time and space. The spatial perturbation can be Fourier decomposed into a complete set of modes. Light scattering involves selection of one mode, so we will consider the temporal evolution of a specific surface mode defined by the interfacial wavenumber (q), given by the projection of the conventional scattering vector on the interfacial plane. The departure of the interface from its equilibrium plane due to a capillary wave of wavenumber q, propagating in the x direction, is given by $\zeta_0 \exp i(q\,x + \omega\,t)$. The wave frequency ω ($= \omega_0 + i\Gamma$) represents the temporal evolution of the surface: the spatial fluctuation oscillates with frequency ω_0 and decays at a rate Γ. The distribution of the intensity of scattered light with q is governed by the mode amplitudes, whereas the temporal evolution of the mode is reflected in the spectrum (or equivalently, as here, the autocorrelation function) of the scattered light.

The scattered light reflects the spectrum of the thermally excited capillary waves. To measure the frequency shifted spectrum of the scattered light it is necessary to employ a heterodyne spectrometer arrangement. A typical experimental correlation function is shown in Fig. 1.

Unbiased estimates of ω_0 and Γ can be derived from such functions[9]. These parameters depend upon those properties of the system which affect them. Thus for the air-water case the capillary waves are affected by four surface properties, whereas for BLM only the tension γ_0 and the corresponding viscosity γ' are involved. In the latter case the two physical properties can be recovered directly from ω, so that data interpretation is simple. However, for a surface film the four surface properties cannot be deduced unambiguously from two measured quantities. Fortunately it has proved possible to extract these four quantities directly from the measured correlation functions[10].

Two points should be made. The surface properties recovered by light scattering are *local* averages over the area illuminated by the laser beam. Further, the values are those appropriate to the capillary wave frequencies *e.g.* $\varepsilon_0(\omega)$.

The intensities of both the scattered light, I_s, and the heterodyne reference beam, I_r, can be determined from the amplitude of the time dependent part of the observed correlation functions together with the background level[11]. The intensity scattered by thermally excited capillary waves is

$$I_s \propto R \frac{kT}{\gamma_0 q^2} \qquad (3)$$

where R is the interfacial reflectivity[7]. This method of determining the scattered intensity is rather direct, as it uses the amplitude of a correlation function of the theoretically predicted form: extraneous scattering processes having different time dependences are automatically excluded. To illustrate the point we show the q and γ_0 dependence of I_s for monolayers of GMO (Fig. 2). The best-fit power-law variations agree well with theoretical predictions (Equation 3): any excess scattering has been ignored.

Some intra-membrane of molecular modes couple to the hydrodynamic modes of interest and may scatter light[3]. In particular molecular splay couples to the capillary waves. The effects on the capillary wave propagation, which arise from $\gamma \rightarrow \gamma + M q^2$, are negligible as the splay modulus M, ~ 10^{-19} J, is many order of magnitude too low to affect experimentally accessible wavenumbers (< 2000 cm^{-1}). However the molecular modes may still be significant. The light which they scatter may interfere with that scattered by the corresponding hydrodynamic modes[3]. Due to the coupling of the capillary and splay modes this affects the amplitude of the measured correlation function and hence the apparent scattered intensity.

One final experimental point has proved vital in certain of the studies to be discussed. Minor changes in the manner of operation of our apparatus enable the very rapid acquisition of statistically adequate heterodyne correlation functions[12]. As expected, this is most useful when the signal is non-stationary.

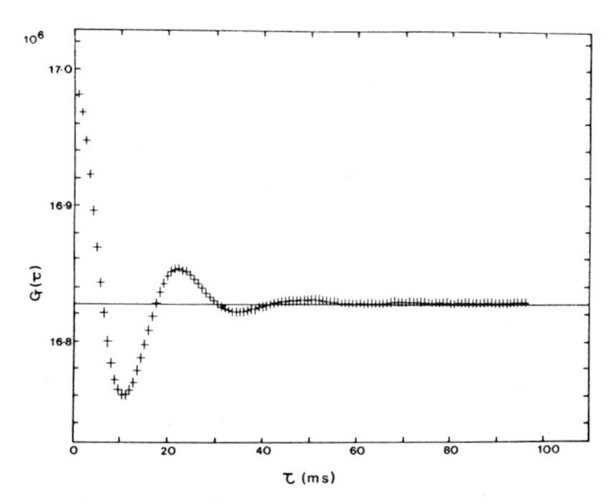

Figure 1. An unnormalised correlation function for light scattered by capillary waves of q = 922.3 cm^{-1} on a BLM. The line is the calculated background level.

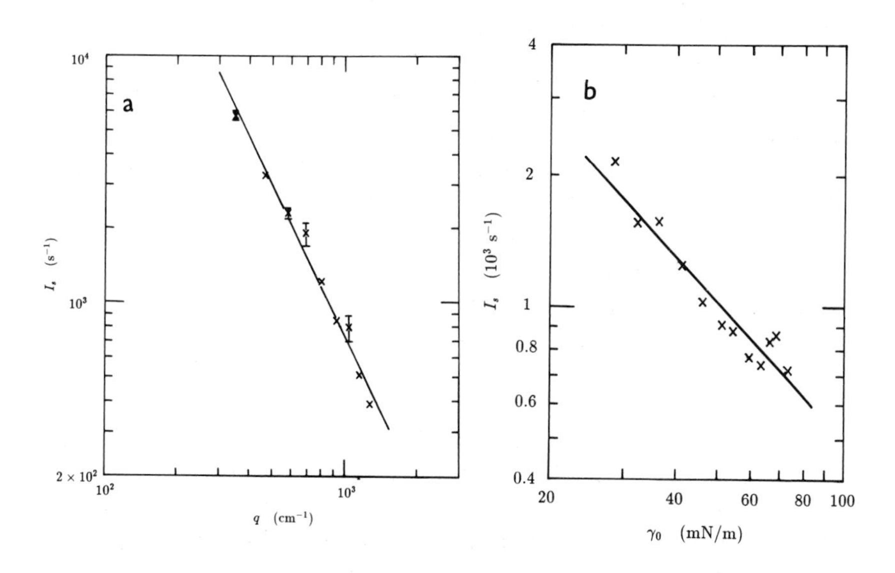

Figure 2. The variation of scattered intensity with (a) q and (b) surface tension. The fitted power-laws shown had exponents of -2.02 ± 0.09 and -1.11 ± 0.09 respectively.

3. STRUCTURAL STUDIES

It is not possible entirely to separate discussion of structural and dynamic aspects of our studies of model membranes, particularly as some of the structural information derives from the time-dependence of the correlation functions. However it simplifies the argument to treat the one before the other.

We first treat examples of structural information derived from non-stationarity of the correlation function, before considering cases where the scattered and reference beam intensities, perhaps in combination with spectral data, yield structural information.

Non-stationarity of the signal

There have been a few reports in the literature of fluctuations in correlation functions recorded in the liquid/vapour phase transition in insoluble monolayers[13,14]. These were ascribed to coexistence of phase separated domains. However in these studies any fluctuations would have been temporally averaged over the data acquisition time. Our rapid data acquisition techniques have proved invaluable in the investigation of such fluctuations, enabling the temporal averaging to be reduced by at least an order of magnitude.

The simplest system to consider is the monolayer at the air-water interface. PDA is a particularly suitable material for such studies as the transitions of the molecular film are simple first order in nature, provided that the materials are sufficiently pure (>99.95%) [5]. It is well established that phase separation occurs in such transitions. We follow the conventional nomenclature for the sequence of phases traversed on compression: vapour, liquid expanded (LE), liquid condensed (LC) and solid.

Monolayers: liquid/vapour coexistence As a PDA monolayer was compressed through the liquid/vapour transition it was noticed that the damping of successive correlation functions fluctuated visibly[15]. Analysis proved that this was due to rather considerable differences between the magnitudes of ε_0 in the pure liquid and vapour phases (observed at opposite ends of the coexistence region). In the vapour phase the surface differed imperceptibly if at all from that of the clean subphase, whereas in the liquid phase $\varepsilon_0 \sim 13$ mN/m, close to the value at which it has its maximum effect upon the capillary wave damping, Γ. This amplified the difference between the two phases, making Γ the natural variable to observe in this case.

The fluctuations of Γ were studied for fixed surface concentrations to gain further insight into the coexistence phenomena. Data for one Γ_s are shown in Fig. 3. The observed values fluctuated between those appropriate to the two pure phases, establishing that the entire illuminated area could be occupied by one of the two phases at a time. The change from one behaviour to the other often occurred within a single interval between the recording of successive correlation functions. The domain speed could be estimated from the fact that the transition time between states was typically less than 100 s. Using this speed the average duration of the periods of high damping in Fig. 3 suggests that the liquid domains were 1.1 ± 0.7 cm in size. The laser beam could be moved about 1 cm across the surface before the damping fell from a high value, confirming this estimate.

Monolayers: Liquid condensed/liquid expanded coexistence Rather smaller fluctuations in the damping of correlation functions were observed in this case[15]. It proved easier to analyse the data in terms of ε_0. While the equilibrium dilational modulus is zero in any coexistence region as the π-A isotherms were flat, this was not the case for the light scattering values. The average light scattering ε_0 rose steadily through the LC/LE transition, as did the fluctuations about these average values. This is the first demonstration of fluctuations in this phase transition, and clearly derives from the lesser degree of temporal averaging involved in our rapid data acquisition methods.

Fig. 4 shows the time course of ε_0 for four different surface concentrations, the lowest of which corresponds to a pure LE state. The rms fluctuations of this last record are similar to the errors expected in determining ε_0 from noisy correlation functions. The other traces, having larger fluctuations than expected, involved phase separation. Analysis of the data suggests that the LC domains were of the order of 500 μm across. This value, although an order of magnitude greater than suggested by fluorescence microscopy methods[16], agrees well with values deduced from literature reports of surface potential fluctuations for PDA monolayers[17]. Impurities, such as fluorophores, in the monolayer appear to modify the domain structure.

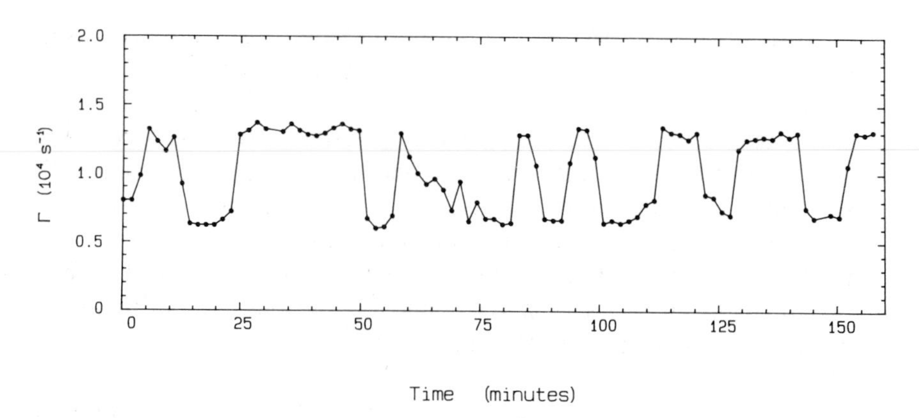

Figure 3. The capillary wave damping as a function of time for a PDA monolayer at 19°C at a surface concentration of 0.017 molecules/Å^2. Data were recorded over 10 s, the intervals between successive observations being 100 s.

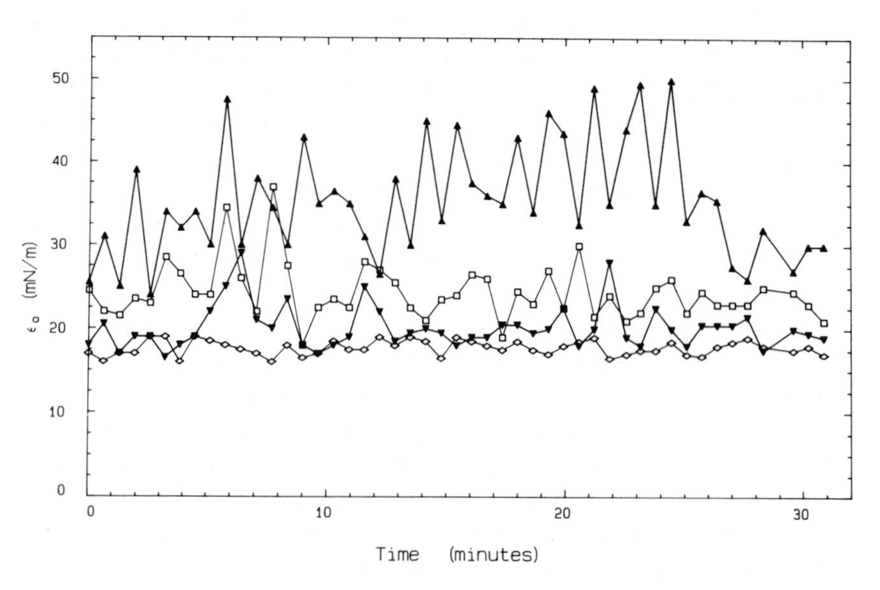

Figure 4. The time dependence of ε_0 for four surface concentrations, three in the LC/LE coexistence region: ▲ 0.025, □ 0.028 and ▼0.031, and one in the LE phase: ◊ 0.036 molecules/Å^2.

Bilayers: cholesterol stabilised domains We have used light scattering to study GMO bilayers. Whilst BLM formed from a solution of GMO in decane contain a high proportion of hydrocarbon in their structure, it is possible to form bilayers which are essentially 'solvent-free'. For the former case the retained solvent fluidised the BLM and the transverse shear viscosity appeared negligibly small at room temperature[18]. However when cholesterol was added to the film-forming solution the surface viscosity acquired a substantial value[19], reaching values similar to those for 'solvent-free' BLM at saturation[18].

When such cholesterol containing BLM were cooled through the GMO phase transition (~16°C) substantial fluctuations were observed in the light scattering signal (Fig. 5)[20]. The variations above the transition presumably indicate the fluid phase behaviour of this cholesterol saturated BLM. The excess wave damping due to γ' is clearly apparent. Below the transition both ω_0 and Γ exhibit large fluctuations, extending right up to extrapolations of the variations above T_t. The lines shown for ω_0 correspond to spline fits to all the data (solid), to all but those below T_t lying well above the trend (chain) and omitting all data below T_t except these high points (dashed). This last line is very similar to that observed for GMO/decane bilayers, which remain fluid down to low T. The data indicate that phase separation has occurred in the BLM: such separation was never observed for either GMO/decane or 'solvent-free' bilayers.

Close inspection of the figure, particularly for ω_0, shows that in several places a few adjacent points are all high. It thus appears that the phase separated regions of gel phase may be as large as the illuminated area (here 300 µm). Such large domains have not previously been reported for lipid bilayers incorporating cholesterol. At present we cannot distinguish between two competing models: a fluidizing action of cholesterol leading to separation into a sterol-rich fluid phase and a pure lipid gel phase below T_t, or a domain edge stabilising influence of cholesterol, leading to a concentration of the sterol at the interfaces between the phases.

Information from the intensities

We turn now to extraction of structural information from the

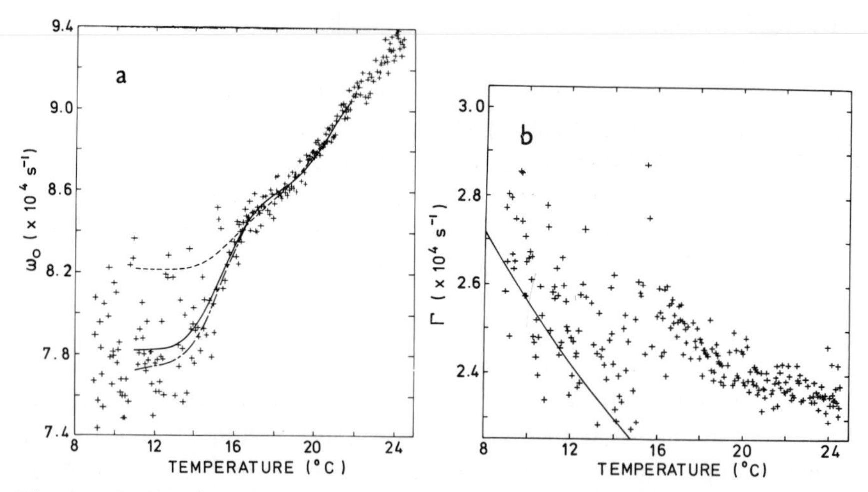

Figure 5. The frequency and damping of capillary waves (q = 1400 cm^{-1}) on a GMO/cholesterol bilayer cooled from 24°C to 9°C. The line with the Γ data indicates the variation expected for damping caused solely by the viscosity of the ambient fluid. The lines for ω_0 are discussed in the text.

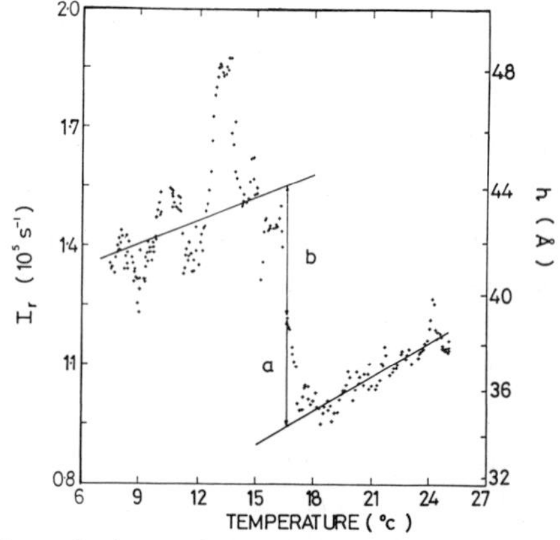

Figure 6. The T variation of I_r. Linear fits above and below T_t are shown (neglecting peak between 12°C and 14°C). Note the non-linear right hand scale indicating r.m.s. bilayer thickness.

intensities of scattered and reference beams. We have seen that the conventional $I_s(q)$ does not directly carry structural information. We have used changes with temperature in I_s and I_r at fixed q to infer changes in structure.

The most illuminating example of this approach is provided by a study of the thermotropic transitions of 'solvent-free' BLM formed of GMO[21]. Both the reference beam intensity and that of the scattered light were deduced from the observed correlation data. Whilst absolute values cannot be found in this way, the temperature variations of the relative intensities carry the physically interesting information.

BLM reflectivity It is simplest to consider I_r first. The data, shown in Fig. 6, display a considerable change about the transition temperature, as well as rather pronounced fluctuations below T_t. The very conspicuous peak between 12°C and 14°C is thought to be unconnected with structural changes, and so will be omitted from further discussion here.

Now I_r depends upon the membrane reflectivity: the heterodyne reference beam involved reflection at the BLM. The reflectivity depends upon the membrane thickness as h^2. Thus, provided that the mean refractive index of the BLM does not vary significantly with T, as reported, the observed variation in I_r can be interpreted directly in terms of the variation of h^2. Using literature data[22] to normalise this variation we can convert the I_r scale to one of h (see Fig. 6). The BLM is 44 Å thick just below the transition, in good accord with expectation for a bimolecular film of fully extended GMO molecules oriented normal to the film. The decrease of 8 Å through the transition is compatible with that expected for some 6 or 7 *gauche* rotations per GMO molecule[23].

Using the fractional change in I_r at a given T as an 'order parameter' (defined as $\phi = a/(a+b)$, *cf.* Fig. 6) permits a comparison with changes in other bilayer parameters. In particular there is a small change in tension over a rather narrow range of T, superposed on a slower variation (see Fig. 9), which exactly parallels the variation of $(1-\phi)$, as shown in Fig. 7. This comparison rather convincingly supports an identification of the transition as involving chain-melting: as the BLM is cooled the acyl chains of the lipid molecules tend to adopt an all-*trans* configuration, oriented normal

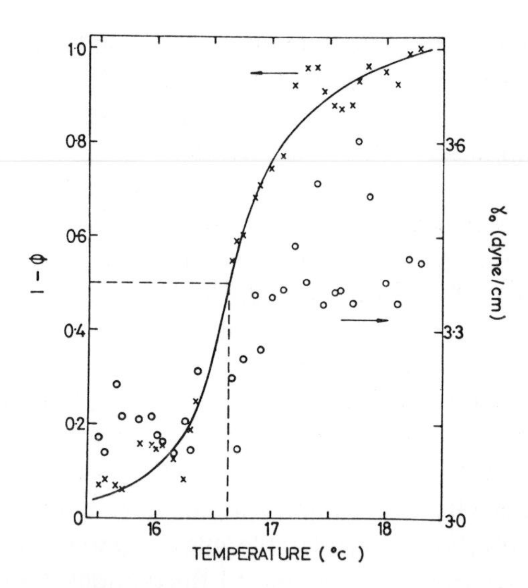

Figure 7. A comparison of the variations of order parameter (x) and bilayer tension (o).

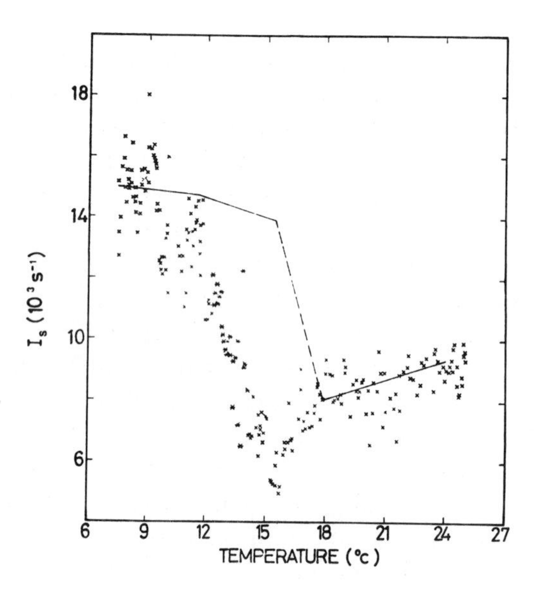

Figure 8. The variation of I_S with T for a 'solvent-free' GMO bilayer. The lines, which are normalised to the data at 24°C, are discussed in the text.

to the bilayer; the consequent increase in packing density at the water/BLM interfaces leads to a reduction in tension.

Further information, of less present concern, can be extracted from the data of Fig. 6[21]. Here we simply note that the linear fits shown in the figure do apparently represent a real increase in bilayer thickness with T. This conflicts with present theoretical ideas[23], but is supported by other experimental evidence[24].

Scattered intensity for BLM The scattered intensity data for this bilayer is shown in Fig. 8. Now, as we have noted in eq. 3 the intensity of light scattered by thermally excited capillary waves depends upon both R and γ_0. The T dependence of both these quantities was measured in our experiments; R *via* I_r and γ_0 from the wave frequencies (Fig. 9). Combining the *forms* of the two variations as in eq. 3 led to the predicted behaviour for I_s shown by the lines in Fig. 8. The prediction in the neighbourhood of T_t is shown dotted because of some uncertainties in the functional dependence of γ_0 in this region.

The general features of the comparisons of the experimental I_s data with the predicted form are clear: above 18°C the agreement is excellent, but below this temperature there are substantial discrepancies. How are these to be explained? We have emphasised that I_s is derived from the amplitude of the time-dependent part of the correlation functions, which agrees excellently with the form expected for capillary waves on the BLM. Any processes invoked to explain the discrepancies of Fig. 8 must thus be closely coupled to the capillary waves. As noted in Section 2, the intra-membrane molecular splay (or membrane bending) modes couple as required. The intensity of light scattered by these modes depends upon the combined dielectric anisotropy and chain orientational order parameter of the BLM (ε_a S_0) [3]. With certain assumptions about instrumental factors (*e.g.* polarisation of incident and detected light) we infer that a rather abrupt increase in this combination (Fig. 10) as T decreases through T_t is needed to explain the observed I_s behaviour. Only the size of the combination $\varepsilon_a S_0$ would be affected by changes in the magnitudes of the instrumental factors: the *form* of the variation shown seems assured.

The scattered intensities thus lead to conclusions which support those drawn from I_r: the transition involves a substantial and abrupt

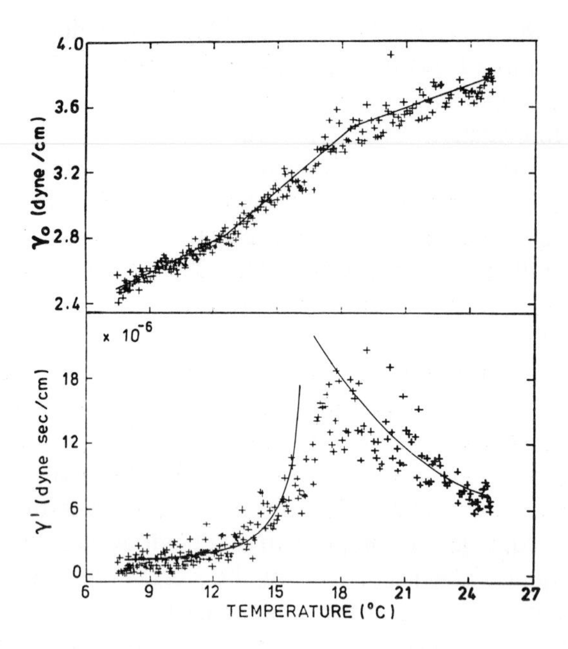

Figure 9. The T variation of γ_0 and γ' for a 'solvent-free' GMO bilayer. The piece-wise linear fit to the γ_0 data was used to compute the expected behaviour of I_S.

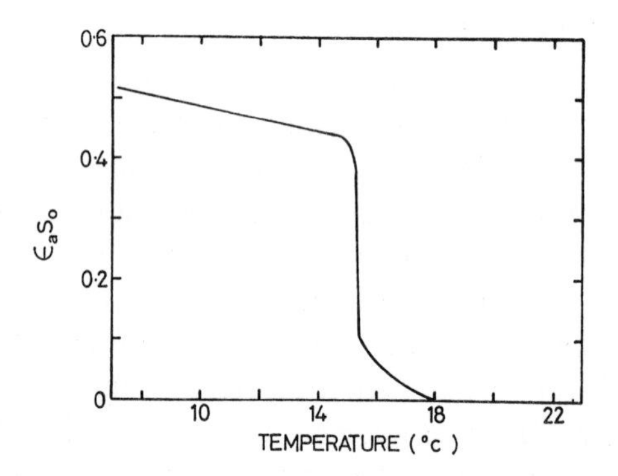

Figure 10. The variation of dielectric anisotropy multiplied by the lipid orientational order parameter, inferred from the data of Fig. 8.

change in the order of the lipid molecules. The transitional change in $\varepsilon_a S_0$ occurs just at that T where the BLM achieves a thickness equal to two fully extended GMO molecules. The entire picture is satisfyingly coherent.

Molecular structure in monolayers Before leaving structural aspects, we briefly note that very similar changes in molecular order to those for the BLM transition have been inferred in the PDA monolayer studies previously discussed.

In the liquid/vapour coexistence region I_s fluctuated synchronously with the fluctuations of wave damping[15]. Fig. 11 shows the correlation of these two quantities. The I_s values were somewhat scattered but on average were 12% greater in the liquid domains than the vapour phase. The reflectivity, as determined from I_r, did change somewhat, but by too little ($\lesssim$ 4%) to account for the observed change in I_s. The tension was the same in both phases. Scattering by thermally excited capillary waves cannot account for the data of Fig. 11. It appears that $\varepsilon_a S_0$ increases in the denser phase, as for the bilayer case. Numerical estimates of the change are difficult due to uncertainties in the polarisation factors.

No noticeable fluctuations in I_s were apparent in the LC/LE monolayer transition.

4. DYNAMICAL STUDIES

In a sense, all information from QLS studies concerns dynamics: we are probing relatively high frequency fluctuations which, by definition, involve dynamic motions. Our present focus must, to keep within reasonable bounds, be narrower - we concentrate upon the recently discovered intra-molecular relaxations of model membranes, and their interpretation in molecular terms. This excludes several fascinating but peripheral aspects, including pre-transitional fluctuations in BLM[21] (*cf.* γ' data of Fig. 9) and the influence of fluctuations of molecular packing upon the average tension in monolayers[25]. Some rather revealing aspects of our studies of transitions in both BLM and monolayers will thus be passed over.

The mechanical properties of the membrane cannot be constant, independent of the frequency of the perturbation involved. If, say, γ' were indeed constant then the dissipative force ($\propto \omega \gamma'$)

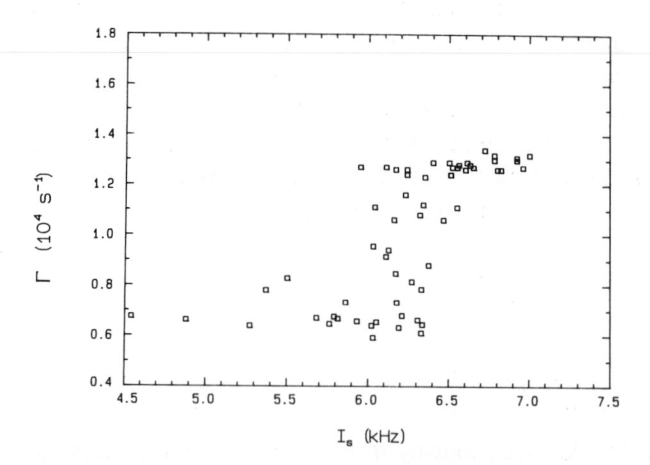

Figure 11. The variation of Γ with I_S for the liquid/vapour transition of the PDA monolayer of Fig. 3.

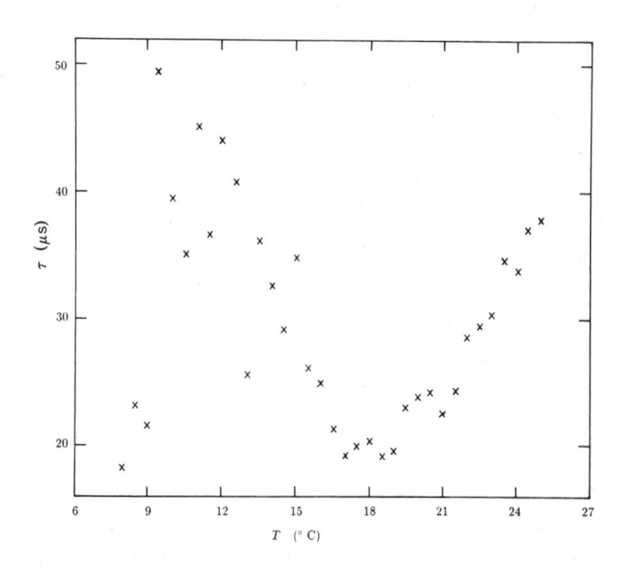

Figure 12. The temperature variation of the relaxation time of the transverse shear modulus in BLM.

would become infinitely large compared to the restoring force ($\propto \gamma_0$) as $\omega \rightarrow \infty$. The membrane moduli must thus be viscoelastic in nature. Now, unlike viscoelastic fluids, there is as yet no theory for viscoelastic relaxation in thin molecular films. However the observed data are compatible with one of the simplest models of relaxation, the Maxwell fluid[26]. This corresponds to a simple exponential relaxation of the stress, with time constant τ. Using γ as an example, the elastic modulus and transverse shear viscosity at a frequency ω (taken as the capillary wave ω_0) would be

$$\gamma_0 = \gamma_e + G \frac{\omega^2\tau^2}{1+\omega^2\tau^2} \tag{4}$$

$$\gamma' = G \frac{\tau}{1+\omega^2\tau^2} \tag{5}$$

where γ_e is the classical, $\omega \rightarrow 0$, limit of the elastic modulus and G is the strength of the relaxation. We note only that at low frequencies ($\omega \ll 1/\tau$) $\gamma_0 = \gamma_e$ and $\gamma' = G\,\tau$, whereas for $\omega \gg 1/\tau$ we expect $\gamma_0 \rightarrow \gamma_e + G$ and $\gamma' \propto \omega^{-2}$. Similar relations can be written for ε.

Transverse shear modulus of bilayers

A careful study of the dispersion behaviour of capillary waves on BLM demonstrated differences between those which contained hydrocarbon solvent in their structure and those which were 'solvent-free'[27]. In the former case the data implied a constant membrane tension and $\gamma' = 0$, whilst both γ_0 and γ' of the 'solvent-free' BLM were apparently q dependent. While somewhat scattered - particularly for γ' - the data were consistent with a Maxwell viscoelastic relaxation of time-scale equal to 37 μs. The equilibrium tension of 3.0 mN/m agreed with literature values. The observed time scale is compatible with values associated with chain-melting in temperature jump studies of bilayer transitions[28].

Now consideration of eqs 4 and 5 suggests that, provided the equilibrium modulus γ_e is known, then τ may be estimated from γ_0 and γ' determined at a single frequency. Using the temperature variation of the interfacial tension of GMO at the oil-water interface (S.H. White, personal communication), normalised to the value of γ_e just quoted (for 24.7°C), the measured variations of γ_0 and γ' through the transition of 'solvent-free' GMO bilayers permitted the T dependence of the relaxation time to be estimated (Fig. 12) [29].

Remarkably τ is lowest just at the transition temperature T_t. If the relaxation of the transverse shear modulus γ is indeed associated with the melting of the lipid acyl moieties this is a significant finding. Such chain-melting leads to bobbing up and down of the lipid molecules, which has been associated with passive ionic permeation through membranes[30]. The quicker relaxation is thus entirely compatible with the established transitional peak in lipid vesicle permeability[31]. It has been suggested that this may involve lipid molecules at the interfaces between gel and fluid regions in the bilayer, which would certainly be compatible with the present data[32].

Monolayers

Transverse shear modulus Monolayers of both GMO[33] and PDA[15] below its triple point display similar phase transitions from an expanded gaseous phase to a condensed phase on compression. Both systems display very similar relaxation phenomena for the modulus γ[33,34], suggesting that these may arise from the monolayer phase, rather than details of the individual molecule.

The viscosity γ' is more precisely determined for GMO, being rather larger than for PDA. It was found for the fully compressed monolayer that the variations of both γ_0 and γ' were entirely compatible with a single Maxwell relaxation of time constant 9 μs [33]. A particularly elegant demonstration involves plotting $\omega_0\gamma'$ *versus* γ_0 as in Fig. 13. The data agree well with the expected form - the semicircle shown. Note particularly that γ_0 tends to the measured γ_e as $\omega \rightarrow 0$, clearly showing that there cannot be any slower relaxation than that cited.

As the monolayer was expanded from the fully compressed state the relaxation time fell, apparently exponentially[33]. A 10% increase in A caused τ to fall by e^{-1}. This decrease appeared to continue throughout the condensed phase into the condensed/vapour coexistence region.

Without going into details we briefly mention that PDA monolayers below the triple point behaved very similarly. The relaxation time was less well defined because of the lower γ' values, but was about 20 μs. As for GMO the relaxation time fell on monolayer expansion[34].

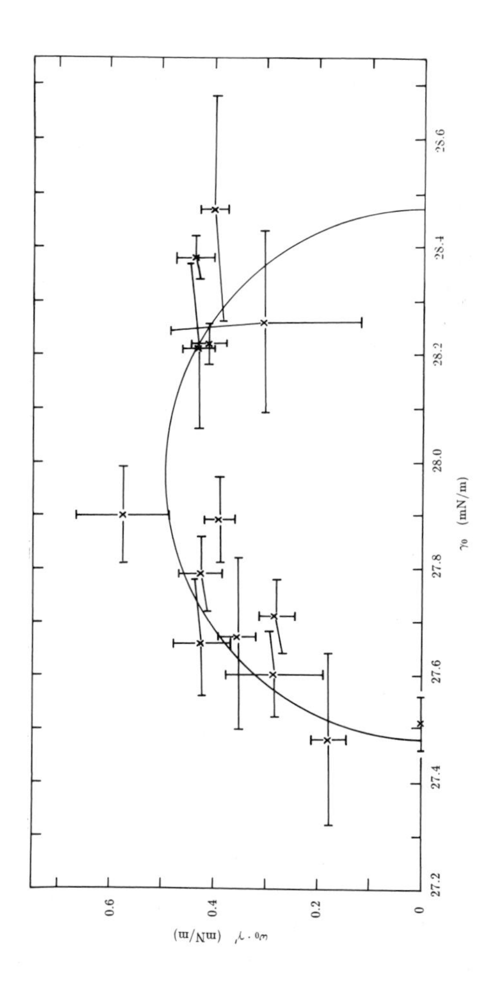

Figure 13. A Cole-Cole plot of the transverse shear modulus γ for a fully compressed GMO monolayer. The semicircle represents the behaviour of a Maxwell relaxation of G = 1 mN/m and $\tau = 9$ μs.

These findings are all consistent with a chain-melting process. In the BLM the apposed lipid films would sterically hinder straightening of the acyl chains, rather naturally explaining the larger τ value in this case. Similarly fully saturated chains would be more closely packed than the unsaturated oleate chains in GMO, inhibiting their chain melting in the case of PDA.

Dilational modulus In all monolayers which have been studied by QLS there are regions of the phase diagram within which relaxation of the dilational modulus ε occurs[33,34]. For example in phase coexistence regions for PDA above the triple point the elastic modulus ε_0 determined by light scattering is non-zero (*cf.* Fig. 4). However the classical isotherms are flat in these regions, so that the equilibrium ε_0 must be zero. The relaxation processes involved appear to be comparatively slow - $\tau \ll 1/\omega_0$ - so that the viscosity ε' is commonly very low at light scattering frequencies. This, together with the indirect action of ε upon the capillary waves, makes it difficult to observe any frequency dependence of either ε_0 or ε' [35]. It has thus been rather more difficult to establish the details of the relaxation for ε than for γ.

A further difficulty is that, as it affects the capillary waves, ε is not a simple quantity[34]. The capillary waves impose a uniaxial dilational stress upon the surface film; the appropriate modulus is the sum of those corresponding to pure shear and to pure 'hydrostatic' compression. It is therefore difficult to be certain that ε_0 and ε' from light scattering correspond to similar combinations of the two moduli. In only one case have comparisons been possible: just at the condensed end of the LC/LE transition of PDA the light scattering ε' seems to comprise the shear component alone, the dilational viscosity having relaxed entirely into the storage modulus ε_0 [34]. The relaxation in this case must involve at least two time scales, one, > 1 s, being visible in the Wilhelmy plate response while the other must be $1/\omega_0 \ll \tau \ll 1$ s. Any relaxation involving the shear viscosity is much faster than $1/\omega_0$.

To establish the details of the relaxation of ε and to determine the processes involved will require experiments with observed correlation functions having much lower noise than to date.

5. CONCLUSIONS

Quasi-elastic light scattering has proven a most useful tool in the investigation of model membranes. Much detail has been determined in studies of the thermotropic transitions of both lipid bilayers and insoluble monolayers. This has included observation of phase separation and of transitional changes in membrane structure. The viscoelasticity of membranes has been shown to be an unexpectedly rich area of study. In particular the observation of viscoelastic relaxation on microsecond timescales in these model systems is completely unexpected. These are, as far as we know, the first observations of the macroscopic manifestations in membrane viscoelasticity of molecular relaxation. The timescales and their dependence upon temperature and molecular packing are strongly suggestive of possible biological significance.

ACKNOWLEDGEMENTS

This work has largely been supported by the S.E.R.C. The author wishes to express his thanks to the many collaborators who have contributed to these studies of model membranes, particularly Drs. J.F. Crilly, G.E. Crawford, R.C. McGivern and P.J. Winch.

REFERENCES

1. *e.g.* A. Milon, J. Ricka, S.-T. Sun, T. Tanaka, Y. Nakatani and G. Durisson, *Biochim. Biophys. Acta*, 1984, *777*, 331.
2. L. Kramer, *J. Chem. Phys.*, 1971, *55*, 2097.
3. C.-P. Fan, *J. Colloid Interf. Sci.*, 1973, *44*, 369.
4. D.H. Martin, *Contemp. Phys.*, 1977, *18*, 81, 193.
5. N.R. Pallas and B.A. Pethica, *Langmuir*, 1985, *1*, 509.
6. D. Langevin, J. Meunier and D. Chatenay, in 'Surfactants in Solution', eds. K.L. Mittal and B. Lindman, Plenum, New York, 1984, Vol. 3, 1991.
7. M.A. Bouchiat and D. Langevin, *J. Colloid Interf. Sci.*, 1978, *63*, 193.
8. F.C. Goodrich, *Proc. Roy. Soc. London A*, 1981, *374*, 341.
9. J.C. Earnshaw and R.C. McGivern, *J. Colloid Interf. Sci.*, 1988, *123*, 36.
10. J.C. Earnshaw, R.C. McGivern, A.C. McLaughlin and P.J. Winch, *Langmuir*, 1990, *6*, 649.
11. G.E. Crawford and J.C. Earnshaw, *J. Phys. D*, 1985, *18*, 1029.

12. P.J. Winch and J.C. Earnshaw, *J. Phys. E: Sci. Instrum.*, 1988, *21*, 287.
13. S. Hård and R.D. Neuman, *J. Colloid Interf. Sci.*, 1981, *83*, 315.
14. Y. Chen, M. Sano, M. Kawaguchi, H. Yu and G. Zografi, *Langmuir*, 1986, *2*, 349.
15. P.J. Winch and J.C. Earnshaw, *J. Phys.: Condens. Matter*, 1989, *1*, 7187.
16. B. Moore, C.M. Knobler, D. Broseta and F. Rondelez, *J. Chem. Soc., Faraday Trans. II*, 1986, *82*, 1753.
17. S.R. Middleton and B.A. Pethica, *Faraday Symposium, Chem Soc*, 1981, *16*, 109.
18. J.F. Crilly and J.C. Earnshaw, *Biophys. J.*, 1983, *41*, 197.
19. J.F. Crilly and J.C. Earnshaw, *Biophys. J.*, 1983, *41*, 211.
20. G.E. Crawford and J.C. Earnshaw, *Eur. Biophys. J.*, 1984, *11*, 25.
21. G.E. Crawford and J.C. Earnshaw, *Biophys. J.*, 1986, *49*, 869.
22. J.P. Dilger, *Biochim. Biophys. Acta*, 1981, *645*, 357.
23. T. Kambara and N. Sasaki, *Biophys. J.*, 1984, *46*, 371.
24. S.H. White, *Biophys. J.*, 1975, *15*, 95.
25. J.F. Crilly and J.C. Earnshaw, *J. de Phys.*, 1987, *48*, 485.
26. J.D. Ferry, 'Viscoelastic Properties of Polymers', John Wiley & Sons, New York, 1980.
27. G.E. Crawford and J.C. Earnshaw, *Biophys. J.*, 1987, *52*, 87.
28. A. Genz and J.F. Holzwarth, *Coll. Polymer Sci.*, 1985, *263*, 484.
29. G.E. Crawford and J.C. Earnshaw, *Biophys. J.*, 1989, *55*, 1017.
30. R.N. Robertson, 'The Lively Membrane', C.U.P., Cambridge, 1983, Chap. 1.
31. D. Papahadjopoulos, K. Jacobsen, S. Nir and T. Isac, *Biochim. Biophys. Acta*, 1973, *311*, 330.
32. L. Cruzeiro-Hansson and O.G. Mouritsen, *Biochim. Biophys. Acta*, 1988, *944*, 63.
33. J.C. Earnshaw, R.C. McGivern and P.J. Winch, *J. de Phys.*, 1988, *49*, 1271.
34. J.C. Earnshaw and P.J. Winch, *J. Phys.: Condens. Matter*, 1990, *2*, 8499.
35. But *cf.* J.C. Earnshaw, R.C. McGivern and G.E. Crawford, in 'Dynamic Properties of Biomolecular Assemblies', eds. S.E. Harding and A.J. Rowe, Royal Society of Chemistry, Cambridge, 1989, p. 348.

27

Laser Light Scattering Characterization of Size and Dispersity of Vesicles and Neurosecretosomes Isolated from Mouse Neurohypophysis

By D.B. Sattelle[1,3], A L. Obaid[2,3] and B.M. Salzberg[2,3]

[1]AFRC LABORATORY OF MOLECULAR SIGNALLING, DEPARTMENT OF ZOOLOGY, UNIVERSITY OF CAMBRIDGE, DOWNING STREET, CAMBRIDGE. CB2 3EJ U.K.

[2]DEPARTMENT OF PHYSIOLOGY, SCHOOL OF MEDICINE AND THE DAVID MAHONEY INSTITUTE OF NEUROLOGICAL SCIENCES, UNIVERSITY OF PENNSYLVANIA, B-400 RICHARDS BUILDING, PHILADELPHIA, PA 19104-6085, U.S.A.

[3]MARINE BIOLOGICAL LABORATORY, WOODS HOLE, MA 02543, U.S.A.

1. INTRODUCTION

Optical methods provide non-invasive measurements of nerve terminal function *in situ* and in isolated subcellular fractions with a high degree of temporal resolution, and a spatial resolution limited only by wavelength and the detection optics. In this way action potentials have been recorded from vertebrate neurosecretory terminals not accesible by other means[1,2]. In addition, two large, rapid changes in intrinsic optical signals have been described, one of which can be attributed to a total intensity light scattering signal intimately linked to the secretory process[2]. Other, slower, light scattering changes have been detected by monitoring intensity fluctuations (quasi-elastic laser light scattering - QLS) of light scattered by invertebrate nerve terminals[3]. These signals are attributed to long-term structural changes in the nerve terminal[4-6] or volume changes[7] following stimulation. Studies on isolated nerve terminal preparations (synaptosomes) have permitted total intensity measurements of volume[8], and QLS determination of size and dispersity[9]. Nerve terminals for which the *in situ* scattering signals have been investigated extensively are those of the mouse neurohypophysis (see ref. 10). In order to define more precisely the source of the *in situ* scattering signals, one experimental approach is to examine the scattering properties of isolated nerve endings (neurosecretosomes) and subcellular organelles purified from neurosecretory terminals. Here, we employ QLS to provide the first optical study of the size and dispersity of neurosecretosomes and neurosecretory vesicles prepared from mouse neurohypophysis.

2. METHODS

Preparation of neurosecretosomes from mouse neurohypophysis

Isolated mouse pituitaries (n = 30) were accumulated in a 10 ml volume of homogenizing medium of the following composition: 0.25 M sucrose containing 10.0 mM HEPES (2-N-2-hydroxyethylpiperazin-N'-yl) ethanesulphonic acid (pH 7.0) and homogenized gently (3-4 passes) in a Thomas BB 15220 homogenizer in 3.0 ml fresh homogenizing medium. The homogenizer was rinsed with a further 2.0 ml homogenizing medium, and the combined homogenate was spun for 2 min at 2,500 r.p.m. (754 x g) using the SS-34 rotor of a Sorvall RV2B centrifuge. The pellet was discarded or rewashed and the supernatant was recentrifuged at 6,500 r.p.m. (5,100 x g) for 30 min. The pellet of crude neurosecretosomes was resuspended in a small volume (50 μl - 100 μl) of oxygenated mouse Ringer's for 45 min prior to optical measurements, in order to allow for restoration of ionic gradients. The whole procedure was performed at room temperature to minimize the discharge of neurohormone that is induced at low temperatures.

Preparation of neurosecretory vesicles from mouse and bovine neurohypophyses

Neurosecretory vesicles were prepared from isolated mouse pituitaries (n = 30) which were accumulated in 10.0 ml homogenizing medium (for composition see previous section). Excess buffer was removed and the tissue was homogenized gently in

1.5 ml homogenizing medium (8 passes). The homogenizer was rinsed with a further 1.0 ml homogenizing medium, and the combined homogenate was centrifuged at 5,000 r.p.m. (3,018 x g) for 30 s at 4°C in the SS-34 rotor of a Sorval RV2B centrifuge. The supernatant (S_1) was collected with a Pasteur pipette, and the pellet (P_1) was discarded. S_1 was recentrifuged at 5,750 r.p.m. (3,991 x g), and its pellet (P_2) discarded after removal by pipette of supernatant (S_2). This supernatant was centrifuged at 17,500 r.p.m. (27,648 x g) in the 50-Ti rotor of a Beckman LS2 ultracentrifuge for 20 min at 4°C. The pellet from this spin (P_3) was very gently resuspended in 50 μl homogenizing medium, taking care to avoid any frothing. Several 3-step sucrose-metrizamide gradients (0.45 M; 0.28 M; 0.13 M) were prepared 16 hr - 18 hr prior to use and stored at 4°C protected from the light. The P_3 suspension was layered onto these cold gradients and spun at 140,000 x g in a Beckman airfuge for 20 min. The 0.13 M band, containing the neurosecretory vesicles was collected. For QLS measurements, vesicles were resuspended in a buffer of the following composition: 0.14 M K^+ glutamate; 1.0 mM EGTA; 0.4 mM $CaCl_2$; 20.0 mM PIPES (Piperazine-N,N'-bis[2-ethanesulfonic acid])-Tris (adjusted to pH 6.5).

Quasi-elastic Laser Light Scattering (QLS)

A Coulter multiangle particle analyzer (model N4MD) equipped with a 1 mW HeNe laser was used for quasi-elastic laser light scattering (QLS) measurements of diffusion coefficient (and hence hydrodynamic diameter), particle size distribution and total scattered intensity from samples of neurosecretory vesicles and neurosecretosomes. The QLS method and its applications have been described in detail elsewhere[11,12] and only a brief account is given here. A schematic representation of the QLS apparatus is shown in Fig. 4a. The N4MD particle analyzer used in this study computes the autocorrelation function of the intensity fluctuations, $g^{(2)}(\tau)$, which, for a heterogeneous population of scatterers could be written as an expansion about $\bar{\Gamma}$ (the mean of the distribution of the decay rates weighted by the intensity scattered from each species present) such that

$$\tfrac{1}{2} \ln \{g^{(2)}(\tau) - 1\} = \bar{\Gamma}\tau + \frac{\mu_2\tau^2}{2!} - \frac{\mu_3\tau^3}{3!} \quad (1)$$

where μ_2 and μ_3 (the second and third cumulants) measure the width and skewness of the weighted decay rate distribution[12]. The quantity

$\mu_2/\bar{\Gamma}^2$ is the normalized variance of the diffusion coefficient distribution and provides a measure of the sample polydispersity[12]. From the scattering vector, K, given by $(4\pi n/\lambda_0)\sin(\theta/2)$, with n equal to the refractive index, and the equation

$$\bar{\Gamma} = \bar{D}K^2 \tag{2}$$

an average diffusion coefficient ($\bar{D}$) can be determined, and, using the Stokes-Einstein equation

$$\bar{D} = \frac{kT}{3\pi\eta\bar{d}} \tag{3}$$

where η is the viscosity of the medium, an average diameter ($\bar{d}$) of an equivalent sphere can be calculated.

The size distribution processor (SDP) of the Coulter N4MD utilizes CONTIN, a computer programme developed at the European Molecular Biology Laboratory (EMBL) in Heidelberg, Germany, for the constrained regularization of linear equations[13-16]. The calculated particle size distribution is approximated by a histogram with the height of each column adjusted by parameters in the fitting programme to achieve the best fit to the observed correlation function. Both SDP and cumulants analysis have been used in the present study. The time required to collect data for a single autocorrelation function was between 10 s and 200 s. Control samples of latex spheres ($\bar{d}$ = 40 nm, 100 nm and 500 nm) were measured (Fig. 4b).

Chemicals

Veratrine sulphate was obtained from Sigma Chernical Co. (St. Louis, Mo.) and purified veratridine was the gift of Prof. G. Holan (CSIRO, Melbourne, Australia). Mast cell degranulating peptide (MCDP) was the gift of Prof. R. Hider, University of London, U.K.

RESULTS

Neurosecretory vesicles

Most vesicle samples prepared from mouse neurohypophysis yielded monotonically decaying, multiexponential, intensity autocorrelation

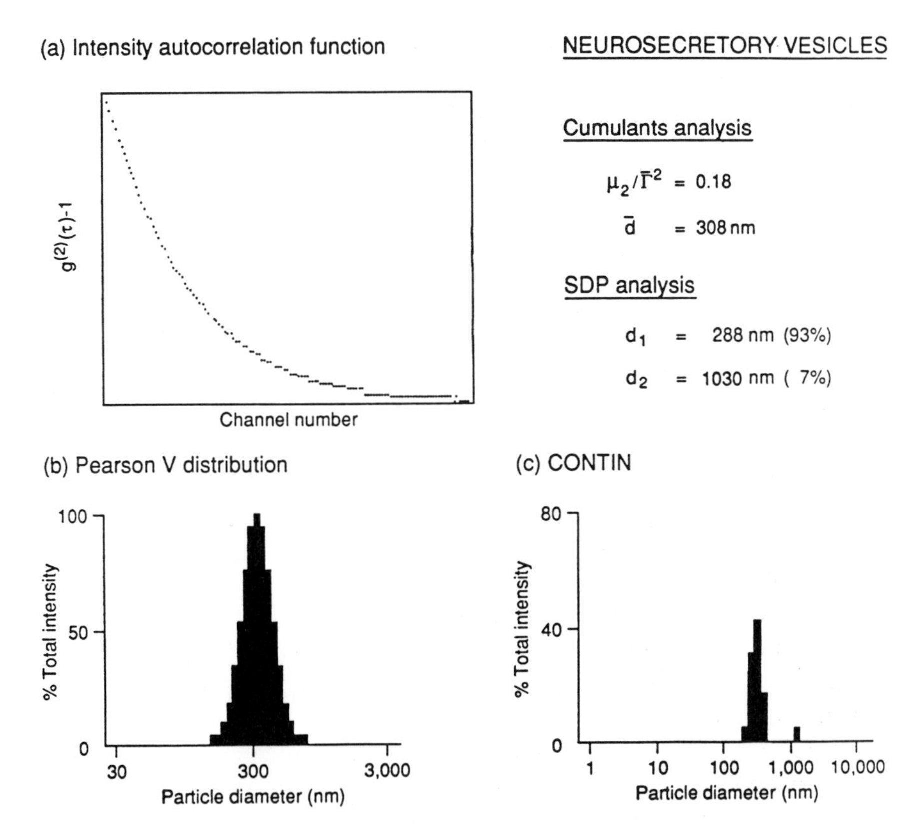

Figure 1. (a) The intensity autocorrelation function illustrated is derived from purified neurosecretory vesicles prepared from mouse neurohypophysis by iso-osmotic density gradient centrifugation. Vesicles are resuspended in buffer (see Methods and ref. 17) at 20°C, and the autocorrelation function is plotted as $g^{(2)}(\tau)$-1 against channel number (each channel representing 34 μs). The polydispersity index ($\mu_2/\bar{\Gamma}^2$) is determined from the decay rate (τ_0=1/$\bar{\Gamma}$) of the correlation function. The average hydrodynamic diameter ($\bar{d}$) of the neurosecretory vesicles is estimated from the Stokes-Einstein equation using $\bar{D}_{20,w}$. (b) Particle size distribution of the neurosecretory vesicles is estimated by fitting the distribution to the Pearson V distribution function[29]. (c) Particle size distribution of neurosecretory vesicles is determined using the CONTIN computer programme[15,16].

functions (Fig. 1), indicating that the neurosecretory vesicles were polydisperse ($\mu_2/\bar{\Gamma}^2$ ranging from 0.13 - 0.36, *cf.* Table 1). Values for $\bar{D}_{20,w}$ ranged from 1.13 x 10^{-8} cm^2s^{-1} to 2.12 x 10^{-8} cm^2s^{-1}, yielding mean hydrodynamic diameters of 202 nm - 367 nm. In all but one of the 8 preparations examined, particle size distribution (SDP) analysis revealed the presence of contaminant particles, which in 4 preparations were of larger diameter than the predominant scattering species, and in the other 3 preparations were smaller. By SDP analysis the most abundant scatterers ranged in size from 195 nm to 375 nm diameter.

Vesicles prepared from bovine neurohypophysis were equally polydisperse, but exhibited smaller diffusion coefficients than those observed for murine vesicles (Table 1). Mean hydrodynamic diameters were in the range 355 nm - 415 nm. The presence of both smaller and larger scattering particles was detected, often in much higher proportions than in mouse preparations. The frequent presence of a high proportion of larger scatterers may account for the higher mean diameter determined for bovine neurosecretory vesicles. Alternatively, vesicles from this source may be of larger average diameter.

Neurosecretosomes

Neurosecretosomes prepared from mouse neurohypophysis also yielded multiexponential intensity autocorrelation functions (Fig. 2). Neurosecretosomes were highly polydisperse ($\mu_2/\bar{\Gamma}^2$ ranging from 0.36 - 0.59). Values for $\bar{D}_{20,w}$ varied by a factor of 3 - 4 between preparations, and thus neurosecretosomes were much more variable in size between preparations than corresponding vesicle samples (Table 1). Mean hydrodynamic diameters for neurosecretosomes ranged from 495 nm - 1776 nm. Normally, SDP analysis revealed the predominant scatterer to be a particle in the size range 2925 nm - 4565 nm, though another particle, of the order of size of neurosecretory vesicles, was often a contaminant.

Veratrine sulphate and pure veratridine (final concentration = 1.0 x 10^{-4} M) failed to generate any consistent change in mouse neurosecretosome size, dispersity, and/or total scattered intensity following application in mouse Ringer's solution (Fig. 3). Mast cell degranulating peptide (MCDP), when tested at a final concentration

Table 1. Diffusion coefficient, size (*i.e.* diameter) and dispersity of neurosecretory vesicles and neurosecretosomes prepared from mouse neurohypophysis (data from bovine neurosecretory vesicles are also included). Values are expresed as ± one standard deviation (n=5). Concentrations (in μg ml^{-1} protein) were as follows: secretosomes, 1-3; secretory vesicles, 30 (mouse), 50 (bovine). All measurements were obtained at a scattering angle (θ) of 90°.

Neurosecretory Tissue	Subcellular Fraction	Preparation	Method of Cumulants			SDP Analysis					
			$10^9 \times \bar{D}_{20,w}$ ($cm^2\ s^{-1}$)	$\bar{d}$ (nm)	$\mu_2/\bar{\Gamma}^2$	d_1 (nm)	%	d_2 (nm)	%	d_3 (nm)	%
Neurohypophysis (mouse)	Neurosecretory vesicles	1	13.0 (± 0.9)	323 ± 20	0.25 ± 0.04	357 ± 109	87	2740 ± 320	13	-	-
		2	12.9 (± 0.3)	324 ± 10	0.13 ± 0.03	375 ± 45	81	156 ± 37	19	-	-
		3	12.4 (± 0.9)	340 ± 26	0.29 ± 0.07	271 ± 31	78	1130 ± 320	22	-	-
		4	11.5 (± 0.6)	367 ± 17	0.30 ± 0.04	274 ± 28	69	1868 ± 598	31	-	-
		5	17.4 (± 0.4)	242 ± 7	0.14 ± 0.06	261 ± 25	91	17 ± 5	9	-	-
		6	21.2 (± 1.2)	202 ± 11	0.26 ± 0.06	195 ± 46	24	3 ± 1	76	-	-
		7	11.3 (± 1.2)	384 ± 47	0.36 ± 0.10	313 ± 72	60	2340 ± 1143	30	13 ± 1	10
		8	15.8 (± 0.4)	266 ± 6	0.25 ± 0.33	286 ± 17	100	-	-	-	-
Neurohypophysis (bovine)	Neurosecretory vesicles	1	10.7 (± 0.1)	399 ± 150	0.26 ± 0.03	268 ± 89	40	768 ± 195	41	93 ± 10	19
		2	11.0 (± 0.2)	389 ± 7	0.31 ± 0.03	394 ± 109	56	865 ± 290	43	26 ± 3	1
		3	10.3 (± 0.4)	415 ± 130	0.25 ± 0.04	233 ± 80	24	749 ± 88	76	-	-
		4	12.0 (± 0.3)	355 ± 10	0.22 ± 0.03	351 ± 82	96	59 ± 9	4	-	-
Neurohypophysis (mouse)	Neurosecretosomes	1	4.48 (± 0.57)	965 ± 113	0.38 ± 0.04	2925 ± 1821	83	221 ± 17	17	-	-
		2	2.58 (± 0.68)	1776 ± 512	0.59 ± 0.10	4292 ± 980	70	375 ± 130	30	-	-
		3	3.53 (± 0.49)	1365 ± 450	0.50 ± 0.10	3350 ± 780	52	394 ± 140	45	44 ± 17	3
		4	8.71 (± 0.69)	495 ± 40	0.36 ± 0.03	4565 ± 2464	30	291 ± 16	70	-	-

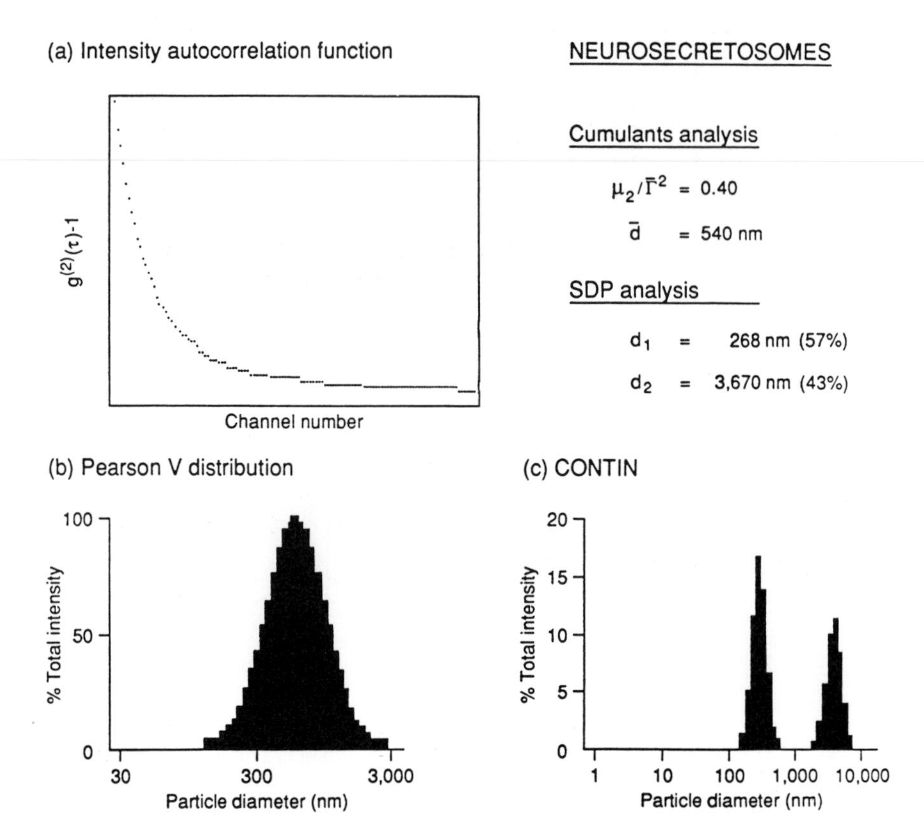

Figure 2. (a) The intensity autocorrelation function shown is derived from neurosecretosomes prepared from mouse neurohypophysis by differential centrifugation. Neurosecretosomes are resuspended in oxygenated mouse Ringer's solution at 20°C, and the autocorrelation function is plotted as $g^{(2)}(\tau)$-1 against channel number (each channel representing 100 μs). The polydispersity index ($\mu_2/\bar{\Gamma}^2$) is determined from the decay rate ($\tau_0=1/\bar{\Gamma}$) of the correlation function. The average hydrodynamic diameter ($\bar{d}$) of the neurosecretosomes is estimated from the Stokes-Einstein equation using $\bar{D}_{20,w}$. (b) Particle size distribution of the neurosecretosomes is estimated by fitting the distribution to the Pearson V distribution function[29]. (c) Particle size distribution of neurosecretosomes determined using the CONTIN computer programme[15,16].

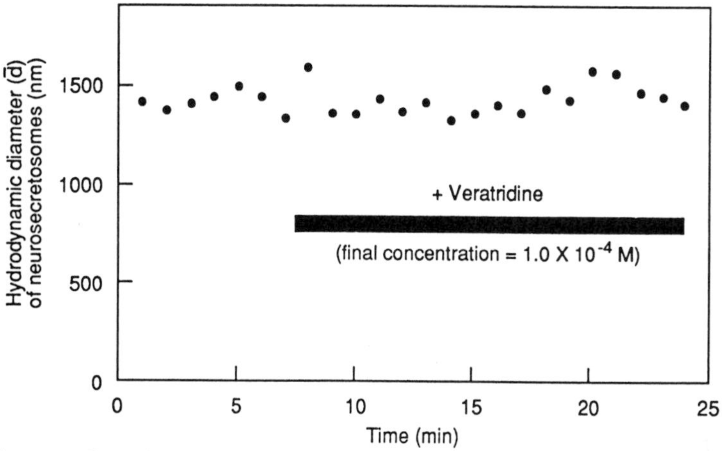

Figure 3. The changes with time of mean neurosecretosome hydrodynamic diameter ($\bar{d}$) are shown before and after the addition of veratridine (1.0 x 10^{-4} M).

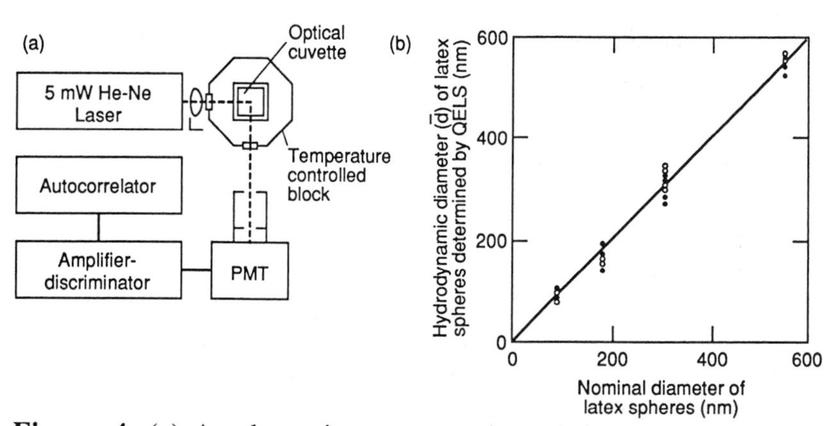

Figure 4. (a) A schematic representation of the main optical and electronic components of the QLS apparatus is shown. By this means hydrodynamic diameters and particle size distributions are obtained for purified neurosecretory vesicles and neurosecretosomes. The cuvette is illuminated by a 5 mW He-Ne laser and light scattered at 90° is collected by a photomultiplier tube (Hamamatsu Model R928). The output of the phototube is led *via* an amplifier-discriminator to a Langley-Ford Instruments autocorrelator (Model 1096). (b) Hydrodynamic diameters (d) of latex spheres (four different sizes) are derived by fitting to a Pearson V distribution function[29], and by using the CONTIN computer programme[15,16].

of 1.0 x 10^{-4} M also failed to modify these same hydrodynamic and scattering parameters derived from mouse neurosecretosomes.

As a check on the stability of the neurosecretosomes used in the present study, the mean hydrodynamic particle diameter ($\bar{d}$) was monitored for extended periods. An increase in $\bar{d}$ 16 hr - 24 hr after preparation of the neurosecretosomes may be attributable to swelling or aggregation, but no change was noted during the period (1 hr - 8 hr) for which observations were normally made after preparation of fresh neurosecretosomes.

4. DISCUSSION

The nonapeptide hormones vasopressin and oxytocin, together with their respective neurophysins are stored in neurosecretory vesicles (≡ granules) in the mammalian neurohypophysis. Neurosecretory vesicle isolation procedures using iso-osmolar density gradients[17], have yielded a convenient preparation of purified neurosecretory vesicles from neurohypophysis. By means of QLS, the hydrodynamic diameter of these purified vesicles is found to agree well with measurements on chromaffin vesicles[18] and islet granules[19] using the same technique. Data from several laboratories (Table 2) reveal hydrodynamic diameters in the range 200 nm - 400 nm, depending on the type of granule and the method of isolation, with the single exception of zymogen vesicles, where a hydrodynamic diameter of 678 ± 12 nm is reported[19]. Zymogen vesicles from pancreatic acinar cells are of larger mean diameter than other secretory vesicles examined by QLS, and also exhibit higher polydispersity[19]. This is consistent with data from stereological electron microscopical studies on acinar cells where a broad size distribution is observed[20]. Some quantitative ultrastructural data for the neurohypophysis of rodents and other vertebrates are available. For example, a size range of 100 nm - 200 nm is reported for neurosecretory vesicles from mouse neurohypophysis nerve endings[21]. A mean diameter of 160 nm has been determined for neurosecretory vesicles from rat neurohypophysis[22], and, in a detailed study on the same tissue using nonparametric statistical methods, 5 subtypes of vesicle have been characterized with mean diameters of 143 nm, 155 nm, 167 nm, 180 nm and 193 nm [23]. Data for a wide range of fish, amphibian, avian and mammalian species yield values ranging from 55 nm - 300 nm for neurosecretory vesicles from the neurohypophysis[24]. In

Table 2. Hydrodynamic diameter and polydispersity of purified secretory vesicles determined by quasi-elastic laser light scattering (QLS).

Tissue Source	Secretory vesicle	Hydrodynamic diameter ($\bar{d}$) (nm)	Polydispersity ($\mu_2/\bar{\Gamma}^2$)	Reference
Bovine adrenal gland	Chromaffin vesicle[a]	324 ± 50 to 360 ± 60	0.25 ± 0.03 to 0.26 ± 0.02	18
Bovine adrenal gland	Chromaffin vesicle[b]	375 ± 70 to 400 ± 80	0.10 ± 0.02 to 0.23 ± 0.02	18
Bovine adrenal gland	Chromaffin vesicle[c]	394 ± 7	0.18 ± 0.01	19
Bovine neurohypophysis	Neurohypophysial vesicle[d]	355 ± 10 to 415 ± 130	0.22 ± 0.03 to 0.31 ± 0.03	This study
Murine neurohypophysis	Neurohypophysial vesicle[d]	202 ± 11 to 384 ± 47	0.13 ± 0.03 to 0.36 ± 0.10	This study
Bovine pancreatic islet cells	Islet vesicle[c]	346 ± 4	0.19 ± 0.01	19
Bovine pancreatic acinar cells	Zymogen vesicle[c]	677 ± 12	0.27 ± 0.02	19
Bovine pancreatic acinar cells	Zymogen vesicle[c]	1225 ± 18	0.20 ± 0.001	25

Secretory vesicles are purified by [a]sedimentation through sucrose/Ficoll, [b]iso-osmotic sucrose, [c]iso-osmotic gradient centrifugation with sucrose gradients, [d]iso-osmotic gradient centrifugation with metrizamide-sucrose gradients.

general it appears that hydrodynamic diameters reported for secretory vesicles from different sources are greater than those detected when electron microscopy is applied to secretory tissues.

There are several factors that may contribute to the observed differences between the light scattering and electron microscopy findings. First, the neurosecretory vesicle fraction is polydisperse, and particle size distributions often reveal a component of much greater diameter than the neurosecretory vesicles. Even when present as only a small proportion of the particles in the fraction, the mean particle diameter ($\bar{d}$) will be distorted toward a higher value. Aggregation and swelling of neurosecretory vesicles and/or the presence of other subcellular contaminants will all yield an overestimate of neurosecretory vesicle diameter by QLS.

Secondly, it is possible that electron microscopical methods can introduce artefactual size changes in neurosecretory vesicles, some of which might be expected to underestimate vesicle diameter. For example, the cross-linking involved in fixation and subsequent preparative stages involving dehydration might be expected to lead to shrinkage of subcellular particles. Another factor that may contribute to the larger particle sizes detected for isolated vesicles has recently been recognized[25]. Using zymogen vesicles purified from rat pancreatic acinar cells and QLS, a vesicle diameter of 1225 ± 18 nm and a polydispersity ($\mu_2/\bar{\Gamma}^2$) of 0.02 ± 0.01 are detected, but in the presence of a cytoplasmic protein factor, a lower mean hydrodynamic diameter (998 ± 15 nm) and a slightly lower polydispersity are recorded. These observations are attributed to an aggregation inhibitory factor which is absent from isolated, highly purified vesicle preparations, thereby promoting aggregation. Deep-etching of frozen tissue may be expected to produce less distortion of vesicle dimensions, but, to date, measurements are not available for the mouse neurohypophysis.

It will be of interest to obtain ultrastructural data on other particles that are detected by QLS in a number of preparations of mouse neurosecretory vesicles (Table 1). Also, it remains to be resolved whether or not the larger hydrodynamic diameter observed for bovine compared to murine neurosecretory vesicles reflects the existence of larger granules in the bovine material or more aggregation or swelling of purified bovine neurosecretory granules.

Alternatively, the larger mean diameter may indicate contamination by a population of larger vesicles and/or non-vesicular material.

Neurosecretosomes prepared from the mouse neurohypophysis facilitate biochemical and biophysical experimental approaches to the processes involved in the regulated discharge of neurohypophysial hormones. The improved access to the neurosecretory membranes and the advantages of working with a reasonably homogeneous suspension of isolated nerve endings permits the application of optical methods for investigating nerve terminal function. Hydrodynamic diameter, polydispersity and particle size distribution analysis reveal a fraction containing a predominant particle of the size expected for isolated neurosecretory nerve terminals ($\bar{d}$ = 0.5 μm - 2.0 μm), though other particles are also present. Quantitative ultrastructural measurements on the rat neurohypophysis have revealed nerve endings corresponding to spheres of mean diameter 1580 ± 40 nm and swellings of mean diameter 2700 ± 100 nm [26].

The constancy of the hydrodynamic diameter observed over the 1 hr - 8 hr period when QLS experiments were performed suggests that the light scattering measurements reported here reflect the properties of stable functional neurosecretosomes. This view is given suppport by recent findings showing that neurosecretosomes prepared by an almost identical procedure can exhibit depolarization-induced release of arginine vasopressin 4 days after the initial preparation (J.T. Russell, personal communication).

A rapid change in total scattered intensity, probably representing an increase in large angle light scattering, has been shown to accompany the calcium-sensitive secretory release process in intact nerve terminals of the mouse neurohypophysis[1,2]. Veratridine[27] and mast cell degranulating peptide (MCDP)[28], both of which depolarize nerve terminals and lead to secretion, do not appear to induce any maintained changes in size, dispersity or total scattered intensity of the neurosecretosome fraction. However, 10 s - 200 s are required to collect each intensity autocorrelation function with the present apparatus. Additional experiments having improved time resolution are needed in order to examine whether or not the light scattering changes detected in the neurohypophysis *in situ* can be reproduced in isolated, resealed nerve terminals *in vitro*.

5. CONCLUSIONS

Quasi-elastic laser light scattering (QLS) was employed to probe the size and dispersity of neurosecretory vesicles and neurosecretosomes prepared from murine (and bovine) neurohypophyses. Vesicle fractions were highly polydisperse ($\mu_2/\bar{\Gamma}^2$ = 0.13 - 0.36) and mean hydrodynamic diameters ($\bar{d}$) ranged from 202 ± 11 nm to 384 ± 47 nm (murine vesicles) and 355 ± 10 nm to 415 + 130 nm (bovine vesicles). Size distribution histograms yielded one predominant scattering particle ($\bar{d}$ = 195 ± 46 nm to 375 ± 45 nm), though many preparations included one or two distinct populations of scatterers which normally contributed less than 20% of the total scattering signal.The mean hydrodynamic diameter detected by QLS was consistently slightly larger than that measured by electron microscopy, a finding probably attributable to preparative procedures, and the different weighting of size distributions inherent to optical and electron microscopical methods. Purified neurosecretosomes yielded more complex particle size distributions. A component with a mean hydrodynamic diameter in the range $\bar{d}$ = 965 ± 113 nm to 1776 ± 512 nm was detected, though other particle sizes were often observed. Particle size distribution analysis reveals a dominant scattering species ($\bar{d}$ = 2925 ± 1821 nm to 4565 ± 2464 nm) with a major contribution from particles of the order of size of neurosecretory vesicles ($\bar{d}$ = 221 ± 17 nm to 394 ± 140 nm). Depolarization by veratridine (1.0 x 10^{-4} M) and exposure to mast cell degranulating peptide (MCDP) did not result in a detectable change in size of the neurosecretosomes.

ACKNOWLEDGEMENTS

The authors acknowledge the support of the Marshall and Orr bequest of The Royal Society and the Summer Fellowships programme of the Marine Biological Laboratory, Woods Hole, Massachusetts (DBS). They are grateful to Dr. James Russell for helpful advice regarding neurosecretosome and neurosecretory vesicle preparations. The support of NIH grant NS 16824 is gratefully acknowledged (BMS, ALO), as is the generous provision by the Langley-Ford Instruments Division of Coulter Electronics of the N4MD particle size analyzer used throughout the study.

REFERENCES

1. B.M. Salzberg, A.L. Obaid, H. Gainer and D.M. Senseman, *Nature (Lond.)*, 1983, *306*, 36.
2. B.M. Salzberg, A.L. Obaid and H. Gainer, *J. Gen. Physiol.*, 1985, *86*, 395.
3. T.I. Shaw and B.J. Newby, *Biochim. Biophys. Acta*, 1972, *255*, 411.
4. D.B. Sattelle, D.J. Green and K.H. Langley, *Biol. Bull. Mar. Biol. Lab. (Woods Hole)*, 1975, *149*, 455.
5. R.W. Piddington and D.B. Sattelle, *Proc. Roy. Soc. B*, 1975, *180*, 415.
6. D.B. Sattelle, *J. Exp. Biol.*, 1988, *139*, 233.
7. D. Englert C. Edwards, *Proc. Natl. Acad. Sci. USA*, 1978 *74*, 5759
8. K. Kamino, K. Inouye and A. Inouye, *Biochim. Biophys. Acta*, 1973, *330*, 39.
9. D.B. Sattelle, K.H. Langley, A.L. Obaid and B.M. Salzberg, *Eur. Biophys. J.*, 1987, *15*, 71.
10. B.M. Salzberg and A.L. Obaid, *J. Exp. Biol.*, 1988, *139*, 195.
11. D.B. Sattelle, W.I. Lee and B.R. Ware, (eds.) 'Biomedical Applications of Laser Light Scattering', Elsevier Biomedical Press, Amsterdam, 1982, p. 1.
12. D.E. Koppel, *J. Chem. Phys.*, 1972, *57*, 4814.
13. S.W. Provencher, J.Hendrix and L. De Maeyer, *J. Chem. Phys.*, 1978, *69a*, 4273.
14. S.W. Provencher, *Makromol. Chem.*, 1979, *180*, 201.
15. S.W. Provencher, *Comput. Phys. Commun.*, 1982, *27*, 213.
16. S.W. Provencher, *Comput. Phys. Commun.*, 1982, *27*, 229.
18. D.J. Green, G.W. Westhead, K.H. Langley and D.B. Sattelle, *Biochim. Biophys. Acta*, 1978, *539*, 369.
19. E.K. Matthews, M.D.L. O'Connor, D.B. McKay, D.R. Ferguson, and A.D. Schuz, "Insulin secretory mechanisms and antidiabetic drug action: an investigation by photon correlation spectroscopy and laser Doppler electrophoresis"in 'Biomedical Applications of Laser Light Scattering' (eds. DB Sattelle WI Lee and BR Ware) p. 311, Elsevier Biomedical Press, Amsterdam, 1982.
20. T.H. Ermak, and S.S. Rothman, *Cell Tissue Res.*, 1981, *214*, 51.
21. M. Murakami, *Z. Zellforsch*, 1962, *56*, 277.
22. J.F. Morris, *J. Endocrinol.*, 1976, *68*, 209.

23. S. Ishii, P. Thomas and T. Nakamura, *Z. Zellforschung und mikros. Anat.*, 1973, *146*, 463.
24. J.C. Sloper and R.G. Bateson, *J. Endocrinol.*, 1965, *31*, 139.
25. J. Rogers, E.K. Matthews and D.B. McKay, *Biochim. Biophys. Acta*, 1987, *897*, 217.
26. J.J. Nordmann, *J. Anat.*, 1977, *123*, 213.
27. M.C.W. Minchin, *J. Neurosci. Methods*, 1980, 2, 111.
28. D. Lagunoff, T.W. Martin, and G. Read, *Ann. Rev. Pharmacol. Toxicol.*, *23*, 331.
29. J.T. Russell, J.T., *Analyt. Biochem.*, 1981, *113*, 2298.
30. R.L. McCally and C.B. Bargeron, *J. Chem. Phys.*, 1977, *67*, 3151.

28

PCS and Zeta Potential Characterization of Model Drug Carriers - Essential Information for the Interpretation of *in vitro* and *in vivo* Data

By R.H. Müller, S. Heinemann, T. Blunk and S. Rudt

UNIVERSITY OF KIEL, DEPARTMENT OF PHARMACEUTICS AND BIOPHARMACEUTICS, GUTENBERGSTR. 76-78, D-2300 KIEL, GERMANY.

1. INTRODUCTION

Colloidal particulates are potential carriers for the intravenous delivery of drugs[1]. Polymeric particles, liposomes and parenteral fat emulsions are possible carrier systems. The organ distribution and the fate of intravenously injected particles are determined by their properties such as particle size, charge, surface hydrophobicity, presence of complement activating groups and the interaction with blood components (opsonisation, adsorption of apolipoproteins). In general, injected particles are opsonized in the blood, recognized by the reticuloendothelial system (RES) and mainly taken up by the liver and spleen macrophages within a few minutes. This 'natural' accumulation in liver (60-80%) and spleen (2-5%) can be used for the site-specific delivery of drugs to these organs (so called natural or passive targeting). Targeting to the liver is of high interest for the treatment of liver metastasis which is the main cause of death in cancer. The modification of the surface properties can alter the organ distribution. Adsorption of block copolymers such as Poloxamer and Poloxamine (Fig. 1) on polymeric particles modifies the surface by the formation of an adsorption layer (coating layer) (Fig. 2). After coating with the polymers the particle charge decreases significantly and the surface possesses an increased surface hydrophilicity. Non-charged particles are cleared more slowly by

Poloxamer: $(EO)_n - (PO)_m - (EO)_n$

Poloxamer 407: n = 98; m = 67; MW = 11.500

Poloxamine:

$$\begin{matrix} (EO)_n\text{-}(PO)_m \\ (EO)_n\text{-}(PO)_m \end{matrix} > N\text{-}CH_2\text{-}CH_2\text{-}N < \begin{matrix} (PO)_m\text{-}(EO)_n \\ (PO)_m\text{-}(EO)_n \end{matrix}$$

Poloxamine 908: 80 % EO; MW = 27.000

Figure 1. Structure of Poloxamer and Poloxamine block copolymers. Note, in particular, that they possess a hydrophobic centre part (polypropylene oxide block, PO) and hydrophilic polyethylene oxide (EO) chains at the end.

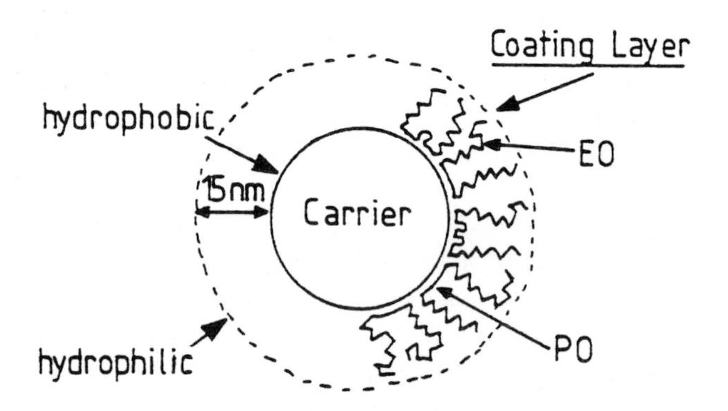

Figure 2. The polymers adsorb on the hydrophobic particle surface by their hydrophobic propylene oxide blocks (anchor part), the hydrophilic EO chains protrude into the dispersion medium forming a new hydrophilic surface. The adsorption layer (coating film) covers the surface charge of the particles and the diffuse layer, the particles are uncharged.

the RES, the increased surface hydrophilicity reduces the adsorption of blood components. Coating of model carriers (polystyrene latex particles) with Poloxamine 908 prevented recognition by the RES and the particles circulated in the blood[1-3]. Such circulating carriers could be employed for the prolonged or sustained release of drugs, *e.g.* peptides or cytotoxics. Adsorption of Poloxamer 407 on small particles (60 nm) led to an accumulation of more than 80% in the bone marrow[4]. The accumulation is possibly due to an uptake by the bone marrow macrophages and/or the endothelial lining cells[1]. The examples demonstrate the potential to achieve a targeted delivery of drugs by modification of the surface properties of colloidal particles. However, to benefit from these major advances in targeting the non-biodegradable polystyrene particles need to be replaced by carriers made from biodegradable polymers such as poly-lactic acid (PLA).

To target to sites other than the liver it is essential to prevent the phagocytosis of the particles by liver and spleen macrophages. Cell cultures were employed as *in vitro* test system to study the phagocytosis reducing effect of coating layers on particles[5]. However, block polymers reducing phagocytosis *in vitro* exhibited very different distribution patterns *in vivo*. The apparent contradiction between *in vitro* and *in vivo* results could be solved by intensive physico-chemical characterization. To allow a better interpretation of cell culture results a thorough characterization is required in terms of:

1. Hydrophobicity of the particle surface (which determines the thickness of the adsorbed block copolymer layer);
2. Thickness of the adsorbed block copolymer layer;
3. Charge reducing effect of the adsorbed layer;
4. Surface hydrophobicity of the coating layer (which determines interaction with blood components);
5. Adsorption of blood components.

Photon correlation spectroscopy (PCS) and laser Doppler anemometry (LDA) were employed to characterize surface-modified model particles and also biodegradable carriers with regard to the points 2, 3 and 5. The aim of this Chapter is to demonstrate the potential of the laser light scattering techniques in the development of drug carriers.

2. MATERIALS AND METHODS

Polystyrene latex particles of various sizes were purchased from Polysciences (Northampton, UK) and used as model carriers, Poloxamer and Poloxamine polymers were obtained from BASF (Wyandotte, USA).

PCS measurements were performed with a Malvern Spectrometer RR102, a 4-bit K7025 correlator and a PET 8000 computer (Malvern Instruments, UK). A Malvern Zetasizer II based on laser Doppler anemometry (LDA) was used for the electrophoresis measurements. The electrophoretic mobility was transformed to a zeta potential using the Helmholtz-Smoluchowski equation. Zeta potential measurements were performed in 0.001 M NaCl (large 4 mm capillary) and in serum (small bore capillary, 1 mm), field strength 20 V/cm.

The surface of the latex particles was modified by adsorption of polymers (coating). The particles were coated with Poloxamer and Poloxamine by mixing equal volumes of latex dispersion (2.5% w/w) and aqueous polymer solution (2.0% in distilled water, w/w) and incubating overnight. The coating was verified by PCS measurement of the coating layer thickness.

Organ distributions of the surface-modified particles were determined in NZW rabbits by gamma scintigraphy (particles labelled with iodine-131, injection *via* the ear vein). The amount in the organs is given as percentage radioactivity of the injected dose.

3. MODEL CARRIERS - *IN VITRO* CELL CULTURE *VERSUS IN VIVO* ORGAN CULTURE RESULTS

Distribution

Polystyrene particles were coated with different Poloxamer and Poloxamine polymers and incubated in cultures of mouse peritoneal macrophages to study the protective effect of the polymers against phagocytosis[5]. A size of 5.25 µm was chosen to enable detection of the phagocytosed particles by light microscopy. From PCS measurements on 60 nm polystyrene latex particles it is known that the polymers form adsorption layers of a thickness between 2 nm and 15 nm. Layers with a similar thickness were assumed on the larger sized particles because the adsorption depends mainly on the nature of the surface rather than on their curvature (particle size)[5]. The phagocytosis was quantified as relative phagocytic uptake, the control (uncoated) particles was 100% uptake. Coating layers of low molecular weight polymers had little protective effect, high molecular weight polymers such as Poloxamer 407 and Poloxamine 908 led to a distinct reduction (Table 1). Surprisingly, small latex particles could be protected by low molecular weight polymers which were non-protective on large particles.

From cell culture results obtained with the 5.25 μm latex particles it was expected that the high molecular weight polymers Poloxamer 407 and Poloxamine 908 would possess the highest potential to protect against uptake by the RES, that means the macrophages in liver and spleen. These two polymers should be able to avoid uptake of large and small particles (60 nm). Low molecular weight polymers (Poloxamer 188) should at least avoid the uptake of smaller particles (« 5 μm). However, intravenous injection of differently sized particles yielded unexpected organ distributions (Table 2).

Table 1. Coating layer thickness of polymers on latex particles (after ref. 1) and relative phagocytic uptake of coated 5.25 μm particles in macrophage cultures (after ref. 5).

Coating Polymer	Molecular Weight	Coating layer on 60 nm latex particles (nm)	Rel. uptake 5.25 μm latex particles	Rel. uptake 60 nm latex particles
None (control)			100	100
Poloxamer 235	4,600	3.5	86.5	none
Poloxamer 188	8,350	7.6	95.4	none
Poloxamer 407	11,500	15.4	21.6	none
Poloxamer 908	27,000	13.4	69.5	none

Table 2. Organ distribution of differently sized latex particles coated with Poloxamer 407 and Poloxamine 908 (NZW rabbits, organ distribution determined by gamma scintigraphy).

Latex size / coating polymer	Organ Distribution after i.v. injection			Ref.
	Blood circulation	Accumulation in bone marrow	Liver and spleen uptake	
60 nm + 908	Yes	-	-	2,3
60 nm + 407	-	Yes	-	4
142 nm + 908	Yes	-	-	1
142 nm + 407	Yes	-	-	1
383 nm + 908	-	-	Yes	1
383 nm + 407	-	-	Yes	1
> 383 nm *with* both coatings	-	-	Yes	1

Small particles coated with Poloxamine 908 were protected against liver/spleen uptake and circulated in the blood. The similar polymer Poloxamer 407 caused an accumulation of the particles in the bone marrow. Increasing the particle size to 142 nm led to particles circulating in the blood, further size increase to uptake by liver and spleen. The *in vivo* results seemed to contradict the cell culture data indicating a protective effect even for large particles against phagocytosis. In addition, the uptake of 60 nm particles coated with Polxamer 407 by the bone marrow was surprising. According to the cell culture data the Poloxamer 407 was thought to be able to prevent phagocytosis by bone marrow macrophages or pinocytosis by endothelial cells. Furthermore low molecular weight polymers such as Poloxamer 188 proved not to be very effective in preventing phagocytosis by the RES *in vivo*[6]. A thorough characterization of the coated model carriers was performed to resolve these apparent contradictions.

4. PCS AND LDA CHARACTERIZATION OF MODEL CARRIERS TO CORRELATE *IN VITRO* AND *IN VIVO* RESULTS

Determination of the Coating Layer Thickness

The thickness of adsorption layers can be easily determined by subtracting the PCS diameters of uncoated and coated latex particles. The method is limited to small particles (below 100 nm - 150 nm) because of the accuracy of PCS (standard deviation of about 1%). The coating layer thickness needs to be sufficiently large (*e.g.* 10 nm) for precise determination. In addition, the particles need to be free of agglomerates. Addition of block copolymers to particle dispersions has a sterically stabilizing effect and removes particle agglomerates. There will be a size increase due to polymer adsorption on the particle surface and simultaneously a size decrease due to the removal of agglomerates. Therefore the coating layer thicknesses given in Table 1 were determined on 60 nm polystyrene particles.

Because it is impossible to measure the coating layer thickness on 5 μm particles by PCS for the time being the assumption was made that the layers are identical to the ones on 60 nm latex. Determination is possible by small angle neutron scattering (SANS)[7] but the technique is of limited availability. In addition, the data analysis is rather complex. Alternatively to SANS zeta potential (ZP) measurements were employed to assess the coating layer thickness on large particles[8]. The zeta potential is the potential at the plane of shear in the diffuse layer (Fig. 3). An adsorbed polymer layer shifts the surface of shear and reduces the zeta potential depending on the thickness of the adsorbed layer. The thickness of the diffuse layer is relatively large in distilled water (> 100 nm). A shift of the plane of shear by an adsorbed layer of 5 nm to 15 nm will therefore cause only a slight reduction of the ZP. The thickness of the diffuse layer decreases with increasing electrolyte concentration. To use this method, the thickness was adjusted to about 10 nm by addition of 0.001 M NaCl, the pH of the water was stabilized by addition of 0.0001 M citrate-phosphate buffer (pH 5.0).

Polymers with an adsorbed layer thickness of about 12 to 15 nm cover most of the diffuse layer and create a reduction of the ZP close to zero. The reduced potential after coating was calculated in

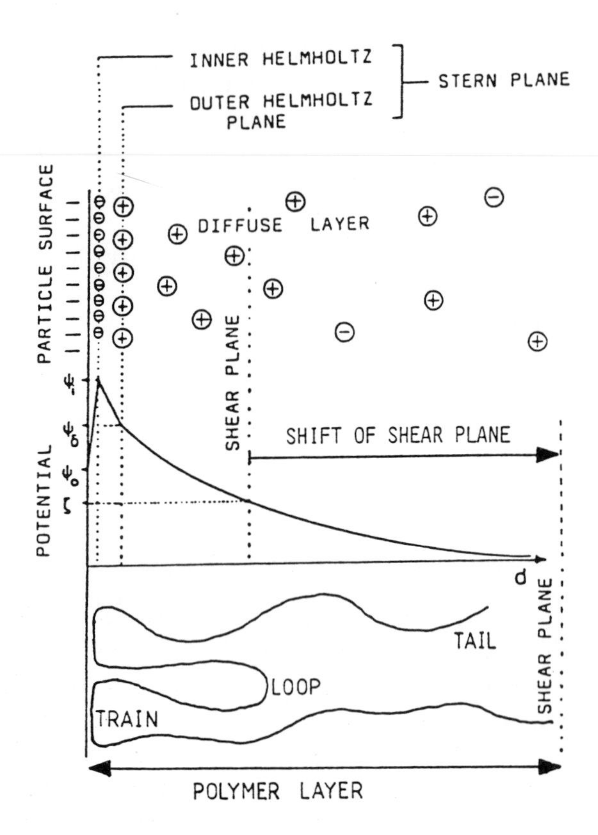

Figure 3. Composition of the Stern plane and the diffuse layer around a particle. The potential of the surface (Nernst potential) increases in the inner Helmholtz plane, is partially compensated by counter ions in the outer Helmholtz plane and decreases exponentially to zero in the diffuse layer. During the movement of the particle in an electric field a part of the diffuse layer is stripped off. The potential at this plane of shear is the measured zeta potential. An adsorbed polymer layer protects partially the diffuse layer against removal and shifts the plane of shear. This reduces the measured zeta potential as a function of the thickness of the adsorbed polymer layer (after ref. 1).

per cent of the original potential of the uncoated latex particle. A calibration was performed by plotting the measured zeta potential reduction *versus* the known coating layer thickness of various Poloxamer and Poloxamine polymers on 60 nm and 145 nm particles (layer determined by PCS). The calibration using small particles is possible because the ZP reduction does not depend on the size of the particles. To obtain a straight line, a logarithmic y-axis was used (Fig. 4).

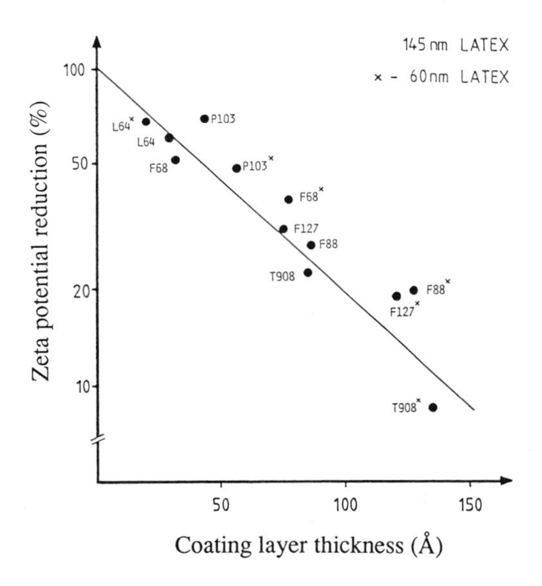

Figure 4. Plot of the ZP of coated 60 nm (x) and 145 nm latex *versus* the coating layer thickness. The ZP of the coated particles are calculated as percentage of the ZP of the uncoated particles (ZP of uncoated particles = 100%). Coating was performed with Poloxamer 184 (L64), 188 (F68), 283 (F88), 333 (P103), 407 (F127) and Poloxamine 908 (T908). A semi-logarithmic plot was used to obtain a straight line (exponential drop of the ZP in the diffuse layer) (after ref. 1).

The thickness of Poloxamer 407 and Poloxamine 908 coating layers were determined on differently sized latex particles using this method (Table 3). The adsorption layers were distinctly thinner on

Table 3. Coating layer thickness on latex particles of different size determined by PCS (60 nm and 145 nm particles) and by LDA *via* zeta potential reduction (large particles)[1].

Latex particle size (nm)	Coating layer thickness (nm) after coating with ...	
	Poloxamine 908	Poloxamer 407
60	13.5	12.0
145	8.0	7.5
383	7.0	5.0
496	8.1	7.1
910	6.2	4.5
5600	7.0	5.5

large particles - despite that they were all made from the same material: polystyrene. This raised two questions:

1. Why were the coating layers thinner on large polystyrene particles?
2. Which effect had the thinner coating layers on the interaction with blood components *in vivo*?

The thinner coating layers could be explained by the surface properties of the polystyrene particles. Driving forces for the adsorption are the hydrophobicity of the particle surface (gain in interaction energy when water is replaced by a hydrocarbon on the surface), the reduction in the number of hydrocarbon-water contacts[9] and the gain in entropy of the water[10]. The surfaces of polystyrene particles larger than 60 nm are less hydrophobic which results in a thinner coating layer. The reduced hydrophobicity could be shown by adsorption studies of hydrophobic dyes[11]. The differences can be explained by variations in polymerisation conditions used for small and larger particles. The effect of this reduced coating layer thickness had to be determined.

Interaction of the Coating Layer with Serum Components

The coating layers of Poloxamer and Poloxamine create a relatively hydrophilic surface and reduce therefore the interaction (adsorption) with blood components after i.v. injection. The conformation of the adsorbed polymers is different in case of thinner coating layers. The relatively hydrophobic parts of the polymers (PO chains) are more exposed. The reduction in coating layer thickness was found to be correlated with an increase in surface hydrophobicity of the coating films[12]. The surface hydrophobicity of coated particles was determined by hydrophobic interaction chromatography (HIC)[13,14]. The particles are passed through hydrophobic columns of alkyl-agaroses. The retention volume and the hydrophobicity of the columns which can be passed are parameters to quantify the hydrophobicity of the coating films (Table 4).

The increased hydrophobicity of a coating film resulted in an increased interaction with blood proteins[12]. Zeta potential measurements of coated particles in serum can be performed as a measure of this interaction[15]. The measurements are based on the assumption that the zeta potential increases with increasing amount of protein absorbed. This assumption was confirmed by studying the adsorption pattern of proteins by polyacrylamide gel electrophoresis

Table 4. Surface hydrophobicity of differently sized polystyrene particles coated with Poloxamer 407 and Poloxamine 908. For simplicity reasons the hydrophobicity is quantified by the number of crosses (details see ref. 1).

Size of coated latex particle (nm)	Surface of particles coated with ...	
	Poloxamine 908	Poloxamer 407
60	+	++
142	+++	+++
383	+++++	+++++
> 383	+++++	+++++

(PAGE)[16]. An increased adsorption of proteins will therefore enhance also the adsorption of charged components and cause subsequently a charge (zeta potential) increase of the particles (Table 5).

Table 5. Zeta potential of uncoated and Poloxamine 908 and Poloxamer 407 coated particles in serum (different particle size, after ref. 1).

Coating polymer	Zeta potential (mV) of 60 nm latex	Zeta potential (mV) of 383 nm latex
No coating	-20.5	-21.4
Poloxamine 908	-5.0	-15.9
Poloxamer 407	-4.9	-18.3

Uncoated latex particles are very hydrophobic, heavily opsonized after i.v. injection and rapidly taken up by the liver and spleen macrophages. This heavy adsorption of blood components leads to a relatively high ZP of about -20 mV in serum. Coating of 60 nm latex with Poloxamine 908 creates a hydrophilic surface, reduces opsonisation and particles escape liver and spleen recognition. The ZP in serum drops to -5 mV. A similar low potential was obtained for the hydrophilic Poloxamer 407 coating. The coating films of Poloxamine 908 and Poloxamer 407 were found distinctly more hydrophobic on particles ≥ 383 nm (Table 4). This results in an increased adsorption of proteins/opsonins as shown by the high ZP which is close to the ZP of uncoated particles.

Consequently the particles are recognized by the RES and taken up by liver and spleen macrophages.

PCS Measurements in Serum

The reduction in the adsorption of serum components by coating can be followed by PCS measurements in serum. Uncoated 60 nm particles adsorbed a protein layer with a thickness of about 38 nm. The adsorption of serum proteins could be distinctly reduced by coating the particles with low molecular weight Poloxamer 184. After coating with Poloxamine 908 a layer could hardly be detected. The PCS measurements were not sensitive enough to distinguish between the effect of medium and high molecular weight polymers (*e.g.* Poloxamer 188 and 407).

Explanation of the *in vivo* Organ Distribution by PCS and LDA Data

From the characterization of coating layer thickness (PCS/LDA), hydrophobicity (HIC) and interaction with serum components (LDA in serum) it could be concluded:

1. The coating layers on larger particles were thinner than on small ones (*e.g.* 60 nm);
2. The thinner coating layers were more hydrophobic as seen by HIC and the ZP in serum;
3. The more hydrophobic layers led to an increased adsorption of serum/blood components (ZP in serum, PCS in serum).

Small particles coated with Poloxamine (60 nm, 142 nm) and Poloxamer 407 (142 nm) are sufficiently hydrophilic to avoid opsonisation and RES clearance. They circulate in the blood. Larger particles coated with Poloxamer 407 and Poloxamine 908 possess a more hydrophobic coating film which interacts strongly with blood components. They are recognized by the immune system and taken up by liver and spleen.

The accumulation of Poloxamer 407 coated 60 nm particles in the bone marrow cannot be explained by the presented PCS and LDA data. However, the HIC detected a slightly increased hydrophobicity for these particles compared to the Poloxamine 908 coating (Table 4). It is suggested that they might adsorb a blood component which leads to specific adherence to bone marrow cells[17]. This bone marrow specific opsonin might be uncharged or present in a low concentration and therefore not detectable by ZP measurements. At present PAGE studies are under way to compare the protein adsorption patterns on the particles to confirm this hypothesis. The uptake of particles by the bone marrow cells is limited to a size below approximately 150 nm. The bone marrow is not accessible for Poloxamer 407 coated particles larger than this size. Thus the 142 nm particles circulate in the blood similarly to the Poloxamine 908 coated ones.

Explanation of the *in vitro* Cell Culture Results by PCS and LDA Data

The requirements to prevent phagocytosis are different *in vivo* and *in vitro*. Less reduction in surface hydrophobicity is required to reduce or eliminate phagocytosis in cell cultures. The thin coating layers (4 nm -5 nm) on 5 μm particles were still too hydrophobic to prevent uptake by macrophages *in vivo*. However, they were sufficiently hydrophilic to reduce phagocytosis *in vitro* (phagocytic index 20-60% for Poloxamer 407, Poloxamine 908). It should be noted that the surface hydrophobicity of Poloxamer and Poloxamine coating films decreases with increasing thickness[15].

The coating layers on small 60 nm particles are thicker, even low molecular weight Poloxamers create a thickness of almost 4 nm[1]. The surface hydrophilicity obtained by such a thickness is sufficient to prevent phagocytosis *in vitro*. This explains why even low molecular weight Poloxamers could prevent phagocytosis of 60 nm particles in cell culture (Table 1).

It was concluded that the reduction of phagocytosis *in vitro* is a function of the coating layer thickness (hydrophilicity), the coating layer thickness is a function of the particle surface hydrophobicity, the latter varies with the particle size. By choosing a particle size between 60 nm and 5 μm it should therefore be possible to optimize the *in vitro* cell culture test system[18]. Optimized means that the uptake is identical *in vitro* and *in vivo*: The phagocytosis is eliminated by Poloxamine 908 and Poloxamer 407, strongly reduced by Poloxamer 188 and hardly effected by Poloxamer 184. This could be achieved with 1000 nm polystyrene particles (Fig. 5). The *in vitro* test system is adjusted to differenciate between the efficiency of polymers to protect small particles (60 nm) against RES uptake *in vivo*. It avoids the confusing results obtained in cultures

(a) with 5 μm particles (only reduction of uptake to 69.5% after coating with Poloxamine 908) and
(b) with 60 nm particles (elimination of phagocytosis with all polymers).

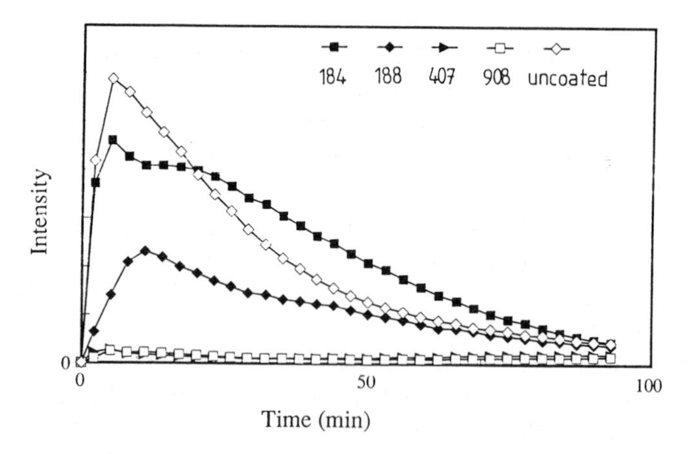

Figure 5. Uptake of 1000 nm particles (uncoated and coated with Poloxamer 184, 188 and 407, Poloxamine 908) by granulocytes in suspension culture. The uptake was determined by chemiluminescence[18] (y-axis: chemiluminescence intensity, x-axis: time (min)).

5. PERSPECTIVES

The future work in the area of controlled drug delivery with colloidal carriers will focus on the search for new polymers eliminating the recognition and the uptake by the RES. These polymers will be used for the coating of polymeric colloidal particles but also for the production of fat emulsion carriers and liposomes. PCS and LDA will be useful techniques in selecting potential new polymers with suitable properties.

Transferring the advances achieved with surface-modified non-biodegradable particles to biodegradable carriers will be of paramount importance. Unfortunately Poloxamer and Poloxamine adsorb only in thin layers on particles made from biodegradable, non-toxic polymers such as poly lactic acid (PLA) and its copolymer with glycolic acid (PLA/GA)[19]. Enhanced adsorption can be achieved by modification of the particle surface properties[19] or selection of coating polymers with higher affinity to the particle surface[15]. PCS and LDA are regarded as basic techniques to characterize the coating films on the biodegradable carriers with regard to thickness, film stability against desorption[20], charge reducing effect and interaction with serum components.

REFERENCES

1. R.H. Müller, 'Colloidal Carriers For Controlled Drug Delivery - Modification, Characterization and *In Vivo* Distribution', Wissenschaftliche Verlagsgesellschaft Stuttgart, 1990.
2. L. Illum, S.S. Davis, R.H. Müller, E. Mak and P. West, *Life Sci.*, 1987, *40*, 367.
3. S.S. Davis, S. Douglas, L. Illum, P.D.E. Jones, E. Mak, R.H. Müller, in 'Targeting of Drugs with Synthetic Systems', G. Gregoriadis, J. Senior and G. Poste, Plenum Press, New York, 1986, p. 123.
4. L. Illum and S.S. Davis, *Life Sci.*, 1987, *40*, 1553.
5. L. Illum, L.O. Jacobsen, R.H. Müller, E. Mak and S.S. Davis, *Biomaterials*, 1987, *8*, 113.
6. L. Illum and S.S. Davis, *J. Pharm. Sci.*, 1983, *72*, 1086.
7. T. Cosgrove, T. L. Crowley and B. Vincent, *Farady Symp. Chem. Soc.*, 1981, *16*, 101.
8. R.H. Müller, T. Blunk and F. Koosha, *Arch. Pharm.*, 1988, *321*, 678.
9. B. Kronberg, *J. Colloid and Interf. Sci.*, 1983, *96*, 55.
10. H. Carstensen, B.W. Müller and R.H. Müller, *Arch. Pharm.*, 1990, *323*, 776.
11. R.H. Müller, S.S. Davis, L. Illum and E. Mak, in 'Targeting of Drugs with Synthetic Systems', G. Gregoriadis, J. Senior and G. Poste, Plenum Press, New York, 1986, p. 239.
12. R.H. Müller and T. Blunk, *Arch. Pharm.*, 1989, *322*, 699.
13. E. Mak, R.H. Müller, S.S. Davis, L. Illum, *Acta Pharm. Technol.*, 1988, *34*, 23S.
14. H. Carstensen, B.W. Müller and R.H. Müller, *Int. J. Pharm.*, 1991, *67*, 29.
15. T. Blunk and R.H. Müller, *Arch. Pharm.*, 1989, *322*, 755.
16. R.H. Müller and T. Blunk, publication in preparation.
17. R.H. Müller and S. Heinemann, in 'Bioadhesion - Possibilities and Future Trends', R. Gurny and H. E. Junginger, Wissenschaftliche Verlagsgesellschaft, Stuttgart, 1990, p. 202.

18. R.H. Müller and S. Rudt, submitted to *Proceed. Intern. Symp. Contr. Rel. Bioact. Mater.*, 1991, *18*.
19. R.H. Müller and K. Wallis, *Arch. Pharm.*, 1989, *322*, 768.
20. K.H. Wallis and R.H. Müller, *Acta Pharm. Technol.*, 1990, *36*, 127.

Index

NOTTINGHAM
UNIVERSITY
AGRICULTURAL
AND FOOD
SCIENCES
LIBRARY